SOIL ACIDITY AND PLANT GROWTH

Sponsored by the Australian Wool Corporation

SOIL ACIDITY AND PLANT GROWTH

Edited by

A. D. Robson,
Soil Science and Plant Nutrition,
School of Agriculture,
The University of Western Australia

ACADEMIC PRESS
Harcourt Brace Jovanovich, Publishers
Sydney San Diego New York Berkeley
Boston London Tokyo Toronto

ACADEMIC PRESS AUSTRALIA
30–52 Smidmore Street, Marrickville, NSW 2204

United States edition published by
ACADEMIC PRESS INC.
1250 Sixth Avenue
San Diego, California 92101–4311

United Kingdom edition published by
ACADEMIC PRESS, INC. (LONDON) LTD.
24/28 Oval Road, London NW1 7DX

Printed in Australia

National Library of Australia Cataloguing-in-Publication Data

---

Soil acidity and plant growth.
Includes bibliographical references and index.
ISBN 0 12 590655 2.

1. Soil acidity – Australia. 2. Acid soils – Australia.
3. Plants, Effect of soil acidity on. I. Robson, A. D.
(Alan David), 1945– .

631.8'21'0994

Library of Congress Catalog Card Number: 88–82638

Contents

Contributors

Numbers in parentheses indicate the pages on which the authors' contribution begins.

L. K. Abbott (139) Soil Science and Plant Nutrition, School of agriculture, University of Western Australia, Nedlands, Western Australia 6009, Australia.

M. K. Conyers (205) Agricultural Research Institute, Department of Agriculture New South Wales, P.M.B., Wagga Wagga, New South Wales 2650, Australia.

D. R. Coventry (103) Rutherglen Research Institute, Victorian Department of Agriculture and Rural Affairs, Rutherglen, Victoria 3685, Australia.

P. D. Cregan (205) School of Agriculture, Riverina-Murray Institute of Higher Education, P.O. Box 588, Wagga Wagga, New South Wales 2650, Australia.

J. Evans (103) Agricultural Research Institute, Department of Agriculture New South Wales, P.M.B., Wagga Wagga, New South Wales 2650, Australia.

J. A. Fisher (167) Agricultural Research Institute, Department of Agriculture New South Wales, P.M.B., Wagga Wagga, New South Wales 2650, Australia.

D. L. Godyn (265) Regional Office, Department of Agriculture New South Wales, P. O. Box 477, Wagga Wagga, New South Wales 2650, Australia.

K. R. Helyar (61) Agricultural Research Institute, Department of Agriculture New South Wales, P.M.B., Wagga Wagga, New South Wales 2650, Australia.

J. R. Hirth (205) Rutherglen Research Institute, Victorian Department of Agriculture and Rural Affairs, Rutherglen, Victoria 3685, Australia.

Z. Hochman (265) North Coast Agricultural Institute, Department of Agriculture New South Wales, P.M.B., Wollongbar, New South Wales 2480, Australia.

W. M. Porter (61) Dryland Research Institute, Western Australia department of Agriculture, Merredin, Western Australia 6514, Australia.

G. S. P. Ritchie (1) Soil Science and Plant Nutrition, School of Agriculture, University of Western Australia, Nedlands, Western Australia 6009, Australia.

A. D. Robson (139) Soil Science and Plant Nutrition, School of Agriculture, University of Western Australia, Nedlands, Western Australia 6009, Australia.

B. J. Scott (167–265) Agricultural Research Institute, Department of Agriculture New South Wales, P.M.B., Wagga Wagga, New South Wales 2650, Australia.

Acknowledgements

I would like to thank the following: Dr W. M. Porter and Dr J. S. Yeates participated in the selection of topics and authors. Ms J. McFarlane assisted with proof reading. Miss Y Wedgwood, Miss G. Heather and Mrs L. Murrish typed the manuscripts. Dr N. J. Barrow, Dr G. P. Gillman, Dr D. C. Edmeades, Dr R. A. Date, Dr M. Bromfield, Dr D. N. Munns, Mr J. Howieson, Dr P. Taylor, Dr S. Smith, Dr J. F. Loneragan, Dr J. W. Bowden, Dr P. Kerridge and Dr I. Barclay refereed the manuscripts. The Australian Wool Corporation on the advice of the Wool Research Trust Fund supported the meeting that led to this monograph.

A. D. ROBSON

Foreword

In Europe and the USA, the need for regular dressings of lime has been accepted for many years as a basic principle of good farming on soils with low lime reserves.

In Australia, the ubiquitous and acute deficiencies of phosphorus and nitrogen, the relatively high cost of liming in the dominantly extensive agricultural systems, and the frequent induction of severe trace element deficiencies by liming have led to a much more cautious approach: much greater emphasis has been given to understanding the nature of lime responses with a view to correcting their primary causes rather than using blanket dressings of lime. This approach had its greatest success in A. J. Anderson's brilliant dissection of pasture responses to lime into primary effects of soil pH on two distinct components of legume nutrition in the system — rhizobial growth prior to nodulation, and release of soil molybdenum (Anderson and Moye 1952). These findings opened the way for replacing the use of heavy dressings of lime with cheaper and more efficient procedures of establishing improved pastures on many problem acid soils. At the same time, limits to these procedures were recognised; on other more acid soils, they were not as effective as heavy dressings of lime, indicating that other factors became limiting at more extreme acidity.

Concurrent with the research seeking to identify the primary factors responsible for lime responses in pastures, C. H. Williams and C. M. Donald were investigating the long-term effect of pasture improvement on soil properties. In 1954, they published results showing that, under continuous improved pastures, soils gradually became more acidic (Williams and Donald 1954). The consequences of this prophetic observation have become evident some 30 years later, as this book testifies. Indeed, the present book has arisen directly from concerns over increasing acidification of soils under improved pastures over wide areas of southern Australia.

While the book has its origin in the problems of acidification of Australian soils under pastures, the authors examine soil acidity within a much broader framework, making their views relevant to all agricultural and natural ecosystems on acid soils. The first two chapters are concerned with the chemistry of soil acidity and the ecological processes leading to it. They are followed by three chapters on biological responses to soil acidity including mineralisation of soil nitrogen, incidence of plant diseases, plant mycorrhizal associations, symbiotic nitrogen fixation in legumes, and genetic variability in plant response to toxicities.

The book concludes with two chapters on the correction of soil acidity probems by liming: Chapter 6 is concerned with the rates of application and effectiveness of liming materials and Chapter 7 with the development and use of computer modelling procedures both for assisting researchers in identifying the effects and interactions of soil pH on component processes and to provide assistance to farmers in the management of long-term subterranean clover pastures.

Throughout the book, the authors emphasise the need to understand the processes involved in soil acidification and in the response of plant growth. This analytical approach is very timely and should help greatly in understanding present problems and in defining profitable lines of future research.

J. F. LONERAGAN,
Murdoch University, Perth.

1

The Chemical Behaviour of Aluminium, Hydrogen and Manganese in Acid Soils

G. S. P. RITCHIE

I. INTRODUCTION

The challenge facing researchers in the field of soil acidity today is no different from the situation 30 years ago. The major problem is still our inability to identify accurately which of the many effects of soil acidity is reducing plant growth. This results partly from our lack of knowledge of how nutrients or toxic substances behave in acid soils. It has become increasingly evident from conflicting reports in the literature that many complex reactions occur. Until we can assess the frequency of occurrence and relative importance of these reactions, we shall not be able to improve our ability to identify and ameliorate problem soils.

SOIL ACIDITY AND PLANT GROWTH
ISBN 0 12 590655 2

Soil acidity decreases plant growth in many ways (e.g. deficiencies of phosphorus, molybdenum, calcium or toxicities of aluminium, manganese or hydrogen ions), but toxicities have been recognised as one of the most common causes of reduced yields. This chapter, therefore, will consider the chemical behaviour of aluminium, manganese and hydrogen ions in relation to their toxic effects on plant growth, but not with respect to their effects on the availability of other nutrients.

The complex nature of the chemical reactions that occur in acid soils often necessitates dividing the soil system into simpler fractions in order to study mechanisms. Even though the method of division may appear somewhat arbitrary in some cases, we cannot expect to understand the overall behaviour without some basic knowledge of the behaviour of the individual soil components. Consequently, the reviewed literature includes research that has involved model systems designed to study the chemical properties of individual soil components as well as research that looks at the overall soil behaviour in relation to plant growth.

II. ALUMINIUM

An adequate understanding of the mechanisms which control the distribution of aluminium between the solid and liquid phases of a soil is necessary for identifying soils with an aluminium toxicity problem and for determining the long-term consequences and effectiveness of liming.

The distribution of aluminium in a soil is the net result of two sets of reactions: (1) competition between ligands for aluminium. The ligands may be soluble species, or ligands on particles that can react with cations (i.e. adsorption), or species that can precipitate aluminium ions. This form of competition decides the *potential* distribution of aluminium and is illustrated in Figure 1; (2) competition between aluminium and other

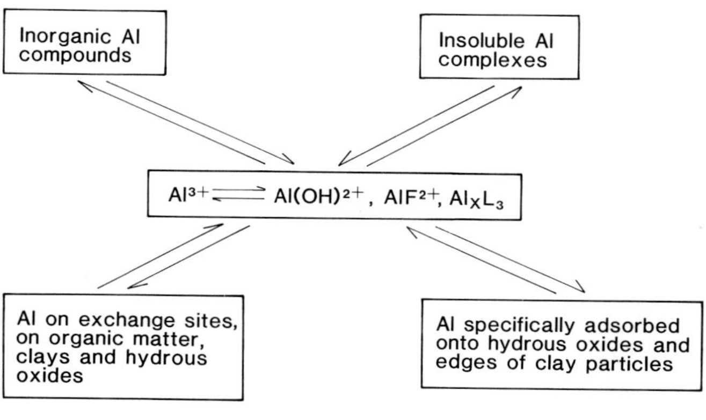

Fig. 1. Aluminium distribution in acid soils.

cations for ligands. This is often overlooked even though it decides the *actual* distribution of aluminium.

This section will consider the chemical behaviour of soluble aluminium, its reactions with other constituents in acid soils before and after liming, and our ability to predict situations where toxicity may arise.

A. Aluminium in the soil solution

Many years of research have been devoted to the complex behaviour of aluminium in pure aqueous solution (e.g. Hem, 1968; Hsu, 1968; Turner, 1968). In soils, however, such work may only be used as a basis for comparison because its direct application to the soil environment is severely limited by the complex composition of the soil solution. In addition, some solution studies (e.g. Hsu, 1968; Adams and Hajek, 1978; Singh, 1982a) have used very high concentrations of aluminium ($> 1000\ \mu M$). The data of Kamprath (1978a), Curtin and Smillie (1983) and Adams and Moore (1983) indicate that total soluble aluminium (Al_T) is usually between 10–350 μM. Levels $> 1000\ \mu M$ occur only in exceptional circumstances such as in acid sulphate soils. The behaviour of soluble aluminium at dilute concentrations has been shown to be very different from that at higher levels (Baes and Mesmer, 1976; Stol *et al.*, 1976; Veith and Sposito, 1979). A greater proportion of aluminium is polymerised at concentrations $> 1000\ \mu M$, which can affect adsorption of aluminium by soil particles and precipitation of aluminium after liming.

There is increasing evidence that the toxicity of aluminium is reduced in the presence of inorganic and organic complexing anions and that the activity of Al^{3+} and/or the Al-hydroxy species are the forms of aluminium that are the most correlated with depressed yields of several plant species (Adams and Lund, 1966; Bartlett and Riego, 1972; Brenes and Pearson, 1973; Helyar, 1978; Pavan and Bingham, 1982; Blamey *et al.*, 1983; Cameron *et al.*, 1986). The use of single ion activities as a measure of ion levels in soils has been questioned on thermodynamic grounds (Sposito, 1984). However, even though such expressions may have no strict thermodynamic meaning, their usefulness in determining critical concentrations of toxic ions cannot be ignored.

The concentration and activity of Al^{3+} in soil solutions will not only depend on the mechanisms by which aluminium is distributed between the solid and liquid phases, but also on its behaviour in solution. Aluminium does not exist in solution just as the 'free' ion, Al^{3+}, but can undergo several reactions depending on the type of ligands (anions) present (Table I). The extent of complexation will depend on the ionic strength and pH of the soil solution and the relative concentrations and complexing abilities of soluble ligands (Sposito, 1981b). An approximate idea of the ability of a ligand to complex aluminium may be gained by considering the reaction of equimolar activities of Al_T and the ligand and the equilibrium constant, K_o. For example, in the reaction:

$$Al^{3+} + SO_4^{2-} \rightleftharpoons Al\,SO_4^+ \qquad \log K_o = 3.2 \tag{1}$$

the relative activities of the different species are given by:

$$(Al\,SO_4^+)/(Al^{3+})(SO_4^{2-}) = 10^{3.2} \tag{2}$$

where round brackets denote activities. This expression may be rearranged in terms of the ratio of the two aluminium species, i.e.

$$(Al\,SO_4^+)/(Al^{3+}) = 10^{3.2}\,(SO_4^{2-}) \tag{3}$$

Equation (3) indicates that for ($Al\,SO_4^+$) to be > 50% of total soluble aluminium (Al_T), the free sulphate activity must be $> 10^{-3.2}$ M. The more free ligand that has to be present to achieve > 50% complexing, the lower the affinity of the ligand for aluminium. In the soil solution, the extent of complexation will be modified further by the total activity of the ligand in comparison to (Al_T) and the presence of competing cations.

Competition between aluminium and other cations is important because it can affect the level of soluble species of aluminium that are thought to be the most toxic to plants, i.e. Al^{3+} and Al-hydroxy species. The extent of competition depends on the equilibrium constant of the reaction of the cation with the ligand relative to that for aluminium and the ratio of aluminium ion:cation in solution.

The influence of these two factors is demonstrated in Figure 2 for aluminium complexing with citrate in the presence of Ca^{2+}, H^+ or Fe^{3+}. Cations that form weaker complexes with the ligand than aluminium (e.g. calcium and magnesium) can only compete effectively with aluminium at aluminium: cation ratios $\ll 1$. On the other hand, cations such as Fe^{3+} that form stronger complexes with the ligand can compete even when aluminium: cation ratios are > 1. The hydrogen ion may also compete with aluminium for a ligand. In the case of citrate, aluminium: H^+ ratios of < 0.1 are necessary to reduce aluminium-citrate complexing by > 15%.

Soluble aluminium species may be broadly divided into two groups: monomers (reactions 1–9, Table I) and polymers (reactions 10–14, Table I). The proportion of soluble aluminium in each category may be estimated theoretically using thermodynamic principles or alternatively by assuming that aluminium species that react quickly (< 2 min) with a coloured complex are monomeric (labile) and aluminium species that

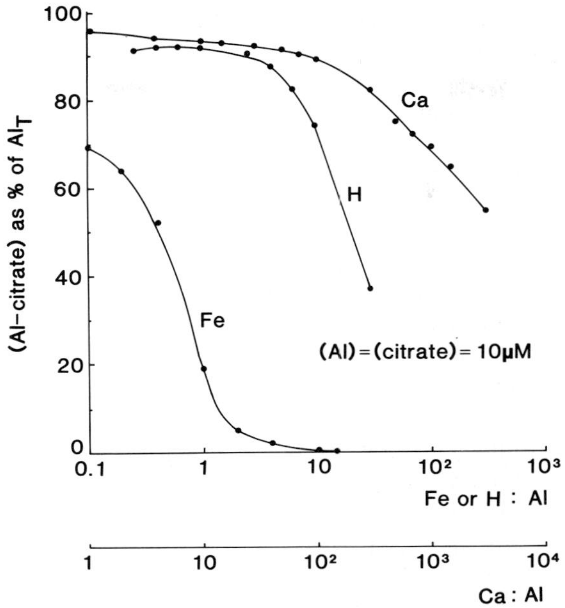

Fig. 2. The effect of pH and of calcium and iron (III) at pH 4 on aluminium-citrate complexing.

Table I. Possible inorganic reactions of aluminium in the soil solution.

1. $Al^{3+} + OH^- \rightleftharpoons AlOH^{2+}$
2. $Al^{3+} + 2OH^- \rightleftharpoons Al(OH)_2^+$
3. $Al^{3+} + 4OH^- \rightleftharpoons Al(OH)_4^-$
4. $Al^{3+} + SO_4^{2-} \rightleftharpoons Al(SO_4)^+$
5. $Al^{3+} + 2SO_4^{2-} \rightleftharpoons Al(SO_4)_2^-$
6. $Al^{3+} + F^- \rightleftharpoons AlF^{2+}$
7. $Al^{3+} + 2F^- \rightleftharpoons AlF_2^+$
8. $Al^{3+} + 3F^- \rightleftharpoons AlF_3^0$
9. $Al^{3+} + 4F^- \rightleftharpoons AlF_4^-$
10. $2Al^{3+} + 2OH^- \rightleftharpoons Al_2(OH)_2^{4+}$
11. $3Al^{3+} + 4OH^- \rightleftharpoons Al_3(OH)_4^{5+}$
12. $13Al^{3+} + 28OH^- \rightleftharpoons Al_{13}O_4(OH)_{24}^{7+} + 4H^+$
13. $(Al)_2{>}OH + Si(OH)_4 \rightleftharpoons (Al)_2{>}O - Si(OH)_3 + H_2O$
14. $(Al)_2{>}OH + H_2PO_4^- \rightleftharpoons (Al)_2{>}O - P(=O)(OH)(OH) + OH^-$

react more slowly (> 30 min) are polymeric or non-labile. The first approach is limited by the lack of reliable equilibrium constants for some aluminium-ligand reactions and the assumptions inherent in their formulation (e.g. equilibrium is achieved, unit activity of solid compounds). The second method assumes that reaction kinetics reflect the extent of polymerisation.

1. Monomeric aluminium

The monomeric hydrolysis of aluminium (reactions 1–3, Table I) usually becomes significant at pH > 4 and by pH 4.9 more than 80 per cent of Al_T is hydrolysed. In the soil solution, the extent of hydrolysis may be reduced by the presence of other anions, particularly nitrate, chloride and sulphate (Singh, 1982c) and organic anions (Kwong and Huang, 1977). At a constant solution concentration of Al_T, increasing the ionic strength increases the *concentration* of Al^{3+} slightly because of a small change in equilibrium constant which does not favour hydrolysis. The activity of Al^{3+}, however, declines sharply because of large decreases in the activity co-efficient of trivalent ions with increased ionic strength.

Sulphate decreases hydrolysis by competing for the Al^{3+} ion (Table I). If the ion pair $AlSO_4^+$ were to be a major species in the soil solution (> 50% of Al_T), free sulphate concentrations would have to be a minimum of 10^{-3} M and concentrations > 10^{-1} M would be necessary for significant formation of other sulphate ion pairs (Lindsay, 1978). Theoretical considerations may suggest that sulphate ion pairing increases with decreasing pH (Lindsay, 1978, p. 44; Singh, 1982b) because aluminium solubility increases. This assumes that a constant, high level of sulphate is present. In soils, however, the opposite often occurs because sulphate concentrations are invariably lower in acid soils due to adsorption and leaching. In four North Queensland soils, total sulphate concentrations in the soil solution varied from $0.02–1.5 \times 10^{-3}$ M (Bell and Gillman, 1978); therefore the sulphate ion may not necessarily be a major complexer of aluminium in unpolluted acid soils.

Fluoride complexes very strongly with aluminium (Lindsay, 1978). Concentrations of soluble fluoride in uncultivated soils are approximately 10^{-5} M (Larsen and Widdowson, 1971; David and Driscoll, 1984) but they may increase in soils fertilized regularly with superphosphate because of its presence as an impurity in the fertilizer.

Aluminium can also form soluble monomeric complexes with organic anions depending on the type and concentration of organic ligand, the proportion of charged sites that are dissociated, pH, ionic strength and the presence of competing cations (Bruckert, 1970; Rashid, 1974; Kwong and Huang, 1977; Ritchie *et al.*, 1982; Stevenson, 1982; Young and Bache, 1985). Most of the work done so far has used specific organic molecules (e.g. citric acid) or purified humic and fulvic acids in simple solutions. The work needs to be extended to more complex systems similar to the soil solution. Aluminium forms stable complexes with simple aliphatic acids (Table II) such as tartaric, oxalic and citric acids (Martell and Smith, 1977; Kwong and Huang, 1977). In solutions containing precipitated aluminium, complexing with citric acid increases Al_T and decreases Al^{3+} at low Al_T concentrations (100 μM). At higher concentrations of aluminium (1000 μM), the same effect has been observed but was attributed to a concurrent drop in pH not observed at lower aluminium concentrations (Kwong and Huang, 1977).

Humic and fulvic acids may also undergo a number of reactions with aluminium (Table II). Young and Bache (1985) have suggested a general model for aluminium speciation in solutions containing an excess of organic complexing ligands similar to humic acid. Complexing increased with pH and the concentration of the dissociated acid. Their data fitted a

Table II. Examples of monomeric complexes formed with hydroxy carboxylic acids.

citrate[a]

$Al^{3+} + L^{3-} \rightleftharpoons AlL$
$Al^{3+} + H^{+} + L^{3-} \rightleftharpoons AlHL^{+}$
$Al^{3+} + L^{3-} \rightleftharpoons Al(H_{-1})L^{-}$

tartrate[a]

$Al^{3+} + L^{2-} \rightleftharpoons Al(H_{-1})L + H^{+}$
$Al^{3+} + L^{2-} \rightleftharpoons Al(H_{-2})L + 2H^{+}$
$Al^{3+} + 2L^{2-} \rightleftharpoons AlL_2^{-}$

gluconic acid[a]

$Al^{3+} + L^{-} \rightleftharpoons AlL^{2+}$
$Al^{3+} + L^{-} \rightleftharpoons Al(H_{-1})L^{+}$
$Al^{3+} + L^{-} \rightleftharpoons Al(H_{-3})L^{-}$

humic and fulvic acids[b]

$Al^{3+} + L^{n-} \rightleftharpoons AlL^{(n-3)-}$
$Al^{3+} + L^{n-} \rightleftharpoons Al(H_{-m})\,L^{(n+m-3)-}$
$Al^{3+} + L^{n-} \rightleftharpoons Al(OH)\,L^{(n-2)-}$

[a] Taken from Motekaitis and Martell (1984). (H_{-1}) signifies the deprotonation of an hydroxyl group.
[b] postulated

model that considered bi- and tri-dentate ligand formation with organic acids alone or in conjunction with the hydroxyl ion. There are two difficulties with their approach. Firstly, Motekaitas and Martell (1984) could not find any evidence for the formation of Al-hydroxy complexes with simple hydroxy carboxylic acids, but showed that aluminium is capable of removing protons from the hydroxyl groups in them. Secondly, Young and Bache (1985) were unable to include the formation of positively charged complexes in their analysis and did not consider competition with other cations for the ligand. Calcium can compete effectively with aluminium for humate ligands if it is present at a much greater concentration, even though it does not form as stable a complex as aluminium (Rashid, 1974). Some organic anions that are capable of complexing aluminium may only be present for short periods because of their susceptibility to microbial breakdown. The easily decomposable, soluble component of alfalfa meal has been found to complex aluminium more strongly than the organic substances that remain after four weeks' incubation of the plant material (Hoyt and Turner, 1975).

2. *Polymeric aluminium*

Aluminium may also form soluble polymers with hydroxyl ions alone or in conjunction with phosphate or silicate ions (e.g. Bache, 1963; Hsu, 1963; 1968; Turner, 1969; Luciuk and Huang, 1974). The polymerisation of aluminium-hydroxy species is well established (Hem, 1968; Turner, 1969; Smith, 1971), but the conditions for the reactions are not fully understood. The occurrence and form of polymers appears to be affected by initial pH, the presence of other ions, degree and method of neutralisation, concentration, temperature and time of ageing (Hem, 1968; Hsu, 1968; Hem *et al.*, 1970; Turner, 1976; Rengasamy and Oades, 1977; Tsai and Hsu, 1984). For example, Hsu (1968) found that 'spontaneous' hydrolysis of Al^{3+} (i.e. the dissolution of an aluminium salt in distilled water) yielded only monomeric species whereas adding OH^- ions to aluminium solutions produced polymers. The kinetics of polymerisation and consequent precipitation and crystallisation are also very susceptible to environmental conditions (Hodges and Zelazny, 1983a). Consequently, it is very difficult to predict precisely the hydrolysis of aluminium in soil solutions. Some of the effects, however, are probably of more importance for understanding liming reactions and will be discussed later.

On the basis of theoretical calculations, Marion *et al.* (1976) considered that only short chain Al-hydroxy polymers existed in the soil solution, whereas Richburg and Adams (1970), Benians *et al.* (1977), and

Rhodes and Lindsay (1978) considered that long chain aluminium polymers could also be present. The latter argument was suggested because this type of speciation had to be considered if the level of aluminium in solution was to be explained adequately by the dissolution of minerals such as kaolinite, gibbsite and alunite. In their calculations, however, they could not include possible complexes with organic anions or mixed polymers containing phosphate and silicate as well as hydroxyl ions. Serious errors in the calculation of Al^{3+} activity can arise if certain species (particularly polyvalent species) are not accounted for or if the ionic strength is not measured accurately. This is probably another reason for the inability of solubility principles to predict solution levels of aluminium.

Aluminium forms polymeric species with phosphate and silicate. Munns (1965), White *et al.* (1976) and Blamey *et al.* (1983) found evidence for soluble Al-hydroxy phosphate polymers in the pH region 4.1–4.8, at aluminium and phosphate concentrations of < 100 and $< 25\,\mu M$ (respectively) and ionic strength $< 910\,\mu M$. Higher concentrations or pH values resulted in precipitation. Hsu (1968) found that phosphate reacted differently with mono- and polymeric aluminium species. Luciuk and Huang (1974) and Farmer *et al.* (1977) have demonstrated that silicate ions can react with Al-hydroxy polymers to produce a soluble complex at pH < 5.5. In the pH range 4.0–5.6, silicate ions appear to react only with polymeric aluminium species, with the level of monomeric aluminium remaining independent of the extent of complexation (Wada and Wada, 1980). Increasing the silicon concentration, hydroxyl:aluminium ratio or time of ageing all increased the incorporation of silicon into the aluminium polymer.

Aluminium humates may also be polymerised. Ritchie and Posner (1982) found that the increase in molecular weight of humic acids with aluminium complexing was far greater than could be attributed to the presence of aluminium in one molecule.

B. Aluminium buffering

The level of soluble aluminium in acid soils may be controlled (i.e. buffered) by the dissolution of inorganic compounds, adsorption onto inorganic minerals or by reactions with solid organic matter (Fig. 1).

1. Dissolution of compounds

Aluminium may exist in soils as mineral oxides or oxyhydroxides (Hsu, 1977; Taylor *et al.*, 1983) and other primary and secondary minerals

(Dixon and Weed, 1977; Lindsay, 1978) which all dissolve to a certain extent, releasing their constituent ions into the soil solution.

Several workers have suggested that the level of soluble aluminium in soils may be explained by the dissolution of gibbsite or kaolinite in equilibrium with quartz (Richburg and Adams, 1970; Marion *et al.*, 1976; Rhodes and Lindsay, 1978) or by the dissolution of halloysite and smectites (Marion *et al.*, 1976). Basaluminite and alunite (aluminium sulphates) have also been considered (e.g. Singh and Brydon, 1969; Adams and Rawajfih, 1977; Adams and Hajek, 1978) as well as aluminium phosphate minerals such as variscite (Lindsay, 1978). However, efforts to relate the solubility products of pure minerals to the solution levels of the appropriate ions have never been completely successful (e.g. Bache, 1974). This is not surprising considering the complex composition of the soil solution from which these minerals would develop. The likelihood of the formation of pure crystalline minerals similar to those that develop in pure systems is not great. The presence of aluminium-substituted goethites and haematites in soils (Norrish and Taylor, 1961; Schwertmann *et al.*, 1977), the retardation of hydrolysis and precipitation by both inorganic and organic ions (Kwong and Huang, 1977; Singh, 1982b) and the detection of amorphous precipitates stable for over 2 years (Turner, 1976; Hsu, 1982) are all evidence of the limited relevance of pure mineral dissolution as a major mechanism controlling aluminium distribution in soils. Also, it is very difficult to distinguish experimentally between adsorption and precipitation in soils. In some cases adsorption may be the initial mechanism followed by precipitation once surface concentrations build up. The latter stage may never be reached in highly adsorbing soils. For example, precipitation of aluminium phosphates and concurrent dissolution of Al-hydroxy sulphates were suggested as the reason for changes in aluminium and sulphate concentrations after the addition of potassium phosphate to a soil (Adams and Rawajfih, 1977). However, the lower aluminium concentration that was observed could also be explained by the concomitant increase in pH, and the higher sulphate level could be due to its displacement from adsorption sites by the phosphate anion.

At aluminium and phosphate levels more commonly found in soils and at pH > 3.2, Hsu (1982) found that amorphous Al-hydroxyl phoshates precipitated and remained amorphous even after 4 years with no evidence of variscite formation. Evidence of the presence of variscite was only observed in solutions containing high concentrations of aluminium and phosphate and with pH < 2.9. Moreover, Bache (1963) showed that variscite dissolves incongruently with the formation of a coating of $Al(OH)_3$ on the variscite surface. Therefore, it would be the new surface that would control the concentration of aluminium in solution rather than variscite. This is not to say that dissolution and precipitation of aluminium minerals does not occur in soils, but rather that the aluminium so released may undergo further reactions with the solid phase by other more dominant mechanisms such as adsorption.

2. Adsorption by inorganic minerals

Aluminium is adsorbed non-specifically (i.e. only electrostatic forces are involved) at negatively charged sites in clay minerals and hydrous oxides of iron, aluminium and manganese. This is more commonly referred to as exchange. At the same time, it may also be specifically adsorbed (i.e. the forces involved are not just electrostatic) at variable charged sites on hydrous oxides and the edges of clay particles. The specific adsorption of aluminium has not been considered much in the literature, possibly because it would not be very extensive at low pH. Therefore, the following discussion refers mainly to non-specific adsorption.

Early work on the adsorption of aluminium onto inorganic minerals in acid soils has been reviewed by Coleman and Thomas (1984). The investigations used systems over-saturated with aluminium (> 500 per cent saturation of the exchange capacity) and artificially high hydroxyl: aluminium ratios (> 3) which often resulted in interlamellar precipitation of aluminium. The hydroxy-Al interlayers were considered a major cause of the aluminium and pH buffering in acid soils. As pointed out by Bloom *et al.* (1977), the high concentrations of aluminium in these systems are uncommon in most acid soils. In other studies, the early literature did not clearly distinguish the form of aluminium preferred by the surface even though there was evidence that both free Al^{3+} and hydrolysed species could be adsorbed (Frink, 1973).

More recent work on the adsorption of aluminium by smectites indicates that Al-hydroxy species are preferred. Veith (1977; 1978), Bloom *et al.* (1977) and Hodges and Zelazny (1983a,b) have studied the adsorption of aluminium onto smectites at low solution aluminium concentrations and low levels of saturation of exchange sites (0–200 per cent). Unfortunately, their work is not directly comparable because of different initial hydroxyl:aluminium ratios in solution and different methods of calculating surface hydrolysis of aluminium. At low pH's, smectites can induce surface hydrolysis of aluminium even though there are negligible levels of hydroxyl species present in solution (Bloom *et al.*, 1977; Veith, 1978). At < 100 per cent aluminium saturation of exchange sites, the ratio of aluminium to Al-hydroxy species is always smaller than in solution even though at low pH values (4–4.5), Al^{3+} would be the predominant species (Bloom *et al.*, 1977; Hodges and Zelazny, 1983a,b). The higher the pH and hydroxyl:aluminium ratio of the original solution, the greater the adsorption of Al-hydroxy species and the smaller the difference in hydrolysis between solution and sorbed aluminium (Hodges and Zelazny, 1983a).

The relative adsorption of Al^{3+} and Al-hydroxy species can be affected by size constraints due to the nature of the adsorbent. Veith (1977; 1978) demonstrated that only free Al^{3+} is exchanged onto vermiculite-type clays because mono- and polymeric Al-hydroxy species could not penetrate the interlayer clay spacings. The lack of penetration of interlayer clay spacings by Al-hydroxy polymers (which are precursors of gibbsite) explained the absence of interlamellar gibbsite formation. Hsu (1968), however, considered the lack of gibbsite precipitation to be due to the high affinity of the clay for aluminium. When aluminium saturation of the surface reaches > 100 per cent, interlamellar precipitation of gibbsite can occur if the aluminium species can penetrate these regions or are present in sufficient quantities. There is also some evidence, over short time periods, that partial dehydroxylation of aluminium species may occur. However, this appeared to be only temporary as an increase in surface hydrolysis was observed with further ageing (Hodges and Zelazny, 1983a).

The effect of aluminium adsorption on pH depends on the type of clay, the hydroxyl:aluminium ratio in solution and the time of ageing. Removal of hydrolysed aluminium species from solution would induce further solution hydrolysis of aluminium and therefore a lowering of the pH. If Al^{3+} was the major species adsorbed then a pH increase would be expected. Accordingly, Veith (1978) observed a pH decrease when aluminium was adsorbed onto smectite and a pH increase when vermiculite was the clay. In addition, the pH change due to Al^{3+} adsorption alone may be confounded by proton uptake by variable charged sites on clays (Bloom *et al.*, 1977). Consequently, the greater the variable anionic charge capacity on the clay, the greater the uptake of protons and so a pH increase may result. However, the pH increase may only be temporary and decrease with time because of buffering by surface reaction with H^+ and OH^- till a new equilibrium is established. In some cases, pH increases due to proton uptake will be counterbalanced by the adsorption of Al-hydroxy species and so only a small change may be observed.

Hodges and Zelazny (1983a) found that adsorption of aluminium onto smectites was accompanied by an increase in pH, whereas Bloom *et al.* (1977) observed a slight decrease in pH (0.1 unit) at a similar percentage saturation of exchange sites. The difference in pH changes could be explained partly by protonation of edge sites on the smectites and partly by dehydroxylation of the more hydrolysed species used by Hodges and Zelazny (1983a); this possibility is supported by the higher pH increase in their initially more hydrolysed solutions. Evidence for the importance of the former mechanism is provided by Bloom *et al.* (1977), who observed pH increase in solution when smectite was left to age for similar periods in solutions not containing aluminium. Hodges and Zelazny (1983a,b) also investigated the adsorption of aluminium onto kaolinite and observed much greater pH increases. This could be attributed to: the greater number of variable charge sites on the edge faces of kaolinite than on smectite; preference for Al^{3+} rather than Al-hydroxy; or dehydroxylation of adsorbed hydrolysed species. The apparently conflicting results of these two groups illustrate the importance of the initial basicity of the aluminium and the effect of time of ageing.

Changes in pH with aluminium desorption will also depend on the environmental conditions and the species involved. If Al^{3+} is exchanged into solution, a pH drop may be observed as some of the Al^{3+} hydrolyses (Bache, 1974; Bolt, 1976). On the other hand, Bloom *et al.* (1977) have pointed out that since the surface aluminium is more hydrolysed than the solution aluminium, exchange of Al-hydroxy into solution would lead to an increase in pH. However, their exchange relative to the exchange of Al^{3+} might be minimal.

3. Reactions with organic matter

The importance of the interaction between aluminium and organic matter in soils is now well recognised, and the inability of certain soil tests to predict toxic levels of aluminium has been attributed to the reactions between aluminium and organic matter (e.g. Adams and Moore, 1983). Aluminium may form soluble or insoluble complexes with organic matter or it may be non-specifically adsorbed onto exchange sites (Fig. 1). Evidence for insoluble organic aluminium complexes comes from studies using infra-red spectroscopy. Vinkler *et al.* (1976) considered that the antisymmetric carboxylate stretching frequency at 1625–1630 cm^{-1} for aluminium humates and fulvates indicated the formation of covalent rather than ionic bonds. Electrostatic adsorption at exchange sites is indicated by the displacement of aluminium from organic matter by non-complexing cations such as potassium. Soluble organic complexes of aluminium have already been discussed. However, changes in pH, ionic strength or the concentration of aluminium may cause their precipitation. For example, aluminium-humate complexes decrease in solubility as pH is decreased and the percentage saturation of carboxyl groups by aluminium approaches 100 per cent (Ritchie *et al.*, 1982). In contrast, aluminium fulvates are more soluble at low pH values.

The extent of both complexation and non-specific adsorption will depend on the same factors that influence the formation of soluble complexes. There has been little distinction between these two mechanisms in published research. Therefore, aluminium will be described as being 'bound' to organic matter when the literature is discussed.

Bloom *et al.* (1979b), Arai and Kumada (1981) and Hargrove and Thomas (1982a) found that aluminium bound with organic matter is more hydrolysed than the soluble aluminium in equilibrium with it. The reactions of aluminium with organic matter can both increase or decrease levels of aluminium in solution, the pH and the pH buffering capacity of the soil. Once more, the apparently conflicting evidence in the literature is due to the sensitivity of aluminium reactions to experimental conditions.

The addition of organic matter to a soil or solution can decrease or increase the level of soluble aluminium depending on the concomitant changes in pH. For example, Bloom *et al.* (1979b) found that if the initial soil pH and the degree of H^+ saturation of the organic matter were high enough, organic matter addition both decreased and increased the level of aluminium in solution. Increases in soluble aluminium could be explained by the high pH of the soil causing some dissociation of H^+ from the organic matter which lowered the soil pH and resulted in the dissolution or release of aluminium. As mentioned earlier, the reactions of aluminium with soluble organic material can also increase the level of aluminium in solution by retarding precipitation of aluminium oxides (Kwong and Huang, 1977). On the other hand, the addition of organic matter can decrease soluble aluminium because the extent of aluminium binding by the organic matter more than counterbalances any increased aluminium dissolution caused by pH decreases (e.g. Bloom *et al.*, 1979b; Hargrove and Thomas, 1981b). The effect of aluminium-organic matter binding on pH appears to depend on the degree of hydrolysis of the aluminium in solution initially. Hargrove and Thomas (1981b, 1982a) and Hodges and Zelazny (1983a) observed a pH increase with aluminium adsorption when the soluble aluminium was present mainly as hydrolysed species initially (Fig. 3). This was probably due to the low pH of the organic matter suspension partially dehydroxylating some of the added aluminium. In contrast, Bloom and McBride (1979), Arai and Kumada (1981) and Hargrove and Thomas (1982b) observed a pH decrease when aluminium was adsorbed by organic matter from solutions at pH < 4. In the latter cases, the aluminium would have been present primarily as Al^{3+} and its adsorption could have involved exchange with H^+ions as well as some surface hydrolysis. Both mechanisms would decrease pH.

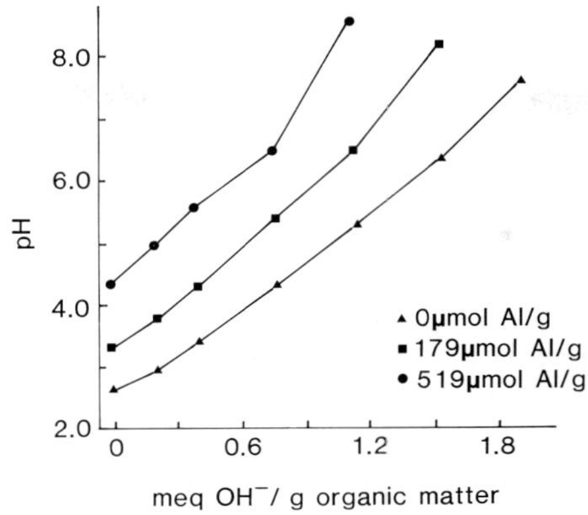

Fig. 3. Potentiometric titration curves for organic matter of various Al contents (adapted from Hargrove and Thomas, 1982b).

Adsorbed aluminium can act as a source of pH buffering by reacting with hydroxyl and thereby decreasing the rise in pH as alkali is added. As pH increases, the extent of hydrolysis of adsorbed aluminium increases, lowering the pH buffering capacity of that surface.

The removal of aluminium from solution by organic matter adsorption and a concomitant increase in pH buffering capacity can have a confounding effect by increasing exchangeable acidity. Hoyt (1977) observed that even though the addition of organic matter reduced concentrations of soluble and exchangeable aluminium there was an increase in exchangeable acidity (i.e. buffer capacity). In some cases, therefore, the benefits from reducing an aluminium toxicity may be counter balanced by an increase in the lime requirement of the soil.

C. Liming and aluminium behaviour in soils

Liming decreases aluminium toxicity by precipitation of hydrolysed polymers and by increasing the amount of specifically adsorbed aluminium at the expense of exchangeable aluminium. The type of precipitate that forms will depend on the environmental conditions, and its nature will affect the amount of aluminium remaining in solution, pH, buffering capacity and the availability of other nutrients.

The earlier discussion as to whether the dissolution of inorganic minerals controls the level of aluminium in the soil solution is also relevant when considering the behaviour of aluminium in the soil in the presence of liming materials. The retarding effect of anions on hydrolysis and precipitation has already been mentioned. Anions also slow down the rate of crystallisation of precipitates and reduce crystal size (Ross and Turner, 1971; Serna *et al.*, 1977; Hsu, 1979; Kwong and Huang, 1979). Hsu (1979) suggested that the effect of inorganic anions on gibbsite crystallisation was in the order: phosphate > silicate > sulphate > nitrate > perchlorate. A smaller crystal size (< 1000 Å) means that the surface area area of the product is much larger and therfore its capacity to adsorb ions such as phosphate is far greater. The type of liming material and time influence the rate of neutralisation which affects the type of aluminium species formed (Turner, 1976). In the absence of soil, the slow release of OH^- from dolomite produced mono- and polymeric aluminium species that did not form a solid phase within two years. On the other hand, at an equivalent level of neutralisation of aluminium, the quicker dissolution rate of calcium carbonate resulted in the formation of a solid phase with a weak X-ray diffraction pattern of gibbsite, which did not change even after two years.

In theory, the presence of different forms of aluminium (soluble and insoluble) leads to differences in pH and buffering capacity (Turner, 1968). Considering solutions of constant aluminium concentration, the more hydrolysed the aluminium species that can form, then the lower the initial pH, the better buffered the system at low pH's and the lower the pH at which precipitation of amorphous $Al(OH)_3$ occurs (Fig. 4). In practice, however, the formation of species will be complicated by the speed of neutralisation.

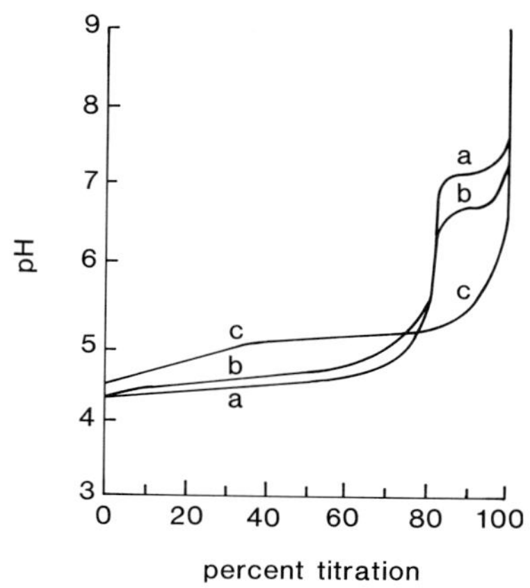

Fig. 4. Theoretical titration curves for the titration of 10^{-4} M $AlCl_3$ with a base. Curve (a): calculations on the basis of the formation of all species except gibbsite and $Al_{13}(OH)_{34}$. Curve (b): calculations on the basis of the formation of mononuclear cations, $Al(OH)_3$ and $Al_6(OH)_{15}$. Curve (c): calculations on the basis of the formation of mononuclear cations and amorphous $Al(OH)_3$ (adapted from Turner, 1968).

Lime application may not increase plant growth on acid soils because of lime-induced deficiencies of phosphorus and/or micronutrients. There are other reasons that may explain the lack of success of liming. Munns (1965) found, that even though liming decreased 0.01 M $CaCl_2$-extractable and soluble aluminium, the effect was only temporary; after 50 weeks the concentration of aluminium had risen to that which existed in the soil after five weeks with no lime added. The increasing aluminium could not be attributed totally to the concomitant decrease in pH of the limed soils because the pH values after 50 weeks were still 0.5–2.5 units above the pH of the unlimed soil.

In a detailed study of the changes in soil solution properties upon liming, Curtin and Smillie (1983) also demonstrated the importance of following the changes through time. As expected, liming initially increased pH and organic matter solubility and decreased Al_T. As time progressed, however, the pH fell due to the production of H^+ by the nitrification of NH_4^+ released during mineralisation of organic matter. Consequently, after 30 weeks of incubation with 2.7 t. ha^{-1} of ground lime (50 per cent of the lime requirement), the soil pH had dropped back to the level of the initial pH measured in the absence of lime and Al_T had increased once more.

The efficacy of liming may also be affected by changes in the concentrations of soluble anions which complex aluminium and therefore change Al^{3+} levels in solution. If liming decreases the concentration of a soluble ligand that complexes strongly with aluminium (e.g. fluoride (Street and Elwali, 1983) or fulvic acids (Stevenson, 1982)), the level of Al^{3+} in solution may not decrease even though Al_T would be lower. As the pH rises the proportion of Al_T that is Al^{3+} increases due to decreased complexing with the lower level of ligand. Provided that this effect is not outweighed by changes in ionic strength, higher pH values may have to be attained before an appreciable drop in Al^{3+} is achieved.

D. Soil tests for aluminium

The previous sections have illustrated the diverse behaviour of aluminium in soils. Not only can it exist in many different pools, but its reactions within the soil solution are also quite complex. Nevertheless, we still need to know which conditions limit plant growth and we also need to estimate the amount of lime required to ameliorate any problems. As with all soil tests, the methods of extraction and analysis must be capable of determining only aluminium that is biologically active as well as reflecting the capacity of the soil to maintain the supply.

In the literature, different extraction methods have been compared to determine if any consistent relationship exists between them, presumably on the premise that they might extract aluminium from specific pools. Methods have also been compared according to their ability to predict lime requirements and plant yield. In this case, the assumption is that they extract aluminium from biologically active pools only. Research of this nature will be considered first, followed by a discussion of the contribution of organic matter and soluble Al^{3+} to aluminium toxicity.

1. Methods of extraction

The most common methods of extracting soil aluminium use 1 M potassium chloride (KCl) and 1 M ammonium acetate (NH_4OAc). Aluminium extracted by these two solutions is commonly referred to as 'exchangeable' and 'extractable' respectively. (Exchangeable aluminium is sometimes expressed as a percentege of the effective cation exchange capacity (ECEC) and is then referred to as the percentage aluminium saturation of the soil). Also, there are methods of measuring the total acidity of a soil using buffered solutions and pH measurements (Kamprath, 1978b) which are discussed in Chapter 2. More recently, the extraction of aluminium by calcium, barium, lanthanum and copper chlorides and total aluminium in the soil solution have received a lot of attention (Hoyt and Nyborg, 1971a; Bloom *et al.*, 1979a; Juo and Kamprath, 1979; Khalid and Silva, 1979; Hargrove and Thomas, 1981a, 1984).

The limited application of exchangeable and extractable aluminium as measures of the intensity and capacity factor of biologically active aluminium is illustrated by the conflicting evidence on the relative ability of KCl and NH_4OAc to extract aluminium. 1 M NH_4OAc has been found to extract more (Gakhamani, 1977), similar amounts (McLean, 1965) and less (Pratt and Blair, 1961) aluminium than 1 M KCl.

Similar inconsistencies have been observed in comparisons of extraction methods. Hoyt and Nyborg (1971a) and Bromfield *et al.* (1983) found 0.01 M $CaCl_2$-extractable aluminium was well correlated with exchangeable aluminium whereas Hoyt and Webber (1974) and Webber *et al.* (1982) provide evidence that this may not always be the case. It has also been considered that percentage aluminium saturation should be related to the total level of aluminium in solution (Nye *et al.*, 1961; Evans and Kamprath, 1970; Kamprath, 1978a), but this is not necessarily so for organic soils (Evans and Kamprath, 1970; Adams and Moore, 1983).

It is difficult to reconcile these results because the amount of aluminium extracted by different solutions primarily depends on the pH, cation type and concentration of the solution and its buffering capacity relative to that of the soil (Bloom *et al.*, 1979a; Oates and Kamprath, 1983a).

A further problem with soil tests for aluminium is highlighted by their inability to predict plant growth or lime requirements even when plant growth is not confounded by concomitant toxicities of manganese or hydrogen ions or deficiencies of calcium and phosphorus. Exchangeable aluminium and percentage aluminium saturation of ECEC have been found to be better predictors of crop response to liming than pH measurements (Kamprath, 1970; Reeve and Sumner, 1970; Juo, 1977; Farina *et al.*, 1980; Webber *et al.*, 1982). However, they are by no means reliable indicators over a wide variety of soil types and environmental conditions. For example, Evans and Kamprath (1970), Adams and Moore (1983) and Bromfield *et al.* (1983) could find no consistent relationship between plant yield and exchangeable aluminium or percentage aluminium saturation. Adams and Lund (1966), Evans and Kamprath (1970) and Adams and Moore (1983) also found that total aluminium in the soil solution was an inadequate predictor of plant yield in the presence of organic matter.

Consequently, the inability of simple measurements such as exchangeable aluminium to predict aluminium toxicity of soils or consistently extract aluminium from biologically active pools has led researchers to consider more closely the retention of Al by organic matter and the behaviour of aluminium in the soil solution.

2. *The contribution of organically bound aluminium to aluminium toxicity*

Organically complexed aluminium appears to be one major cause of the observed inconsistencies. Under certain circumstances, complexing with organic matter lowers the solution concentration and therefore aluminium toxicity does not limit growth even in some very acid soils (Thomas, 1975; Hargrove and Thomas, 1981b).

Effects of organic matter on soluble and exchangeable aluminium vary considerably. When comparing different soils, Evans and Kamprath (1970) found a general increase in soluble aluminium and a decrease in exchangeable aluminium as the organic matter content decreased. Juo (1977) observed the opposite trend for both measurements of aluminium when comparing different horizons from the same soil. In more detailed studies, Thomas (1975) and Hargrove and Thomas (1981b) noted that the changes in exchangeable aluminium with organic matter content depended on pH and the level of organic matter present. They observed a decrease in exchangeable aluminium with increasing organic matter which was more pronounced at low pH and organic matter contents.

Apparently, aluminium bound by organic matter is not extracted by 1 M KCl but is capable of reacting with lime (Juo and Kamprath, 1979; Hargrove and Thomas, 1981a, 1984). Consequently, Hoyt (1977) observed that for soils of similar pH, even though the exchangeable aluminium decreased with increasing organic matter content, there was a concomitant increase in the exchangeable acidity due to the buffering ability of the organically complexed aluminium.

It has been found that copper and lanthanum chlorides are capable of extracting organically bound aluminium (Bloom *et al.*, 1979b; Juo and Kamprath, 1979; Hargrove and Thomas, 1981b, 1984) as well as interlayer aluminium associated with smectites and may therefore be a better measure of biologically active aluminium. In fact, aluminium extracted from an aluminium-saturated peat by $CuCl_2$ or HCl was identical to that determined by ashing and elemental analysis (Hargrove and Thomas, 1981a).

More recently, Oates and Kamprath (1983b) have suggested that $CuCl_2$ may not be suitable for determining the aluminium toxicity status of limed soils because of its apparent ability to dissolve recently-precipitated aluminium. They found that the lower lime requirements predicted by $LaCl_3$ extraction were sufficient to reduce the percentage saturation to < 15 per cent in all soils, whereas liming to neutralise aluminium in KCl extracts only achieved a similar reduction in the percentage saturation for soils with no organically bound aluminium.

E. The relative toxicities of soluble aluminium species

The forms of aluminium in solution appear to have different relative toxicities to plants (Adams and Lund, 1966; Bartlett and Riego, 1972; Helyar, 1978; Pavan and Bingham, 1982; Blamey *et al.*, 1983; Cameron *et al.*, 1986). Current evidence suggests that Al^{3+} and monomeric Al-hydroxy species are more toxic to plants than other forms. The presence of both organic and inorganic complexing ions apparently detoxify aluminium by reducing the percentage of total aluminium that is Al^{3+} and/or $Al(OH)_n^{3-n}$. In addition, there is increasing evidence of the toxicity of H^+ *per se* and the mitigation of aluminium toxicity by calcium (Alva *et al.*, 1986; Cameron *et al.*, 1986). Until now the mitigating effect of calcium on aluminium toxicity has been considered to be a plant effect. However, the effect of calcium on aluminium complexing discussed in Section II,A does not rule out the possibility of it being a soil effect when complexing ligands such as citrate are present.

The effect of ionic strength on aluminium toxicity was considered first by Adams and Lund (1966). In a series of experiments investigating the effects of aluminium toxicity on elongation of cotton roots, they found that neither pH nor exchangeable aluminium were suitable indicators of aluminium toxicity. If their data were expressed as activities rather than concentrations they found a more consistent relationship between soluble aluminium and relative root length (Fig. 5). This is effectively correcting the aluminium concentrations for ionic strength (I) effects. In their experiments, calcium was the major cation in solution and so one cannot distinguish between an ionic strength effect or an effect due to calcium itself. Also, they assumed that Al^{3+} was the only species in solution even though other forms of aluminium may have been present and ion pairing with sulphate may have occurred in some of their nutrient solutions.

Phosphate, sulphate and fluoride are inorganic anions that have been found to detoxify aluminium (Helyar, 1978; Blamey *et al.*, 1983; Alva *et al.*, 1986; Cameron *et al.*, 1986). At constant total aluminium added to nutrient solutions, Blamey *et al.* (1983) found that increasing the phosphate concentration and the aluminium:hydroxyl ratio decreased monomeric aluminium and increased the root growth of soyabean (*Glycine max*). The formation of polymers or precipitation with phosphate and hydroxyl appeared to be the cause of the reduced aluminium toxicity. Cameron *et al.* (1986) found that aluminium complexed by sulphate and fluoride did not inhibit root elongation of barley (*Hordeum Vulgare*) seedlings grown in nutrient solutions (Fig. 6).

Aluminium toxicity has also been reduced by the presence of organic anions such as citrate, tartrate, EDTA and an alkaline soil organic matter extract (Bartlett and Reigo, 1972; Hue *et al.*, 1986). The ability of organic acids to detoxify aluminium appears to increase with concentration and the proximity of hydroxyl and carboxylic groups within the molecule (Hue *et al.*, 1986), and follows the order citrate > oxalic > tartaric > malic > malonic > salicylic which is the same as the order of their ability to complex aluminium.

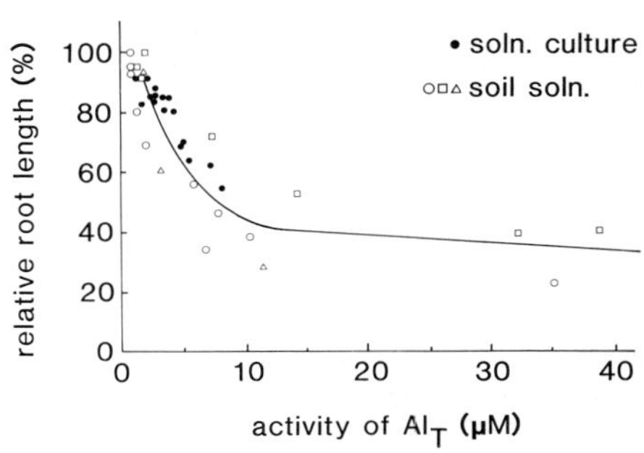

Fig. 5. Effect of molar activity of Al_T in subsoil solutions (○, □, △) and in nutrient solutions (●) on growth of primary roots of cotton (taken from Adams and Lund, 1966).

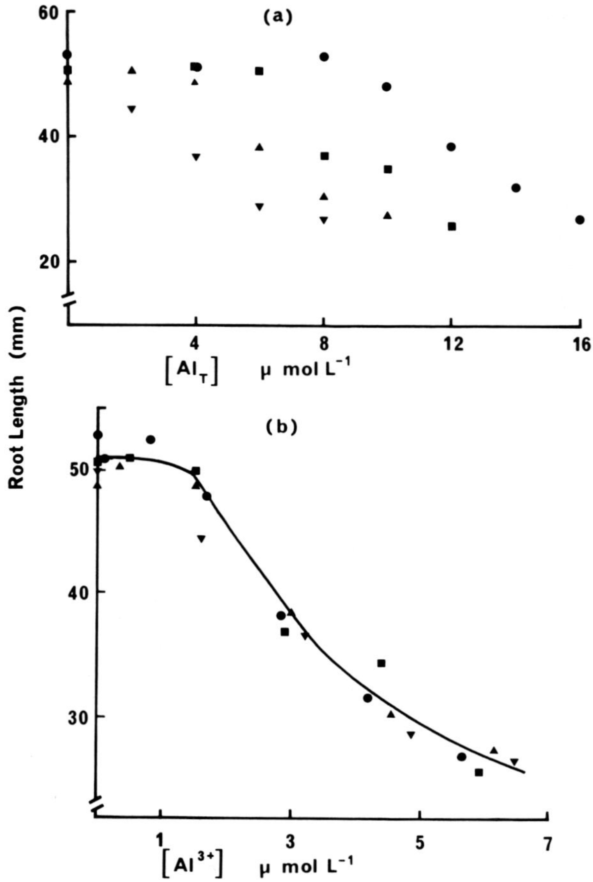

Fig. 6. The variation in root length of barley with (A) [Al_T] and (B) [Al^{3+}] in the presence of 0 (▼), 2.5 (▲), 5 (■) and 10 (●) μM fluoride (taken from Cameron *et al.*, 1986).

The extrapolation of results from solution culture experiments to the soil system must be approached with caution because it may be limited by infinite buffering due to continual changing of solutions, uniform availability of nutrients along the root and the frequent use of much higher nutrient concentrations than found in soils. Additionally, there are the extra difficulties of distinguishing between the confounding effects of pH and aluminium and the competition between cations for complexing ligands. Even so, there is evidence in the literature that observations from solution culture experiments do bear some relevance to plant growth in soils (Adams and Lund, 1966; Prenes and Pearson, 1973; Pavan *et al.*, 1982; Sheppard and Floate, 1984). Pavan *et al.* (1982) found that the relative root growth of coffee seedlings in four acid soils (amended with various rates of different liming materials) was highly correlated with the activity of free aluminium in solution (Fig. 7). An activity of 4×10^{-6} M was required to produce a 10% reduction in growth which was in reasonable agreement with the critical activity (12×10^{-6} M) determined by solution culture techniques (Pavan and Bingham, 1982). In their calculations of (Al^{3+}), no allowance was made for the presence of organic-Al complexes in the soil solutions. If they were present, one would expect the critical (Al^{3+}) of the soil solution to be greater than that of the pure solution cultures. This was not the case, however, and could be due to a concurrent reduction in root growth due to H^+ ions. Relative root growth is usually calculated by expressing root growth in the presence of a toxic ion as a percentage of root growth in the absence of the toxic ion while holding all other soil properties constant. Pavan *et al.* (1982) expressed root growth as a percentage of growth by plants receiving a 'twice-lime equivalent' treatment which not only had a negligible (Al^{3+}) but also had a higher pH. Therefore, their calculations of relative root growth also included the toxic effects of H^+ ions and so the percentage relative root growth (% RRG) at each Al_T is lower than can be attributed to aluminium alone. The error in % RRG increases as the pH of each treatment decreases relative to the pH of the treatment used to assess root growth in the absence of aluminium. Consequently, the toxic effect of aluminium in the soil solutions appears to occur at a lower Al^{3+} activity because of a concomitant H^+ ion effect on root growth.

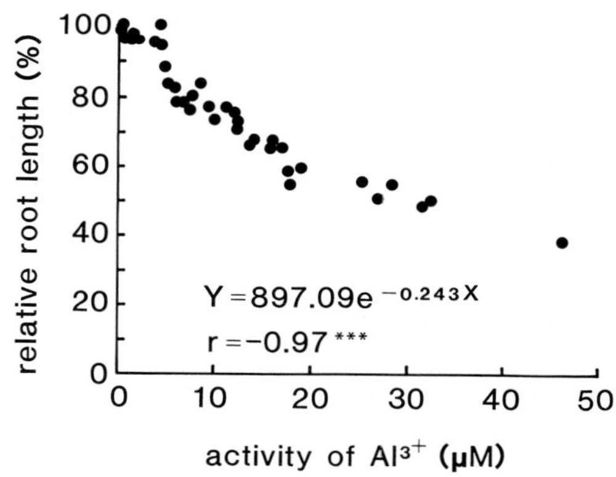

Fig. 7. Relative root growth of coffee in relation to activity of Al^{3+} in soil solution (***: significant at the 0.001 level) (taken from Pavan *et al.*, 1982)

F. Current directions in assessing the aluminium toxicity status of soils

The results of the research discussed in the previous two sections illustrate why no reliable soil test for toxic aluminium levels has been developed yet. The accuracy of soil tests will not improve until the effects of organic matter in both the solid and liquid phases and the role of Al^{3+} are considered more carefully.

Methods which appear to take into account aluminium bound with organic matter (such as the $LaCl_3$ extract) need to be investigated over a wider range of soils and the aluminium extracted related to plant growth. Also, techniques of determining the speciation of aluminium in solution need to be improved to enable a more accurate measurement of Al^{3+} or monomeric aluminium. The correlation of a solution measurement of aluminium (i.e. an intensity factor) with plant growth may seem fortuitous because it does not take into account the buffering capacity of the soil (i.e. the capacity factor). However, this could be due to the capacity factor being in equilibrium with Al^{3+} rather than with A_T.

As previously mentioned, soluble Al^{3+} or monomeric aluminium may be measured experimentally or calculated theoretically. The former approach appears to lack specificity whereas the latter is limited by a lack of reliable thermodynamic data and the need to identify and measure the concentration of all the complexing ligands in solution.

The analytical measurement of soluble Al^{3+} or monomeric aluminium is in fact a measurement of 'kinetically labile' aluminium. This is defined as the proportion of soluble aluminium that reacts to form a coloured complex or is exchanged by a resin in a short time interval (10 s – 30 min). Both methods are based on the assumption that the colouring reagent or resin will only react quickly with aluminium species that form weaker complexes than with the reagent or resin, i.e. the reaction depends solely on the binding constant. Not only is the binding constant for the reaction important, but also the relative concentrations of the competing ligands. An excess of complexing agent or exchange resin may be able to exchange aluminium that is very tightly bound to a very small amount of a soluble ligand. Therefore, kinetic lability does not necessarily indicate toxicity. Hoyt and Turner (1975), Turner and Sulaiman (1971) and James *et al.* (1983) have found that aluminium complexed with salicylate and citrate reacts very quickly with aluminon and 8-hydroxyquinoline, the two most common colouring reagents that are used. In addition, the conditions of the analysis (e.g. pH, ionic strength) may alter the true speciation of aluminium in undisturbed soil solution.

In the case of exchange resins, Campbell *et al.* (1983) have overcome some of the problems with changes in speciation caused by large pH changes by having a mixture of H^+ and Ca^{2+} on the resin that would be in equilibrium with solution levels in the soils under investigation. They found that 95 per cent of Al^{3+}, monomeric and small polymeric Al-hydroxy and Al-fluoride species exchanged onto resin within 30 min and the rates of exchange for the different species were indistinguishable from each other. Aluminium complexed with fulvic and humic acids was not readily exchanged and > 97 per cent remained in solution after 30 min. The long contact time between the sample and exchange resin limits the use of this method because aluminium speciation may change during the analysis as aluminium is released from complexes which are non-labile but are responding to the disturbance of the original equilibrium (i.e. the removal of Al^{3+}). Evidence for this may be seen from the inability of the method to distinguish between monomeric and small, polymeric Al-hydroxy species.

James *et al.* (1983) have developed a technique considered capable of distinguishing between labile aluminium and most of the aluminium complexed by fluoride, carbon, silicon and hydroxyl ions by measuring the aluminium that reacts with 8-hydroxyquinoline after 15 s. The short reaction time decreases disturbances of the natural equilibrium between aluminium species caused by pH changes or removal of Al^{3+}. However, Adams and Hathcock (1984) were unable to find a relationship between Al^{3+} activity measured by this method and plant growth. It would appear that further work is required to ascertain the specificity of this method for aluminium in soil solutions.

Even though current developments are opening up new pathways for research it is necessary to keep the importance of the more recent findings in perspective. The effect of soluble complexing ligands has certainly been established but it is probably only of relevance in soils of low Al_T (< 10–50 μM). At higher concentrations, complexing would be unable to lower the Al^{3+} level below the critical concentration at which toxicity occurs. The ability of different complexing ligands to affect Al_T varies quite widely (Section II,A), but not enough is known about their relative importance in soils. Similarly, the effect of organic matter on aluminium toxicity will no doubt vary from one situation to another as evidenced by the sensitivity of aluminium reactions to environmental conditions (Section II,B,3).

III. HYDROGEN

The pH of the soil solution is of fundamental importance in understanding nutrient availability since plants take up their nutrients from the soil solution. In acid soils, the solution pH is ultimately the result of the distribution of H^+ ions between soil surfaces and the soil solution. This section will consider the behaviour of H^+ ions in the soil; the following chapter will discuss the production and sources of H^+. After a brief consideration of soil pH *per se*, the discussion will concentrate on the different mechanisms of soil pH buffering.

A. Soil pH

The concept of a soil having a certain hydrogen ion activity in solution which is measurable and representative of the soil as a whole is an understatement of reality. The limitations of the meaning of soil pH have been discussed adequately by other authors (e.g. Russell, 1973) and so it will only be discussed briefly here. The deceptively simple process of determining soil pH has been the centre of several controversies for over 50 years and still shows no signs of retiring from the limelight (Sparks, 1984).

Variations in the pH of a soil can be attributed not only to soil variability and seasonal changes (Russell, 1973) but also to the soil: liquid ratio, type and concentration of the ions in the extracting solution (White, 1969) and the position of pH electrodes in a soil-liquid suspension (Sparks, 1984). The latter is usually referred to as the 'suspension-effect' and is manifested by the recording of a higher pH when the electrodes are placed in the supernatant solution than when immersed in the sedimented soil particles. In acid soils, the pH differences observed according to the type and concentration of ions in the extracting medium arise from H^+ exchange and the displacement and hydrolysis of aluminium from soil surfaces (Bache, 1974).

No universal method of pH determination has evolved because of the great diversity of soils. However, it appears that there is general agreement that the most suitable medium for measuring soil pH is that which most closely simulates the environment encountered by a plant root (Schofield and Taylor, 1955; Peech, 1965; White, 1969; Russell, 1973 Bolt *et al.*, 1976; Gillman, 1981). Even though the effect of ionic strength on soil pH and buffering is well documented, there has been little published work to establish if the ionic strength of the solution used for pH measurement bears any relation to the average ionic strength of the soils under investigation. For example, soil pH is commonly measured either in distilled water or 0.01 M $CaCl_2$ (ionic strength, $I = 0.03$) in Australia. Gillman (1981) and Dolling and Ritchie (1985) have found that neither solution approximated the ionic strength of the soils they studied and that soil pH values determined in the two solutions differed from the pH measured at the average ionic strength of the soils at field capacity. Greater effort should be made to ascertain the characteristics of the soil solution in order to select the most appropriate technique for pH measurement.

B. Soil pH buffering

In the presence of a given number of H^+ ions, the pH buffering capacity is the major soil property that determines the soil solution pH and our ability to manipulate the pH of acid soils easily and economically. The many different origins of buffering will now be considered separately.

1. *Buffering by the soil solution*

The soil solution not only contains a certain quantity of H^+ ions, but also has the ability to resist pH changes. According to the Bronsted-Lowry theory, the acidic nature of any element in an aqueous environment may be represented by its ability to repel H^+ from the ordered water molecules immediately surrounding it in solution (i.e. an acid is a proton donor and a base is a proton acceptor). The greater the ability of an element to repel H^+ (hydrolyse), the stronger its acidity:

$$P(H_2O)_4^{5+} \leftrightharpoons PO_2OH(H_2O) + 5H^+$$
$$Al(H_2O)_6^{3+} \leftrightharpoons AlOH(H_2O)_5^{2+} + H^+$$
$$S(H_2O)_4^{4+} \leftrightharpoons SO_3(H_2O) + 4H^+$$

For the examples given above, their ability to repel H^+ is in the order $S > Al > P$. All cations are acids of varying strength according to the pH at which hydrolysis becomes significant. Aluminium hydrolyses at pH ~ 4 and is a stronger acid than copper which hydrolyses at pH ~ 6.5. The buffering ability of strong acids increases with concentration and is a minimum at pH 7. It is only significant at low pH values (< 3) and would be a negligible contribution to the total soil buffering at the low concentrations of strong acids found in the soil solution. For weak acids, buffering increases with the total concentration of the acid, [HA] and the conjugate base, $[A^-]$. Their ability to buffer pH is only significant in the pH range in the immediate vicinity of the pH equal to the negative log of the constant (pK) describing the proton donating reaction (i.e. pH range $\sim$ pK $\pm$ 1), e.g.:

$$H_2CO_3 \leftrightharpoons HCO_3^- + H^+ \qquad pK = 6.31$$

(approximate range of pH buffering: 5.3–7.3)

Therefore, the ability of soluble soil components to buffer pH changes will depend on the initial pH of the soil solution, the total concentration of the cation involved and the pK of dissociation. The greater the number of soluble weak acids with different pK values, the wider the pH range over which buffering will be observed.

In the soil solution the two most important buffers are dissolved CO_2 and organic acids. The buffering ability of dissolved CO_2 is only appreciable at pH > 5.5 and therefore would be insignificant in most acid soils unless waterlogging had built up CO_2 levels. Soluble organic acids have been shown to have a range of pK values with an average value of ~ 5.5 (Posner, 1964) and therefore could contribute a significant proportion of the total soil buffering capacity, particularly in sandy soils. However, there is evidence that complexing with metal ions can reduce the buffering ability of organic acids (van Breemen and Wielemaker, 1974a; Hargrove and Thomas, 1982a).

2. *Buffering by mineral dissolution*

pH buffering by the solid phase may arise from the dissolution of minerals such as kaolinite, smectites or calcite (van Breemen and Wielemaker, 1974a,b; van Breemen *et al.*, 1983). For example, protons added to a gibbsite system are consumed by the dissolution of the mineral. This buffering may be counterbalanced by hydrolysis of the soluble cations *if* the H^+ produced is not removed by leaching. Consequently, the buffering ability of gibbsite and kaolinite (which contain the fairly strong acid, Al^{3+}) is far less than for minerals containing a greater proportion of weak acid cations (e.g. smectites, illites and feldspars which contain Ca^{2+}, Mg^{2+}, Na^+ and K^+). The buffering ability of a mineral increases with its solubility product. The more soluble the mineral, the greater the number of cations released and H^+ ions consumed during dissolution.

In the case of multi-component systems, a situation may arise in which the system has infinite buffering. According to the phase rule: $F = C + 2 - P$, the more phases (P) within a system, the fewer independent variables (F) (e.g. temperature, CO_2 partial pressure) that have to be held constant so that the removal of a component (C) does not disturb the equilibrium, i.e. change the pH. The soil consists of many different components and phases but the likelihood of these being in equilibrium simultaneously is quite low. Therefore, the conditions required for infinite buffering will probably only occur in soils with very high mineral contents such as calcareous or acid sulphate soils (van Breemen, 1976).

3. *Buffering by permanent charged surfaces*

The role of permanently charged sites on clay minerals in pH buffering has been a major area of research for many years, presumably because of the high incidence of these types of clay minerals in soils of the temperate zones where the work was initiated. Several comprehensive reviews of the research have been published (e.g. Russell, 1973; Bolt *et al.*, 1976; Coleman and Thomas, 1984) so it will be dealt with only briefly here. The main mechanism of pH buffering by permanent charge sites is exchange between the surface and solution with other cations. Therefore buffering will depend on the type and concentration of the ions on the surface and in solution as well as the type, structural characteristics and relative affinity of the surface for different ions (Russell, 1973).

In the presence of excess electrolyte, the titration curve of a H^+-saturated clay titrated with a strong base can be similar to that of a strong acid if all the H^+ ions on exchange sites are displaced by the cation of the electrolyte. However, if aluminium still remains on some exchange sites it will titrate as a weak acid because of the hydrolysis of the adsorbed aluminium. Also, if the solution has a low ionic strength, then the diffuse layers of positive edge sites and some of the negative planar surface sites will overlap, mutually neutralising their attractive forces so that ion adsorption is decreased. The addition of base neutralises the positively charged sites thereby releasing negative sites to adsorb H^+ ions. Consequently, clay appears to have better buffering and a weak acid character at $pH < 7$ (Schofield, 1949).

H^+-saturated clays are unstable and revert to an Al^{3+} saturated form (Russell, 1973; Coleman and Thomas, 1984) by partial dissolution of the clay lattice followed by adsorption of the released Al^{3+}. Therefore their incidence and contribution to total buffering in soils will be relatively low.

4. Buffering by variable charge surfaces

The recognition of the importance of variable charge surfaces in the last 10 years has led to a considerable amount of research into the adsorption of H^+ and OH^- by hydrous oxides, the edges of clay particles and by organic matter. The emphasis of research has tended to be concerned with the role of variable charge surfaces in nutrient availability (particularly phosphorus). However, the behaviour of these sites in relation to soil pH and buffering is no less important and is an integral part of our overall understanding of variable charge soils.

Several comprehensive reviews have been published on the behaviour of variable charge systems (Arnold, 1978; Bell and Gillman, 1978; Greenland and Mott, 1978; Theng, 1980; Uehara and Gillman, 1981; Bolt and Van Riemsdijk, 1982; Barrow, 1985).

a. Hydrous oxides and the edges of clay particles: The development of pH-dependent, positively and negatively charged sites (Fig. 8) and the contribution of different hydrous oxides and clay minerals to the total

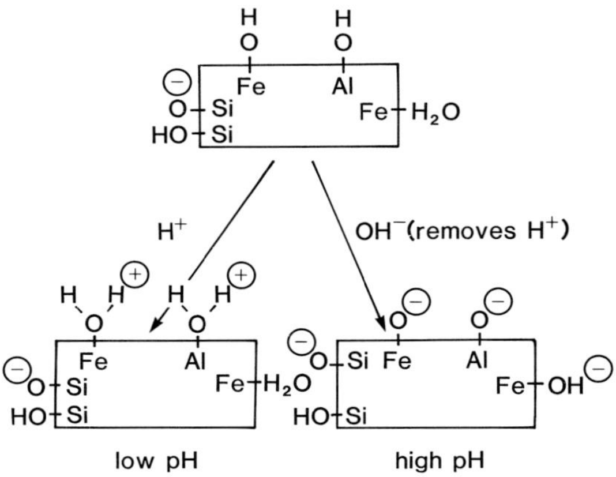

Fig. 8. Schematic representation of the development of positively and negatively charge sites on surfaces of soil constituents, at low and pH (taken from parfitt, 1980)

variable charge in a soil are well documented (Greenland and Mott, 1978; Bowden *et al.*, 1980; Stumm and Morgan, 1981) and so this discussion will concentrate on the soil properties that affect how much H^+ or OH^- is adsorbed and how this affects pH and soil buffering.

Charge development is due to protonation or deprotonation of hydroxyl, oxide or H_2O groups exposed at the surface of hydrous oxides and the edges of clay minerals. It may also be looked upon as the adsorption of OH^- and H^+ ions. Given a certain amount of a particular hydrous oxide or clay mineral, the extent to which H^+ and OH^- ions are adsorbed will depend on the pH, type and concentration of ions in the soil solution and the difference in affinity of the surface for H^+ and OH^- ions. Considering surface properties first, the pH at which a surface adsorbs equal amounts of H^+ and OH^- is termed the isoelectric point (iep) (Parks, 1965). If a surface had equal affinities for H^+ and OH^- then the iep would be at pH 7.0. However, if its affinity for H^+ is greater than that for OH^- (i.e. it repels OH^-), the iep would be at pH values > 7. Therefore gibbsite, which has a higher affinity for H^+ than OH^-, has an iep of ~ 9.0. On the other hand, silica has an iep of 2.0 because it has a more acidic nature than gibbsite (repels H^+ more easily) and therefore has a greater affinity for OH^- than for H^+. The iep will also be affected by the degree of hydration and crystallinity of the hydrous oxide (Parks, 1967).

The effect of solution properties on H^+ and OH^- adsorption by surfaces is quite complex (Bowden *et al.*, 1980). Briefly, at a constant pH

$<$ iep, the adsorption of H^+ increases with ionic strength and the charge of the ions in the soil solution. The same applies for OH^- adsorption at pH $>$ iep. Ions such as HPO_4^{2-} and F^- are capable of displacing surface groups (Parfitt, 1978) and in doing so usually make the surface more negative, e.g.:

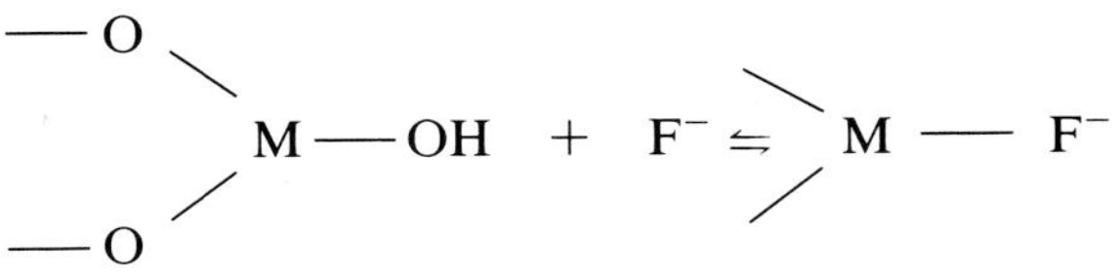

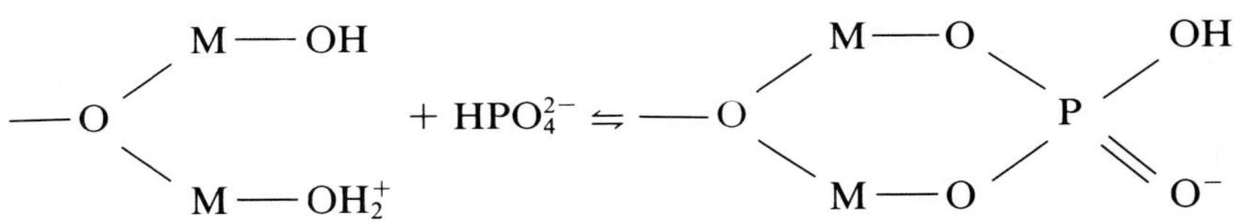

When this type of reaction occurs (usually termed ligand exchange), the pH at which the net surface charge is zero may be referred to as the point of zero charge (pzc) rather than the iep (Sposito, 1981a).

The buffer capacity of a variable charge surface is at a minimum at the pzc and increases at more distant pH values on either side of the pzc (Uehara and Gillman, 1981). Therefore the amount of buffering by a particular surface under constant solution conditions will depend on the difference in pH of the soil solution and the pzc.

At constant pH, buffering will increase with ionic strength and valence of the ions in the soil solution. A typical way of depicting the variation in H^+ and OH^- adsorption in soils is to draw several curves of H^+/OH^- adsorbed vs. pH for different constant ionic strengths on the same axes (Fig. 9). The crossover point of the curves is sometimes referred to as the point of zero salt effect (pzse) and represents the pH at which H^+/OH^- adsorption does not change with ionic strength. If adsorption of other ions does not occur (unlikely in soils) or their adsorption is independent of ionic strength (e.g. adsorption has reached saturation) then pzse = pzc (Sposito, 1981a).

Constant pH and ionic strength are not common phenomena in soils, however, and so no single curve at constant ionic strength will represent soil behaviour in a natural environment. Rather, the curve and point that represent the H^+/OH^- adsorption at a certain moment will change in response to inputs and outputs to the soil system that affect pH, ionic strength and the concentration of specifically adsorbed ions (e.g. lime and fertilizer applications). Uehara and Gillman (1981) give an example of the effect of the addition of fertilizer on variable surface charge which also illustrates pH buffering (Fig. 9). A temporary increase in the ionic strength (by the addition of potassium chloride) at pH > pzc increases OH^- adsorption or deprotonation of surface water molecules. This leads to a drop in solution pH which, in itself, results in a decrease in OH^- adsorption, i.e. if it were possible to observe each step in sequence ($A \rightarrow B \rightarrow C$, Fig. 9) the OH^- adsorption would change from $c_1 \rightarrow c_2 \rightarrow c_3$. If the fertilizer were superphosphate, the pH changes observed would be different and more difficult to interpret because phosphate is specifically adsorbed. In this case surface charge may not only be due to OH^- and H^+ ions but also due to phosphate ions. Phosphorus appears to react with hydrous oxides by displacement of OH^- and water molecules in the surface (Parfitt, 1978; White, 1980). If phosphate ions displace hydroxyl ions they tend not to increase the surface charge as much as when they displace water molecules (Bolan and Barrow, 1984). Consequently, phosphorus adsorption can lead to a decrease in the pzc (due to increased negative charge) and an increase in solution pH (due to hydroxyl displacement). If this is accompanied by a temporary increase in ionic strength, a similar sequence of changes in hydroxyl adsorption will occur as described for potassium chloride. However, in the case where OH^- is displaced by phosphate ions, the net pH drop will not be so great. Also, pH increases due to adsorption may be counterbalanced by the acidic nature of the phosphorus source (e.g. superphosphate) or deprotonation of $H_2PO_4^-$ (Bowden *et al.*, 1980).

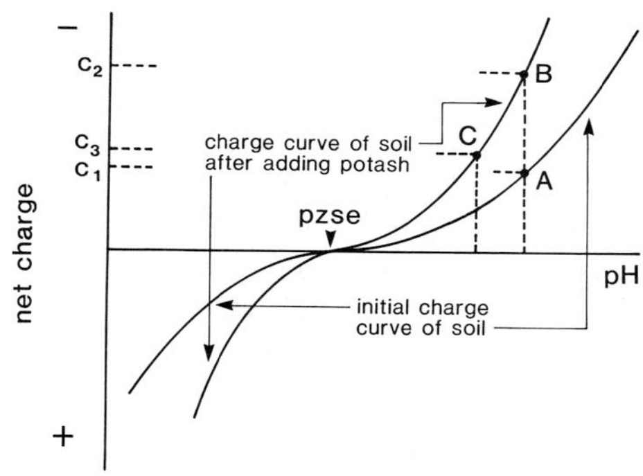

Fig. 9. Changes in pH and the net charge of soil particles (i.e. OH^-/H^+ adsorption) after fertilizing with potash (adapted from Uehara and Gillman, 1981).

b. Organic matter: Buffering by organic matter in the solid phase is similar to its behaviour in solution (Section III BI). Its ability to affect soil pH and buffering will depend upon whether it accumulates in the soil as an acid or a conjugate base (Helyar, 1976; Ritchie and Dolling, 1985). The factors which affect this are discussed more fully in the following chapter.

5. The relative importance of soil components in pH buffering

The buffering ability of different soil components has been discussed separately; now it is necessary to consider their relative importance in the soil as a whole.

The concepts presented so far are those which have been identified from studying individual minerals in pure systems where pH and ionic strength may be controlled easily. In the soil, it is far more difficult to distinguish between these effects and predict responses to changes brought about by fertilizer applications or liming. Soils contain many different variable charge minerals, and H^+/OH^- adsorption is continually changing because of the natural dynamic variation in the factors that affect it. Therefore, one must be cautious in the application of these concepts to the soil in its natural environment. However, we can observe general, overall trends when considering the effects of surface and solution properties on H^+/OH^- adsorption with respect to soil pH buffering.

In general, the lower pzc of a soil with respect to its soil pH, the better buffered the system will be *against pH rises*. The pzc values of most soils appear to be at pH values < 7 (Barber and Rowell, 1972; van Raij and Peech, 1972; Gallez *et al.*, 1976; Fox, 1982; Iñiguez and Val, 1982, Hendershot and Lavkulich, 1983; Wada and Okamura, 1983) even for soils which have reasonably high contents of minerals which in the pure state have pzc values > 8 (e.g. gibbsite and goethite). This is because of the presence of minerals with permanent negative charges and organic matter which increase the total negative charge of the soil. Higher H^+ concentrations are required to neutralise the extra negative charge and therefore the pzc occurs at a lower value than would be anticipated by considering the hydrous oxide content. In addition, the presence of impurities, surface hydration or amorphous coatings of silica can lower the pzc of iron and aluminium oxides (Parks, 1965; Pyman, 1978). Complex mixtures or co-precipitates of silica with alumina or goethite which have pzc values between those of the two pure oxides (depending on the Si:Fe/Al ratio) could also be present (Greenland and Mott, 1978; Pyman, 1978).

The previous discussion has indicated that the buffering ability of different types of surfaces varies with pH, i.e. pH is not a linear function of the amount of alkali added. This is not always the case, however, when the surfaces are present as a complex mixture in soils. Even though Fox (1982) found that buffering of three Hawaiian soils changed markedly with pH, others have observed very little variation in buffering with pH (e.g. Coleman and Thomas, 1964; Bolt *et al.*, 1976; Ritchie and Dolling, 1985), particularly at pH values < 6 and low ionic strength (< 0.005). Also, it is not always possible to correlate the amount of buffering with the content of a particular type of surface (e.g. percentage clay or organic matter) because exchange sites on clays and organic matter which are capable of adsorbing H^+ ions may be blocked by iron and aluminium complexes or oxide coatings (Coleman and Thomas, 1964; Bhumbla and McLean, 1965; Hendershot and Lavkulich, 1983).

H^+-saturated clays will not be a major source of pH buffering because of their instability whereas aluminium-saturated clays are important (Russell, 1973; Coleman and Thomas, 1984). The latter was discussed in the section on aluminium.

IV. MANGANESE

In soils, manganese behaves similarly to aluminium in that it may be adsorbed onto the surface of hydrous oxides, clay particles and organic matter or exist as discrete manganese compounds. On the other hand, it differs from aluminium because it may exist in more than one oxidation state under the conditions naturally found in soils. Given a certain total manganese content of the soil, the level of manganese in the soil solution will not only depend on its interaction with soil surfaces but will also be the net result of the rates of oxidation and reduction. This extra dimension to manganese distribution in soils is complicated further by the fact that the redox reactions may be chemically and/or biologically mediated. Consequently, concentrations of soluble manganese are subject to even more dynamic variation than ions such as aluminium.

Since plants take up manganese from the soil solution in the form of Mn^{2+}, this section will consider the factors affecting concentrations of soluble Mn^{2+}. Mn^{2+} levels in solution are governed by the distribution of Mn^{2+} between the solid and liquid phases, the formation of soluble complexes and the extent and kinetics of redox reactions.

A. Manganese in the soil solution

Manganese (II) is the only oxidation state that has been identified in the soil solution even though under certain conditions Mn(III) could exist (Geering *et al.*, 1969). Manganese (II) may react with several inorganic

Table III. Possible reactions of manganese in the soil solution.

1.	$Mn^{2+} + H_2O \rightleftharpoons MnOH^+ + H^+$
2.	$Mn^{2+} + 3H_2O \rightleftharpoons Mn(OH)_3^- + 3H^+$
3.	$Mn^{2+} + 4H_2O \rightleftharpoons Mn(OH)_4^{2-} + 4H^+$
4.	$Mn^{2+} + SO_4^{2-} \rightleftharpoons MnSO_4^0$
5.	$Mn^{2+} + Cl^- \rightleftharpoons MnCl^+$
6.	$Mn^{2+} + CO_2\,(G) + H_2O \rightleftharpoons MnHCO_3^+ + H^+$
7.	$Mn^{2+} + CO_2\,(G) + H_2O \rightleftharpoons MnCO_3^0 + 2H^+$
8.	$2Mn^{2+} + H_2O \rightleftharpoons Mn_2OH^{3+} + H^+$

and organic ligands to form soluble complexes as well as exist in the free form, Mn^{2+} (Table III). In comparison to aluminium, however, it forms much weaker complexes, e.g. hydrolysis does not become significant until pH > 9. In acid soils, therefore, the inorganic complexes will not be major forms of manganese unless there are unusually high levels of sulphate (e.g. acid sulphate soils) or chloride (e.g. saline soils). The importance of soluble organic complexes is not so clear. Laboratory studies have shown that manganese only forms weak complexes with purified fulvic acids (isolated from the acid soluble component of soil organic matter) in comparison to other metals (Schnitzer and Skinner, 1967; Khan, 1969; Schnitzer and Hansen, 1970). Sanders (1983) also found that > 70% of soluble manganese was as the free ion, Mn^{2+}, in five English soils (pH 5.5–7.05). On the other hand, Geering *et al.* (1969) found that > 90 per cent of soluble manganese was complexed with organic ligands in 13 American topsoils and 39–73 per cent was complexed in subsoils. The differences could be due to the low pH and high ionic strengths used in the fulvic acid studies and the apparent lack of control of pH (Geering *et al.*, 1969; Sanders, 1983) and ionic strength (Geering *et al.*, 1969) during measurement of the soluble, complexed manganese by the resin exchange method. Schnitzer and Hansen (1970) found that as the pH increased from 3.5 to 5 and ionic strength decreased from 0.15 to zero, complexes that were more stable formed which contained greater amounts of manganese per mole of organic ligand. The effect of pH is probably due to decreased competition with H^+ (Page, 1962; Stevenson, 1982) and a concomitant increase with ionic strength in the stability constant for the reaction. It would appear that there is now more evidence that manganese in acid soil solutions is mainly in an uncomplexed form.

B. Manganese buffering

Under natural conditions, it is often difficult to distinguish between adsorption and redox reactions as the cause of a change in manganese availability because both may be affected by the same factor (e.g. pH). Nevertheless, soluble manganese levels are ultimately controlled by buffering from the solid phase (Fig. 10) and only modified by redox reactions. Therefore the two types of reactions will be considered in separate sections and the discussion on solid phase reactions assumes the soil is well aerated.

1. Dissolution of compounds

Manganese-containing minerals which have been identified in soils have been reviewed by McKenzie (1977) and Taylor *et al.* (1983). Lindsay (1978) has also considered several compounds which could control levels of manganese in solution.

Until now, birnessite (γ-MnO_2) and lithiophorite ($[Al, Li]\ MnO_2(OH)_2$) have been identified as the most common minerals (Taylor *et al.*, 1983) in which manganese (IV) is the major oxidation state. The minerals are usually very finely grained or poorly crystalline and non-stoichiometric in composition. Therefore it is not surprising that Mn^{2+} levels found in soil solutions have not coincided with those predicted by thermodynamic solubility data of pure compounds even when redox conditions are also considered (Geering *et al.*, 1969; Ponnamperuma *et al.*, 1969; Bohn, 1970). This does not mean that oxides are not an important source of manganese but reflects the difficulty of predicting the availability of manganese. As with all plant nutrients, a high level of soluble Mn^{2+} is not necessary for adequate plant growth, just a source that will readily revert to Mn^{2+}.

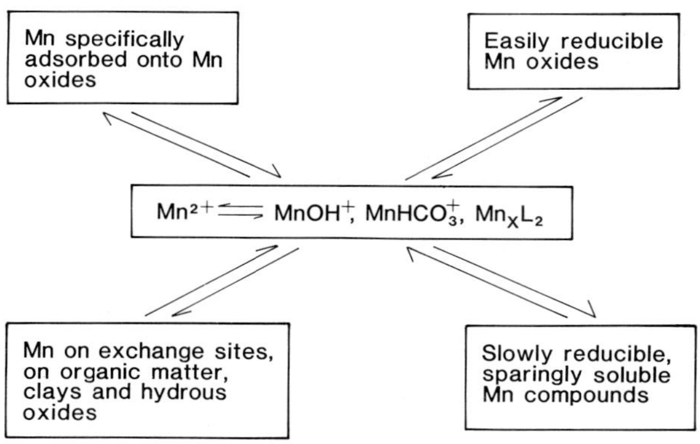

Fig. 10. Manganese distribution in acid soils (pH < 5.5).

2. Adsorption by inorganic minerals

In contrast to aluminium and other metal ions, Mn^{2+} is only weakly adsorbed onto clay minerals (Tiller, 1983). This low adsorption affinity has been attributed to the inability of Mn^{2+} to hydrolyse under acidic conditions. The action of drying can change the mechanism of adsorption onto clays. Reddy and Perkins (1976) found that if clays were subjected to several wetting and drying cycles, then it was more difficult to remove the Mn^{2+} by exchange with I M ammonium acetate than if they were kept moist. They suggested that oxidation/precipitation or interlamellar entrapment upon drying could be responsible for the increased retention. McBride *et al.* (1975) also observed a broadening of the electron spin resonance (ESR) signal of a Mn^{2+}-saturated smectite after drying and attributed it to either partial dehydration of the ion or interactions between neighbouring Mn^{2+} ions.

Regardless of the exact mechanisms, the above research indicates that wetting and drying affects not only redox reactions of manganese but also retention reactions; if wetting and drying were to increase soluble Mn^{2+} levels by reduction of Mn^{4+}, it may be counterbalanced by increased retention by clay minerals.

Similar behaviour has been observed for its adsorption onto aluminium and ferric oxides. McBride (1978) found there was no significant retention of Mn^{2+} by alumina in the pH range 4.6–5.5. At initial concentrations of 10^{-3} M, no Mn^{2+} was adsorbed by haematite and goethite at pH < 5.5 and pH < 5, respectively (McKenzie, 1980). Above these pH values, adsorption occurred and increased quite rapidly with pH (0-20 $\mu mol.g^{-1}$ in approximately 1.5 pH units). On the other hand, manganese oxides have been found to have a much higher affinity for Mn^{2+}. Manganese adsorption onto birnessite increased linearly with pH from 200 $\mu mol.g^{-1}$ at pH 2 to 1600 $\mu mol.g^{-1}$ at pH 6 (McKenzie, 1980). The low pzc ($\sim$ 1.8) of the manganese oxides in comparison to the pzc of ferric oxides ($>$ 6) means that the manganese surfaces will be negatively charged in the acid pH range, making it much easier for hydrolysable metals to be adsorbed onto manganese oxides than onto the positively charged surfaces of ferric oxides.

The high affinity of manganese oxides for Mn^{2+} in comparison to more hydrolysable metals could not be readily explained by just specific adsorption of Mn^{2+} alone or combined with surface induced hydrolysis. Another contributing factor could have been the accumulation of Mn^{2+} and OH^- at the surface resulting in oxidation of Mn (II) to Mn (IV) which would have been catalysed by the oxide surface (Pankow and Morgan, 1981).

Mn^{2+} may also be adsorbed onto calcite surfaces (McBride, 1979), but it is negligible in comparison to adsorption onto hydrous oxides (0.75 μmol.g^{-1} from initial solution concentrations of 10^{-3} M). As might be expected, the mechanism appeared to involve precipitation as well.

3. Adsorption by organic surfaces

Nuclear magnetic and electron spin resonance spectroscopy (NMR and ESR) have shown that under acidic conditions Mn (II) is bound to organic matter by electrostatic forces only (Gamble *et al.*, 1976; McBride, 1978; Bloom and McBride, 1979). Apparently organic matter shows little preference for Mn^{2+} relative to other divalent ions (Passioura and Leeper, 1963a; Bloom and McBride, 1979).

C. Redox reactions of manganese

Manganese (II) may be oxidised to Mn (IV) by both chemical and microbial pathways. The chemical pathway applies to reactions that take place outside plants or microorganisms and are not affected by microbial activity. Microbial pathways are those reactions that rely on microbial activity. Both pathways require an electron sink for oxidation and an electron source for reduction.

$$Mn^{2+} + H_2O \leftrightharpoons Mn^{4+} + 2e^-$$

Microbial oxidation and reduction are two totally separate reactions, whereas chemical oxidation and reduction represent the two extremes of one reaction. Also it is important to note that in soils, even though oxidation may be predominantly biological, it does not mean that reduction is necessarily by the same pathway. Indeed, the inability to explain or predict Mn^{2+} levels in the soil solution probably arises mainly from the oxidation process being microbially controlled, whereas the reduction reaction is mainly chemically controlled.

Superimposed on this complication is the fact that redox reactions are affected by changes in soil properties that can also govern the distribution of manganese between soil surfaces and the solution. Plants, soil pH, moisture content, organic matter, temperature, redox potential and the presence of manganese oxides have all been shown to affect the extent and kinetics of redox reactions (Leeper, 1947; Ponnamperuma, 1972; Pankow and Morgan, 1981). Further, factors that can affect microbial processes may have exactly the opposite effect when the same process is chemically mediated. Some controlling factors can increase both reduction and oxidation, depending on their nature. The factor may act by either changing the *extent* of a reaction, i.e. favouring the oxidised or reduced form or by altering the *kinetics* of oxidation or reduction.

For example, the presence of soluble organic ligands that can complex manganese may retard the *rate* of chemical oxidation of Mn^{2+} (Pankow and Morgan, 1981). On the other hand, it may increase the *extent* of microbial oxidation because of increased microbial activity in the presence of *easily reducible* substrates. Alternatively, increasing organic matter content of soils may increase the extent of manganese reduction (by chemical or microbial pathways) because of an increased source of *easily oxidisable* material.

Consequently, a careful assessment of the major pathway of oxidation and reduction is required before one can assess the effects of changing pH and moisture content, while taking into account the nature of the factor (e.g. type of organic matter, moisture contents below or above field capacity). The amount of manganese available to plants in a soil will depend on manganese distribution between the soil surfaces and the soil solution and how that is modified by the difference in the rates of oxidation and reduction which are governed by environmental conditions, soil properties and microbial activity.

D. Reduction

The chemical reduction of Mn (IV) requires the presence of H^+ ions and a source of readily oxidisable material, usually organic matter (Zende, 1954):

$$MnO_2 + 4H^+ + 2e^- \rightleftharpoons 2H_2O + Mn^{2+}$$

Chemical reduction is favoured by low pH, high levels of oxidisable organic material, increasing temperature and the presence of moisture. Unlike microbial reduction, it can still continue in the absence of moisture, albeit more slowly (Leeper, 1947; Zende, 1954).

Microbial reduction occurs when there is a lack of a more readily oxidisable material than manganese oxides and an anaerobic microbial population. It is enhanced by soil moisture contents > field capacity, low redox potentials and sufficient substrate (organic matter) to maintain microbial activity (Ponnamperuma, 1972).

The increase in available manganese with air-drying and decreasing pH indicates that chemical reduction is usually the more important pathway in soils (Leeper, 1947; Zende, 1954; Jones, 1957b; Harter and McLean, 1965; Grasmanis and Leeper, 1966; Ingols and Enginun, 1968;

Patrick and Henderson, 1981; Goldberg and Smith, 1984; Inubushi *et al.*, 1984; Sparrow and Uren, 1984). Microbial reduction would be limited by lack of moisture and low pH values whereas chemical reduction could continue slowly given a supply of H^+ ions and electrons (i.e. organic matter). The release of Mn (II) oxides by dehydration of mixed Mn (II) and (IV) oxides has also been postulated as a cause of increased manganese availability with drying (Fujimoto and Sherman, 1948). While this mechanism may seem unnecessarily complicated (Zende, 1954), it cannot be ignored completely because of the high incidence of mixed oxides in soils (Taylor *et al.*, 1983). The biological pathway for reduction may become equally important under certain environmental conditions (Pal *et al.*, 1979; Mäntylahti, 1981), but will be of an inconsistent nature.

E. Factors affecting reduction

Both microbial and chemical reduction may be enhanced by an increase in readily oxidisable organic matter, moisture content or temperature and a decrease in redox potential.

1. Redox potentials

Theoretically, redox potentials of 200 mV are required for Mn (IV) to be reduced (Ponnamperuma, 1972). In soils, however, manganese reduction has been observed at 500 mV and it is unclear whether redox potentials are the cause or the result of manganese reduction (Mantylahti, 1981; Patrick and Henderson, 1981). Research into redox effects on soluble manganese is limited by difficulties in controlling and maintaining given reducing conditions whilst achieving a constant pH (Gotoh and Patrick, 1972). Even though increases in soluble manganese with decreasing redox potentials have been observed, the levels always tend to be lower than those predicted by thermodynamic principles (Bohn, 1970; Gotoh and Patrick, 1972). As already discussed, the rare occurrence of pure minerals in soils, the kinetics of redox reactions and the microbial as well as chemical pathways for reduction are all contributing factors to the lack of agreement between practical observations and theoretical predictions.

Mn^{2+} released into solution by mineral dissolution may also be adsorbed on to exchange surfaces (Ponnamperuma, 1972; Sims and Patrick, 1978). Such a phenomenon could help explain the gradual increase and then decrease in soluble manganese after flooding a soil (Ponnamperuma, 1972) or the apparently slow equilibration of manganese levels after a change in reducing conditions. The duration of reducing conditions affects soluble manganese levels because the rates of oxidation and reduction in soil systems are not very fast (Mantylahti, 1981) Patrick and Henderson (1981) found that equilibrium was only achieved after two weeks at a constant redox potential and that the rate of oxidation was slower than the rate of reduction. The likelihood of reducing conditions remaining constant over such a period is only possible in flooded soils such as used for rice growing. Patrick and Henderson (1981) attempted to distinguish between chemical reduction and a combination of chemical plus biological reduction by comparing manganese levels in manganese-water and manganese-soil-water systems. They concluded that chemical reduction depended on pH but not the redox potential, Eh, whereas the combination of chemical and microbial reduction was pH and Eh dependent. However, the soil-free system was not aseptic and therefore their experiments could well have been comparing chemical plus microbial reduction in the absence and presence of soil constituents which may act as catalysts for redox reactions or sinks for the products of the reactions.

2. Organic matter

The presence of organic matter has been shown to increase manganese availability (Leeper, 1947; Christensen *et al.*, 1950; Zende, 1954; Miller and Ohlrogge, 1958; Sanchez and Kamprath, 1959; Lamm, 1960; Ghanem *et al.*, 1971). It is not always clear whether the effect is due to increased solubility by complexing, due to the organic matter acting as a chemical reducing agent or due to stimulating microbial reduction. So far, the evidence indicates that organic matter acting as a chemical reducing agent is probably the major mechanism because soluble organic complexes are probably not as prevalent in acid soils as originally thought and chemical reduction is more common than microbial reduction. However, increasing organic matter does not always indicate increasing manganese availability because the nature of the material is also important i.e. it must be easily oxidisable. For example, Katyal (1980) found that the addition of farmyard manure to soils did not increase manganese availability to plants. Also, more Mn^{2+} may be released by increased reduction, but its availability to plants may be limited by its adsorption onto soil surfaces.

The ability of soils low in organic matter to provide sufficient manganese for plant growth (Bromfield, 1958) led to research which has shown that plants may also act as a source of organic materials by excreting organic reducing agents (Bromfield, 1958; Godo and Reisenauer, 1980; Uren, 1981). The solubility of γ-MnO_2 is enhanced by the

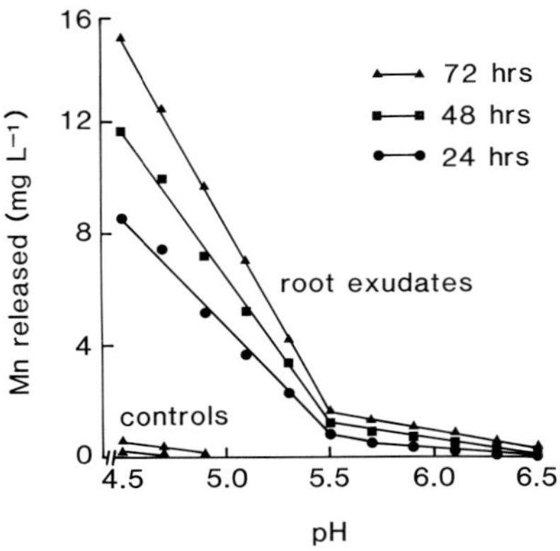

Fig. 11. Dissolution of γ-MnO_2 by root exudates and control solutions at different acidities and times (taken from Godo and Reisenauer, 1980).

presence of root exudates (Fig. 11), the effect being greater at low pH values and after long reaction times (Bromfield, 1958; Godo and Reisenauer, 1980). In some soils, however, the effect is not as great because of concurrent adsorption onto surfaces.

The increase in solution levels of Mn^{2+} by reduction depends on the breakdown pathway of the root exudate. Jauregui and Reisenauer (1982) have suggested that if the root exudate is adsorbed onto soil particles, the intermediate organic acids produced during breakdown of the exudate are more easily oxidised than those produced in the absence of adsorption. As a result, higher levels of soluble Mn^{2+} are observed when root exudates are adsorbed. The effect was enhanced at low pH values but it could not be established whether the increases were due to the low pH favouring reduction or increasing adsorption of breakdown products. The addition of MnO_2 to an iron-coated sand increased soluble manganese and was attributed to iron adsorption onto MnO_2 displacing Mn^{2+}. However, McKenzie (1980) found that very little Mn^{2+} is released when metals are adsorbed on manganese oxides (released manganese was $< 0.05\%$ of the amount of metal adsorbed). Therefore, the increased soluble Mn^{2+} levels observed by Jauregui and Reisenauer (1982) probably arose from increased adsorption of breakdown products leading to production of more easily oxidisable carboxylic acids and hence increased manganese reduction.

The importance of root exudates in increasing manganese availability is difficult to ascertain and may not be consistent. For example, the ability of exudates to increase manganese availability will depend on the incidence of contact between the manganese and the root exudate.

Leeper (1947), Passioura and Leeper (1963b) and Goldberg *et al.* (1983) have all found evidence that compaction can lead to an increased possibility of contact and therefore increased manganese availability. If the exudates are to be effective in increasing nutrient availability they must come into contact with the manganese and do so reasonably quickly before they are broken down by ordinary microbial activity.

3. *pH*

The importance of the effect of pH on chemical reduction has already been mentioned but its real effect in soils can be difficult to establish because of the many concomitant effects of pH on manganese availability (e.g. decreasing pH lowers adsorption onto hydrous oxides). Often, pH effects on manganese availability are initially manifested indirectly as effects of agricultural practices. For example, Goldberg *et al.* (1983) found that ammonium sulphate increased manganese availability by decreasing pH and the effect was accentuated in regions where the soil was compacted or shallow seeding increased plant contact with the acidifying fertilizer. The pH decrease was attributed to the acidifying effect of ammonium sulphate *per se* and to NH_4^+ uptake leading to H^+ excretion. However, Uren (1981, 1982) found that the pattern of acid production by plant roots did not coincide with the extent of reduction of MnO_2 and therefore was not a major contributor to increasing manganese availability.

4. *Moisture content*

The moisture regime of a soil can have an important effect on chemical and microbial reduction of Mn. Manganese toxicity is most commonly attributed to increased moisture contents causing a lowering of reduction potentials and hence an increase in soluble Mn^{2+}.

However, it is not always well appreciated that low reduction potentials only occur if the moisture content is greater than field capacity and exists for an appreciable time. Low redox potentials and excess moisture are not prerequisites for manganese reduction. Also, a previously flooded soil may not necessarily have high levels of soluble Mn^{2+} if it is well drained because the released Mn^{2+} is lost by leaching (Leeper, 1947; Sparrow and Uren, 1984).

5. *Temperature*

Increasing the temperature will increase the rate of chemical and microbial reduction according to the Arrhenius' equation (Castellan,

1970) up to approximately 30°C. At higher temperatures microbial reduction would be affected by the ability of the organisms to survive. The effects of temperature and moisture are closely interrelated. At the same temperature, reduction during a wet period would be greater than reduction during a dry period because both microbial and chemical reduction would be occurring in wet conditions whereas only chemical reduction would occur during the dry period. If the temperature during the dry period is raised, however, the rate of chemical reduction would be increased and could approach the combined rate of chemical and microbial reduction that occurred during a colder, wet period.

F. Oxidation

Microbial oxidation requires the presence of manganese oxidising microbes (which may be autotrophs, i.e. use CO_2 as their source of carbon; Gray and Williams, 1971) and suitable conditions for the active oxidation of Mn (II) (Bromfield, 1974).

Chemical oxidation of manganese is the reverse reaction to reduction and is therefore favoured by high pH values and low organic matter contents. The reaction is catalysed by the presence of manganese oxides and retarded by the presence of a complexing ligand if it is present at high enough concentrations or forms a particularly stable complex (Pankow and Morgan, 1981). Even so, the rate of chemical oxidation at $pH > 7$ is much quicker than microbial oxidation (hours rather than days). At lower pH values, this may not be so (Collins and Buol, 1970).

In acid soils, it would appear that microbial and chemical oxidation are both limited by low pH. Even so, microbial oxidation is the major pathway (Leeper, 1947; Grasmanis and Leeper, 1966; Uren and Leeper, 1978).

G. Factors affecting oxidation

Oxidation is also affected by pH, redox conditions, moisture and temperature regimes and the presence of organic matter.

Many of the organisms involved are apparently pH sensitive, being active between pH 5.8 and 7.8 and then only if the decomposition products from the organic substrate do not alter the pH to outside the range (Bromfield, 1974). Some manganese oxidising bacteria are quite active outside the pH range 5.8–7.8 but use alternative sources of energy/oxygen. Therefore, the manganese oxidising ability of a soil may be under- or over-estimated if it is not measured at the natural soil pH (Bromfield, 1974).

There is some evidence that autotrophic organisms may play a part in manganese oxidation since CO_2 has been shown to stimulate microbial oxidation of manganese in addition to the effect of lowering the pH of alkaline soils into the optimum range for microbial activity (Meek *et al.*, 1973; Uren and Leeper, 1978). This means that oxidation can continue in soils with low redox potentials provided an autotrophic microbial population exists and conditions are suitable for active Mn (II) oxidation.

Drying out of the soil would result in a decrease in microbial oxidation whereas increased temperatures (up to 30°C) in the presence of sufficient moisture would increase microbial activity. The presence of organic matter can have the indirect effect of increasing oxidation if it is a suitable microbial substrate.

H. Liming and manganese behaviour in soils

Liming has been shown by numerous workers to decrease manganese availability (Christensen *et al.*, 1950; Sanchez and Kamprath, 1959; White, 1970; Gupta, 1972; Mantylahti, 1981). In soil, liming may affect the availability of manganese by changing the rates and extent of oxidation and reduction in favour of the former and by increasing reactions of manganese with the solid phase. The magnitude of the effect depends on the amount and nature of soil organic matter, moisture content and the change in pH.

In contrast, Jones (1957a) actually observed an increase in manganese availability with one soil after liming and attributed the effect to a greater retardation in the rate of oxidation in comparison to the reduction rate. It was suggested that reduction was enhanced to such an extent by a large source of oxidisable organic matter that changing the pH had little effect on the reduction rate. Christensen *et al.* (1950) made similar observations after adding lime and organic matter (readily oxidisable sugar) to a soil. The organic matter catalysed the reduction processes to a greater extent than the lime increased the oxidation rate and so an increase in exchangeable manganese was noted. However, it was a temporary effect, only lasting until all the organic matter had been oxidised. Similarly, the addition of manganese to an acid soil high in organic matter increases exchangeable manganese unless lime is added as well (Sanchez and Kamprath, 1959).

In some soils, Mn^{2+} levels at pH 7 have been found to be lower than at pH 8 (Jones, 1957a) because Mn^{2+} was rapidly oxidised at pH 7, whereas the organisms were inactive at pH 8 (Bromfield, 1978). In air dry soils, however, Mn^{2+} was greater at pH 7 than at pH 8 because chemical reduction continued but microbial oxidation was at a minimum due to the lack of water.

I. Soil tests for manganese

The previous discussion has concentrated on the different forms and chemical behaviour of manganese in soils in relation to soluble Mn^{2+} levels and ultimately manganese availability. The majority of research into suitable soil tests for predicting manganese availability to plants has concentrated on deficiency rather than toxicity problems. Nevertheless, discussion of the work is still relevant in the search for an adequate test to predict manganese toxicity. If a soil test is a good indicator of adequate levels of manganese for plant uptake, then the assumption that it will also be good at indicating toxic levels is a better criterion for its choice for further studies than the arbitary choice of any extraction method.

As with aluminium, the more direct studies concerned with manganese availability may be divided into two broad categories: those which have attempted to ascertain the forms of soil manganese by using individual extracts or a series of chemical extracts; and those which have related the manganese extracted by a certain method to plant growth or manganese uptake. The literature has previously been reviewed by Leeper (1947), Viets and Lindsay (1973) and Uren and Sparrow (1983).

Uren and Sparrow (1983) pointed out that the available sources change with pH and therefore the most suitable extractant may also vary with pH. Assuming that total soil manganese is not naturally low, sufficient manganese usually exists in the soil solution at pH < 5 but at less acid pH values (5–6), soluble manganese decreases and the plant must rely more upon exchangeable manganese as well, i.e. the soil's ability to maintain an adequate supply. As the pH increases further (> 6), easily reducible forms of manganese become an important source. In addition, different plant species may be able to influence the availability of manganese in the same soil to various extents by their ability to increase manganese reduction by exudation of reducing agents or to indirectly decrease manganese oxidation in the rhizosphere by decreasing soluble Mn^{2+} levels. Finally, manganese uptake may be affected by interactions with, or deficiencies of, other nutrients (Shuman *et al.*, 1980). These factors are illustrated by: (1) the change in distribution of manganese in soil pools with pH (Sims *et al.*, 1979); (2) the variation in correlation between manganese content of soyabeans and manganese extracted by different methods from soils of varying pH (Shuman and Anderson, 1974); (3) near neutral soils that can supply adequate manganese for plant growth even though very little exists in a soluble form (Bromfield, 1978); and (4) the inability of the same soil test to predict manganese uptake by both wheat and soyabean (Shuman and Anderson, 1974).

Even though soil preparation (i.e. sieving and crushing the soil) limts the ability of all soil tests to account for the available surface area of certain nutrients, Uren and Sparrow (1983) have suggested that preparation could be a particularly important determinant for manganese because of the major role of redox reactions in manganese availability and the presence of most manganese oxides as non-stoichiometric fine grains, low in crystallinity. This is substantiated by the work of Jones and Leeper (1951) who found that the crystallinity and surface structure of synthetic oxides affected the availability of manganese to plants. Manganese was more available from less ordered oxides and those with large surface areas, but availability did not depend on the degree of oxidation.

Several different extractants have been used to determine the amount of manganese in particular forms (e.g. Hammes and Berger, 1960b; Randall *et al.*, 1976; Salcedo and Ellis, 1979). However, the specificity of some of these extractants is questionable (Shuman, 1983), particularly when used in a sequence as a fractionation scheme. In some cases it would appear that the amount of manganese extracted by a particular extractant in a fractionation scheme is more related to its position in the extraction sequence than to a particular form of manganese. For example, Shuman (1979; 1983) pointed out that hydrogen peroxide extracts manganese from manganese oxides and potassisum pyrophosphate dissolves iron oxides (which could release adsorbed Mn^{2+}) even though these two extractants are commonly used to determine organically bound manganese (e.g. Sims and Patrick, 1978; Sims *et al.*, 1979).

In assessing the ability of different soil tests to predict the manganese status of soils, it has frequently been found that several methods may extract manganese levels that give statistically significant correlations with uptake (e.g. Hoyt and Nyborg, 1971b; Shuman and Anderson, 1974; Salcedo and Ellis, 1979). On this premise, most methods may be considered 'successful' at predicting soil manganese status but some of them will have quite a low accuracy. Consequently, it is important to distinguish between tests that give an accurate correlation with uptake and those that just establish a significant relationship.

Originally, water soluble and exchangeable manganese were the two major forms of manganese investigated in soils. The main extractant of exchangeable manganese has been 1 M ammonium acetate (NH_4OAc) (Sherman *et al.*, 1942; Walker and Barber, 1960; Shuman and Anderson, 1974; Sims *et al.*, 1979), but magnesium chloride (Shuman, 1979), magnesium nitrate (Boken, 1955; Lamm, 1960) and NaCa HEDTA (Uren and Sparrow, 1983) have also been used. The ability of both water soluble and exchangeable manganese to predict manganese availability has proved to be somewhat variable (Shuman and Anderson, 1974; Randall *et al.*, 1976). Gupta (1972) and Roorda van Eysinga *et al.* (1978) found water soluble manganese to be a good predictor of available manganese whereas Hoyt and Nyborg (1971b) and Salcedo and Ellis (1979) found it to be inadequate.

Possibly because of the above inconsistencies, a method was developed to measure the soil's capacity to reduce manganese quickly and therefore render it available. 'Easily reducible' manganese is usually extracted in 1 M NH_4OAc containing 0.2% hydroxyquinone at pH 7 (e.g. Hammes and Berger, 1960a,b; Shuman and Anderson, 1974; Randall *et al.*, 1976). Unfortunately, results are not necessarily comparable as extraction times vary from 10 min (Hammes and Berger, 1960b) to 8 h (Goldberg and Smith, 1984) even though it has been pointed out that extraction times > 30 min probably extract more than just easily reducible manganese (Leeper, 1947). For this reason, it is not surprising that 'easily reducible' manganese has not been found to be as well correlated with manganese uptake as other soil tests (Hammes and Berger, 1960a; Hoyt and Nyborg, 1971b; Shuman and Anderson, 1974) but there are some exceptions (e.g. Randall *et al.*, 1976).

Diethylenetriaminepentaacetic acid (DTPA) has also been a popular soil test for manganese even though it was originally designed only for neutral or alkaline soils and the initial screening test did not include any manganese responsive soils (Lindsay and Norvell, 1978). Consequently, it has not proved a more reliable indicator of manganese availability than other soil tests when applied to a wide variety of soil types and plants (Shuman and Anderson, 1974; Randall *et al.*, 1976; Salcedo and Ellis, 1979; Salcedo *et al.*, 1979; Goldberg and Smith, 1984). Lopez and Graham (1972) assumed that DTPA extracted manganese from the labile pool because when Mn^{54} was added to the soil, all the radioactivity remained in the solution phase. However, if DTPA removed more than just the labile pool, one would still find 100 per cent activity in solution. Sims *et al.* (1979) found DTPA extracted manganese from the water soluble, exchangeable and reducible fractions of manganese. Others have shown that DTPA extracts manganese from water-soluble, exchangeable and reducible forms (Sims *et al.*, 1979) and from manganese oxides (Shuman, 1983).

Several authors have found that manganese extracted by 0.033 M H_3PO_4 gave the best correlation with manganese uptake for a number of different plants (oats, Hammes and Berger, 1960b; Randall *et al.*, 1976;

soybean, Salcedo *et al.*, 1979; sudangrass, Salcedo and Ellis, 1979; ryegrass, Randall *et al.*, 1976) even though it is buffered at quite a low pH (2.6). On the other hand, Hoyt and Nyborg (1971b) found that 0.01 M $CaCl_2$-extractable manganese was superior to 0.033 M H_3PO_4 for predicting manganese status of 40 acid soils (pH < 5.6). Bromfield *et al.*, (1983) also found that 0.01 M $CaCl_2$-extractable manganese was well correlated with the manganese concentration of rape and subterranean clover grown on acid soils. The manganese concentration of oats has been shown to be well correlated with 0.01 M $CaCl_2$-extractable manganese (Goldberg *et al.*, 1983).

In conclusion, 0.033 M H_3PO_4 appears to have been the most consistent at accurately predicting the manganese status of acid soils; 0.01 M $CaCl_2$ shows promise but requires further testing with more plant species and soil types. Also, the predictive ability of several soil tests may be improved markedly by using multiple regression with the inclusion of soil properties such as pH, manganese oxide content, organic matter, the base ratio (Ca + Mg)/K (Randall *et al.*, 1976; Salcedo and Ellis, 1979; Salcedo *et al.*, 1979; Shuman *et al.*, 1980) and plant properties such as zinc uptake (Dolar *et al.*, 1971).

REFERENCES

Adams, F. and Hajek, B. F. (1978). Effects of solution sulfate, hydroxide and potassium concentrations on the crystallization of alunite, basaluminite and gibbsite from dilute aluminium solutions. *Soil Science* **126**, 169–173.

Adams, F. and Hathcock, P. J. (1984). Aluminium toxicity and calcium deficiency in acid subsoil horizons of two Coastal Plains Soil Series. *Soil Science Society of America Journal* **48**, 1305–1309.

Adams, F. and Lund, Z. F. (1966). Effect of chemical activity of soil solution aluminium on cotton root penetration of acid subsoils. *Soil Science* **101**, 193–198.

Adams, F. and Moore, B. L. (1983). Chemical factors affecting root growth in subsoil horizons of coastal plain soils. *Soil Science Society of America Journal* **47**, 99–102.

Adams, F. and Rawajfih, Z. (1977). Basaluminite and alunite: a possible cause of sulfate retention by acid soils. *Soil Science Society of America Journal* **41**, 686–692.

Alva, A. K., Edwards, D. C., Asher, C. J. and Blamey, F. P. C. (1986). Effects of phosphorus/aluminium molar ratios and calcium concentration on plant response to aluminium toxicity. *Soil Science Society of America Journal* **50**, 133–137.

Arai, S. and Kumada, K. (1981). Factors controlling stability of Al-humate. *Geoderma* **26**, 1–12.

Arnold, P. W. (1978). Surface-electrolyte interactions. *In* 'The Chemistry of Soil Constituents'. (D. J. Greenland and M. H. B. Hayes, eds.), pp. 355–404. Wiley Interscience, Chichester.

Bache, B. W. (1963). Aluminium and iron phosphate studies relating to soils. I. Solution and hydrolysis of variscite and strengite. *Journal of Soil Science* **14**, 113–123.

Bache, B. W. (1974). Soluble aluminium and calcium-aluminium exchange in relation to the pH of dilute calcium chloride suspensions of acid soils. *Journal of Soil Science* **25**, 320–332.

Baes, C. F. and Mesmer, R. E. (1976). 'The Hydrolysis of Cations'. Wiley Interscience, New York.

Barber, R. G. and Rowell, D. L. (1972). Charge distribution and the cation exchange capacity of an iron-rich kaolinitic soil. *Journal of Soil Science* **23**, 135–146.

Barrow, N. J. (1985). Reactions of anions and cations with variable-charge soils. *Advances in Agronomy* **38**, 183–230.

Bartlett, R. J. and Riego, D. C. (1972). Effect of chelation on the toxicity of aluminium. *Plant and Soil* **37**, 419–423.

Bell, L. C. and Gillman, G. P. (1978). Surface charge characteristics and soil solution composition of highly weathered soils. *In* 'Mineral Nutrition of Legumes in Tropical and Subtropical Soils'. (C. S. Andrew and E. J. Kamprath, eds.), pp. 37–57. CSIR0, Melbourne.

Benians, G., Scullion, P. and Fitzhugh, G. R (1977). Concentrations and activities of ions in solutions displaced from basaltic soils. *Journal of Soil Science* **28**, 454–461.

Bhumbla, D. R. and McLean, E. O. (1965). Aluminium in soils. VI. Changes in pH-dependent acidity, cation exchange capacity, and extractable aluminium with additions of lime to acid surface soils. *Soil Science Society of America Proceedings* **29**, 370–374.

Blamey, F. P. C., Edwards, D. G. and Asher, C. J. (1983). Effects of aluminium, OH:Al and P:Al molar ratios, and ionic strength on soybean root elongation in solution culture. *Soil Science* **136**, 197–207.

Bloom, P. R. and McBride, M. B. (1979). Metal ion binding and exchange with hydrogen ions in acid-washed peat. *Soil Science Society of America Journal* **43**, 687–692.

Bloom, P. R., McBride, M. B. and Chadbourne, B. (1977). Adsorption of aluminium by a smectite. I. Surface hydrolysis during $Ca^{2+}-Al^{3+}$ exchange. *Soil Science Society of America Journal* **41**, 1068–1077.

Bloom, P. R., McBride, M. B. and Weaver, R. M. (1979a). Aluminium organic matter in acid soils: salt-extractable aluminium. *Soil Science Society of America Journal* **43**, 813–815.

Bloom, P. R., McBride, M. B. and Weaver, R. M. (1979b). Aluminium organic matter in acid soils: buffering and solution aluminium activity. *Soil Science Society of America Journal* **43**, 488–493.

Bohn, H. L. (1970). Comparisons of measured and theoretical concentrations in soil suspensions. *Soil Science Society of America Proceedings* **34**, 195–197.

Boken, E. (1955). On the effect of ferrous sulphate on the available manganese in the soil and the uptake of manganese by the plant. *Plant and Soil* **6**, 97–112.

Bolan, N. S. and Barrow, N. J. (1984). Modelling the effect of adsorption of phosphate and other anions on the surface charge of variable charge oxides. *Journal of Soil Science* **35**, 273–281.

Bolt, G. H. (1976). Surface interaction between the soil solid phase and the soil solution. *In* 'Soil Chemistry. A. Basic Elements'. (G. H. Bolt and M. G. M. Bruggenwert, eds.), pp. 43–53. Elsevier, Amsterdam.

Bolt, G. H. and Van Riemsdijk, W. H. (1982). Ion adsorption of inorganic variable charge constituents. *In* 'Soil Chemistry'. (G. H. Bolt, ed.), pp. 459–504. Elsevier, Amsterdam.

Bolt, G. H., Bruggenwert, M. G. M. and Kamphorst, A. (1976). Adsorption of cations by soils. *In* 'Soil Chemistry. A. Basic Elements'. (G. H. Bolt and M. G. M. Bruggenwert, eds.), pp. 54–90. Elsevier, Amsterdam.

Bowden, J. W., Posner, A. M. and Quirk, J. P. (1980). Adsorption and charging phenomena in variable charge soils. *In* 'Soils with Variable Charge'. (B. K. G. Theng, ed.), pp. 147–166. New Zealand Society of Soil Science, Lower Hutt.

Brenes, E. and Pearson, R. W. (1973). Root responses of three Gramineae species to soil acidity in an oxisol and an ultisol. *Soil Science* **116**, 295–302.

Bromfield, S. M. (1958). The properties of a biologically formed manganese oxide, its availability to oats and its solution by root washings. *Plant and Soil* **9**, 325–337.

Bromfield, S. M. (1974). Bacterial oxidation of manganous ions as affected by organic substrate concentration and composition. *Soil Biology and Biochemistry* **6**, 383–392.

Bromfield, S. M. (1978). The effect of manganese-oxidising bacteria and pH on the availability of manganous ions and manganese oxides to oats in nutrient solutions. *Plant and Soil* **49**, 23–39.

Bromfield, S. M., Cumming, R. W., David, D. J. and Williams, C. H. (1983). The assessment of available manganese and aluminium status in acid soils under subterranean clover pastures of various ages. *Australian Journal of Experimental Agriculture and Animal Husbandry* **23**, 192–200.

Bruckert, S. (1970). Influence des composés organiques solubles sur la pédogenèse en milieu acide. I. Etudes de terrain. *Annals d'Agronomie* **21**, 421–452.

Cameron, R. C., Ritchie, G. S. P. and Robson, A. D. (1986). The relative toxicities of inorganic aluminium complexes to barley (*Hordeum vulgare* L.). *Soil Science Society of America Journal* **50**, 1231–1236.

Campbell, P. G. C., Bisson, M., Bougie, R., Tessier, A. and Villeneuve, J-P. (1983). Speciation of aluminium in acidic freshwaters. *Analytical Chemistry* **55**, 2246–2252.

Castellan, G. W. (1970). 'Physical Chemistry'. Addison-Wesley, Massachusetts.

Christensen, P. D., Toth, S. J. and Bear, F. E. (1950). The status of soil manganese as influenced by moisture, organic matter and pH. *Soil Science Society of America Proceedings* **14**, 279–282.

Coleman, N. T. and Thomas, G. W. (1964). Buffer curves of acid clays as affected by the presence of ferric iron and aluminium. *Soil Science Society of America Proceedings* **27**, 187–190.

Coleman, N. T. and Thomas, G. W. (1984). The basic chemistry of soil acidity. *In* 'Soil Acidity and Liming'. (N. T. Coleman and R. W. Pearson, eds.), pp. 1–41. American Agronomy Society, Madison.

Collins, J. F. and Buol, S. W. (1970). Patterns of iron and manganese precipitation under specified Eh-pH conditions. *Soil Science* **110**, 157–162.

Curtin, D. and Smillie, G. W. (1983). Soil solution composition as affected by liming and incubation. *Soil Science Society of America Journal* **46**, 701–707.

David, M. B. and Driscoll, C. T. (1984). Aluminium speciation and equilibria in soil solutions of a haplorthod in the Adirondack mountains. *Geoderma* **33**, 297–318.

Dixon, J. B. and Weed, S. B. (1977). 'Minerals in the Soil Environment'. American Soil Science Society, Madison.

Dolar, S. G., Keeney, D. R. and Walsh, L. M. (1971). Availability of Cu, Zn and Mn in soils. III. Predictability of plant uptake. *Journal of the Science of Food and Agriculture* **22**, 282–286.

Dolling, P. J. and Ritchie, G. S. P. (1985). Estimates of soil solution ionic strength and the determination of pH in West Australian soils. *Australian Journal of Soil Research* **23**, 309–314.

Evans, C. E. and Kamprath, E. J. (1970). Lime response as related to percent Al saturation, solution Al and organic matter content. *Soil Science Society of America Proceedings* **36**, 893–896.

Farina, M. P. W., Sumner, M. E., Plank, C. O. and Letzsch, W. S. (1980). Exchangeable aluminium and pH as indicators of lime requirement for corn. *Soil Science Society of America Journal* **44**, 1036–1041.

Farmer, U. C., Fraser, A. R. and Tait, J. M. (1977). Synthesis of imogolite: a tubular aluminium silicate polymer. *Journal of the Chemical Society, Chemical Communications* 462–463.

Fox, R. L. (1982). Some highly weathered soils of Puerto Rico. 3. Chemical properties. *Geoderma* **27**, 139–176.

Frink, C. R. (1973). Aluminium chemistry in acid sulphate soils. Proceedings of the International Symposium on Acid Sulphate Soils, Wageningen, pp. 131–168.

Fujimoto, C. K. and Sherman, G. D. (1948). Behaviour of manganese in the soil and the manganese cycle. *Soil Science* **66**, 131–145.

Gakhamani, A. B. (1977). Extractable and exchangeable aluminium in synthetic compounds, minerals and subtropical and tropical soils. *Pochvovedeniye* **11**, 45–53.

Gallez, A., Juo, A. S. R. and Herbillon, A. J. (1976). Surface and charge characteristics of selected soils in the tropics. *Soil Science Society of America Journal* **40**, 601–608.

Gamble, D. S., Langford, C. H. and Tong, J. P. K. (1976). The structure and equilibria of a manganese (II) complex of fulvic acid studied by ion exchange and nuclear magnetic resonance. *Canadian Journal of Chemistry* **54**, 1239–1245.

Geering, H. R., Hodgson, J. F. and Sdano, C. (1969). Micronutrient cation complexes in soil solution. IV. The chemical state of manganese in soil solutions. *Soil Science Society of America Proceedings* **33**, 81–85.

Ghanem, I., El-Gabaly, M. M., Hasan, M. N. and Tadros, V. (1971). Effect of organic materials addition on transformation of added manganese dioxide to alkali calcareous soils. *Plant and Soil* **34**, 653–661.

Gillman, G. (1981). Analytical Methods. *Varisol* No. 1, 7–8.

Godo, G. H. and Reisenauer, H. M. (1980). Plant effects on soil manganese availability. *Soil Science Society of America Journal* **44**, 993–995.

Goldberg, S. P. and Smith, K. A. (1984). Soil manganese: E values, distribution of manganese among soil fractions, and effects of drying. *Soil Science Society of America Journal* **48**, 559–564.

Goldberg, S. P., Smith, K. A. and Holmes, J. C. (1983). The effects of soil compaction, form of nitrogen fertilizer, and fertilizer placement on the availability of manganese to barley. *Journal of the Science of Food and Agriculture* **34**, 657–670.

Gotoh, S. and Patrick, W. H. (1972). Transformation of manganese in a waterlogged soil as affected by redox potential and pH. *Soil Science Society of America Proceedings* **36**, 738–742.

Grasmanis, V. O. and Leeper, G. W. (1966). Toxic manganese in near-neutral soils. *Plant and Soil* **25**, 41–48.

Gray, T. R. G. and Williams, S. T. (1971). 'Soil Micro-Organisms'. Oliver and Boyd, Edinburgh.

Greenland, D. J. and Mott, C. J. B. (1978). Surfaces of soil particles. *In* 'The Chemistry of Soil Constituents'. (D. J. Greenland and M. H. B. Hayes, eds.), pp. 321–353. Wiley Interscience, Chichester.

Gupta, U. C. (1972). Effects of manganese and lime on yield and on the concentrations of manganese, molybdenum, boron, copper and iron in the boot stage tissue of barley. *Soil Science* **114**, 131–136.

Hammes, J. K. and Berger, K. C. (1960a). Manganese deficiency in oats and correlation of plant manganese with various soil tests. *Soil Science* **91**, 239–244.

Hammes, J. K. and Berger, K. C. (1960b). Chemical extraction and crop removal of manganese from air-dried and moist soils. *Soil Science Society of America Proceedings* **24**, 361–364.

Hargrove, W. L. and Thomas, G. W. (1981a). Extraction of aluminium from aluminium-organic matter complexes. *Soil Science Society of America Journal* **45**, 151–153.

Hargrove, W. L. and Thomas, G. W. (1981b). Effect of organic matter on exchangeable aluminium and plant growth in acid soils. American Society of Agronomy Special Publication No. 40, 151–166.

Hargrove, W. L. and Thomas, G. W. (1982a). Titration properties of Al-organic matter. *Soil Science* **134**, 216–225.

Hargrove, W. L. and Thomas, G. W. (1982b). Conditional formation constants for aluminium-organic matter complexes. *Canadian Journal of Soil Science* **62**, 571–575.

Hargrove, W. L. and Thomas, G. W. (1984). Extraction of aluminium from aluminium-organic matter in relation to titratable acidity. *Soil Science Society of America Journal* **48**, 1458–1460.

Harter, R. D. and McLean, E. D. (1965). The effect of moisture level and incubation time on chemical equilibria of a Toledo clay loam soil. *Agronomy Journal* **57**, 583–588.

Helyar, K. R. (1976). Nitrogen cycling and soil acidification. *Journal of the Australian Institute of Agricultural Science* **42**, 217–221.

Helyar, K. R. (1978). Effects of aluminium and manganese toxicities on legume growth. *In* 'Mineral Nutrition of Legumes in Tropical and Subtropical Soils'. (C. S. Andrew and E. J. Kamprath, eds.), pp. 207–231. CSIRO, Melbourne.

Hem, J. D. (1968). Aluminium species in water. *Advances in Chemistry* **73**, 98–114.

Hem, J. D., Russo, E. J., Bahal, S. M. and Levi, R. S. (1970). Effect of pH of precipitation on antacid properties of hydrous aluminium oxide. *Journal of Pharmaceutical Science* **59**, 317–321.

Hendershot, W. H. and Lavkulich, L. M. (1983). Effect of sesquioxide coatings on surface charge of standard mineral and soil samples. *Soil Science Society of America Journal* **47**, 1252–1260.

Hodges, S. C. and Zelazny, L. W. (1983a). Interactions of dilute, hydrolyzed aluminium solutions with clays, peat and resin. *Soil Science Society of America Journal* **47**, 206–212.

Hodges, S. C. and Zelazny, L. W. (1983b). Influences of OH : Al ratios and loading rates on aluminium-kaolinite interactions. *Soil Science Society of America Journal* **47**, 221–225.

Hoyt, P.B. (1977). Effects of organic matter content on exchangeable Al and pH-dependent acidity of very acid soil. *Canadian Journal of Soil Science* **57**, 221–222.

Hoyt, P. B. and Nyborg, M. (1971a). Toxic metals in acid soil. I. Estimation of plant available aluminium. *Soil Science Society of America Proceedings* **35**, 236–240.

Hoyt, P. B. and Nyborg, M. (1971b). Toxic metals in acid soil. II. Estimation of plant available manganese. *Soil Science Society of America Proceedings* **35**, 241–244.

Hoyt, P. B. and Turner, R. C. A. (1975). Effects of organic materials added to very acid soils on pH, aluminium, exchangeable NH_4^+ and crop yields. *Soil Science* **119**, 227–237.

Hoyt, P. B. and Webber, M. D. (1974). Rapid measurement of plant-available aluminium and manganese in acid Canadian soils. *Canadian Journal of Soil Science* **54**, 53–61.

Hsu, P.H. (1963). Effect of initial pH, phosphate and silicate on the determination of aluminium with aluminon. *Soil Science* **96**, 230–238.

Hsu, P. H. (1968). Interaction between aluminium and phosphate in aqueous solutions. *Advances in Chemistry Series* **73**, 115–127.

Hsu, P. H. (1977). Aluminium hydroxides and oxyhydroxides. *In* 'Minerals in the Soil Environments'. (J. B. Dixon and S. B. Weed, eds.), pp. 99–143. Soil Science Society of America, Madison, Wisconsin.

Hsu, P. H. (1979). Effect of phosphate and silicate on the crystallization of gibbsite from OH-Al solutions. *Soil Science* **127**, 219–226.

Hsu, P. H. (1982). Crystallization of variscite at room temperature. *Soil Science* **133**, 305–313.

Hue, N. V., Craddock, G. R. and Adams, F. (1986). Effect of organic acids on aluminium toxicity in subsoils. *Soil Science Society of America Journal* **50**, 28–34.

Ingols, R. S. and Enginun, M. E. (1968). Biological studies of manganese solution from its dioxide. *Advances in Chemistry Series* **73**, 143–148.

Iñiguez, J. and Val, R. M. (1982). Variable charge characteristic of andosols from Navarre, Spain. *Soil Science* **133**, 390–396.

Inubushi, K., Wada, H. and Takai, Y. (1984). Easily decomposable organic matter in paddy soil. IV. Relationship between reduction process and organic matter decomposition. *Soil Science and Plant Nutrition* **30**, 189–198.

James, B. R., Clark, C. J. and Riha, S. J. (1983). An 8-hydroxyquinoline method for labile and total aluminium in soil extracts. *Soil Science Society of America Journal* **47**, 893–897.

Jauregui, M. A. and Reisenauer, H. M. (1982). Dissolution of oxides of manganese and iron by root exudate components. *Soil Science Society of America Journal* **46**, 314–317.

Jones, L. H. P. (1957a). The effect of liming a neutral soil on the uptake of manganese by plants. *Plant and Soil* **8**, 301–314.

Jones, L.H.P. (1957b). The effect of liming a neutral soil on the cycle of manganese. *Plant and Soil* **8**, 315–327.

Jones, L. H. P. and Leeper, G. W. (1951). The availability of various manganese oxides to plants. *Plant and Soil* **3**, 141–153.

Juo, A. S. R. (1977). Soluble and exchangeable aluminium in ultisols and alfisols in West Africa. *Communications in Soil Science and Plant Analysis* **8**, 17–35.

Juo, A. S. R. and Kamprath, E. J. (1979). Copper chloride as an extractant for estimating the potentially reactive aluminium pool in acid soils. *Soil Science Society of America Journal* **43**, 35–38.

Kamprath, E. J. (1970). Exchangeable aluminium as a criteria for liming leached mineral soils. *Soil Science Society of America Proceedings* **34**, 252–254.

Kamprath, E. J. (1978a). Lime in relation to Al toxicity in tropical soils. *In* 'Mineral Nutrition of Legumes in Tropical and Subtropical Soils' (C. S. Andrew and E. J. Kamprath, eds.), pp. 233–245. CSIRO, Melbourne.

Kamprath, E. J. (1978b). The role of soil chemistry in the diagnosis of nutrient disorders in tropical situations. *In* 'Mineral Nutrition of Legumes in Tropical and Subtropical Soils' (C. S. Andrew and E. J. Kamprath, eds.), pp. 313–327. CSIR0, Melbourne.

Katyal, J. C. (1980). Influence of farmyard manure on the chemical kinetics and growth of rice in a submerged vertisol. *Field Crops Research* **3**, 137–145.

Khalid, R. A. and Silva, J. S. (1979). A study of soil aluminium extraction methods in relation to plant aluminium and yield in tropical soils. *Tropical Agriculture (Trinidad)*, **56**, 53–63.

Khan, S. U. (1969). Interaction between the humic acid fraction of soils and certain metallic cations. *Soil Science Society of America Proceedings* **33**, 851–854.

Kwong, N. K. and Huang, P. M. (1977). Influence of citric acid on the hydrolytic reactions of aluminium. *Soil Science Society of America Journal* **41**, 692–697.

Kwong, N. K. and Huang, P. M. (1979). Surface reactivity of aluminium hydroxides precipitated in the presence of low molecular weight organic acids. *Soil Science Society of America Journal* **43**, 1107–1113.

Lamm, C. G. (1960). Some investigations of the chemistry and plant uptake of manganese in soils by use of radioactive manganese[54]. Transactions of the 7th Congress of the International Soil Science Society, Commission II (Madison), 223–229.

Larsen, S. and Widdowson, A. E. (1971). Soil fluorine. *Journal of Soil Science* **22**, 210–221.

Leeper, G. W. (1947). The forms and reactions of manganese in the soil. *Soil Science*, **63**, 79–94.

Lindsay, W. L. (1978). Chemical reactions affecting the availability of micro-nutrients in soils. *In* 'Mineral Nutrition of Legumes in Tropical and Subtropical Soils'. (C. S. Andrew and E. J. Kamprath, eds.), pp. 153–167. CSIRO, Melbourne.

Lindsay, W. L. and Norvell, W. A. (1978). Development of a DTPA soil test for zinc, iron, manganese and copper. *Soil Science Society of America Journal* **42**, 421–428.

Lopez, P. L. and Graham, E. R. (1972). Labile pool and plant uptake of micronutrients. I. Determination or labile pool of Mn, Fe, Zn, Co and Cu in deficient soils by isotopic exchange. *Soil Science* **114**, 295–299.

Luciuk, G. M. and Huang, P. M. (1974). Effect of monosilicic acid on hydrolytic reactions of aluminium. *Soil Science Society of America Proceedings* **38**, 235–244.

McBride, M. B. (1978). Retention of Cu^{2+}, Ca^{2+}, Mg^{2+} and Mn^{2+} by amorphous alumina. *Soil Science Society of America Journal* **42**, 27–31.

McBride, M. B. (1979). Chemisorption and precipitation of Mn^{2+} at $CaCO_3$ surfaces. *Soil Science Society of America Journal* **43**, 693–698.

McBride, M., Pinnavaia, T. J. and Mortland, M. M. (1975). Electron spin relaxation and the mobility of manganese (II) exchange ions in smectites. *American Mineralogy* **60**, 66–72.

McKenzie, R. M. (1977). Manganese oxides and hydroxides. *In* 'Minerals in the Soil Environment'. (J. B. Dixon and S. B. Weed, eds.), pp. 181–193. Soil Science Society of America, Madison, Wisconsin.

McKenzie, R. M. (1980). The adsorption of lead and other heavy metals on oxides of manganese and iron. *Australian Journal of Soil Research* **18**, 61–73.

McLean, E. O. (1965). Aluminium. *In* 'Methods of Soil Analysis, Part 2' (C. A. Black, ed.), pp. 978–998. American Agronomy Society, Madison, Wisconsin.

Mäntylahti, V. (1981). Determination of plant-available manganese in Finnish soils. *Journal of the Scientific Agricultural Society of Finland* **53**, 391–508.

Marion, G. M., Hendricks, D. M., Dutt, G. R. and Fuller, W. H. (1976). Aluminium and silica solubility in soils. *Soil Science* **121**, 76–85.

Martell, A. E. and Smith, R. M. (1977). 'Critical stability constants. Other organic ligands, Vol. 3'. Plenum Press, New York.

Meek, B. D., Page, A. L. and Martin, J. P. (1973). The oxidation of divalent manganese under conditions present in tile lines as related to temperature, solid surfaces, microorganisms and solution chemical composition. *Soil Science Society of America Proceedings* **37**, 542–548.

Miller, M. J. and Ohlrogge, A. J. (1958). Water-soluble chelating agents in organic materials. I. Characterization of chelating agents and their reactions with trace metals in soils. *Soil Science Society of America Proceedings* **22**, 228–231.

Motekaitas, R. J. and Martell, A. E. (1984). Complexes of aluminium (III) with hydroxy carboxylic acids. *Inorganic Chemistry* **23**, 18–23.

Munns, D.N. (1965). Soil acidity and growth of a legume. *Australian Journal of Agricultural Research* **16**, 757–766.

Norrish, K. and Taylor, R. M. (1961). The isomorphous replacement of iron by aluminium in soil goethites. *Journal of Soil Science* **12**, 294–306.

Nye, P., Craig, D., Coleman, N. T. and Ragland, J. L. (1961). Ion exchange equilibria involving aluminium. *Soil Science Society of America Proceedings* **25**, 14–17.

Oates, K. M. and Kamprath, E. J. (1983a). Soil acidity and liming. I. Effect of the extracting solution cation and pH on the removal of aluminium from acid soils. *Soil Science Society of America Journal* **47**, 686–689.

Oates, K. M. and Kamprath, E. J. (1983b). Soil acidity and liming. II. Evaluation of using aluminium extracted by various chloride salts for determining lime requirements. *Soil Science Society of America Journal* **47**, 690–692.

Page, E. R. (1962). Studies in soil and plant manganese. II. The relationship of soil pH to manganese availability. *Plant and Soil* **16**, 247–257.

Pal, S. S., Sudhakar-Barik and Sethunathan, N. (1979). Effects of benomyl on iron and manganese reduction and redox potential in flooded soils. *Journal of Soil Science* **30**, 155–159.

Pankow, J. F. and Morgan, J. J. (1981). Kinetics for the aquatic environment. *Environmental Science and Technology* **15**, 1306–1313.

Parfitt, R. L. (1978). Anion adsorption by soils and soil materials. *Advances in Agronomy* **30**, 1–50.

Parfitt, R. L. (1980). Chemical properties of variable charge soils. *In* 'Soils with Variable Charge'. (B. K. G. Theng, ed.) , pp. 167–194. New Zealand Society of Soil Science, Lower Hutt.

Parks, G. A. (1965). The isoelectric points of solid oxides, solid hydroxides and aqueous hydroxo complex systems. *Chemical Reviews* **65**, 177–198.

Parks, G. A. (1967). Aqueous surface chemistry of oxides and complex oxide minerals. *In* 'Equilibrium Concepts in Natural Water Systems'. Advances in Chemistry Series 67. (R. F. Gould, ed.), pp. 121–160. American Chemical Society, Washington, D.C.

Passioura, J. B. and Leeper, G. W. (1963a). Available manganese and the X-hypothesis. *Agrochimica* **8**, 81–89.

Passioura, J. B. and Leeper, G. W. (1963b). Soil compaction and manganese deficiency. *Nature (London)* **200**, 29–30.

Patrick, W. H. Jr. and Henderson, R. E. (1981). Reduction and re-oxidation cycles of manganese and iron in flooded soil and in water solution. *Soil Science Society of America Journal* **45**, 855–859.

Pavan, M. A. and Bingham, F. T. (1982). Toxicity of aluminium to coffee seedlings grown in nutrient solution. *Soil Science-society of America Journal* **46**, 993–997.

Pavan, M. A., Bingham, F. T. and Pratt, P. F. (1982). Toxicity of aluminium to coffee in ultisols and oxisols amended with $CaCO_3$, $MgCO_3$ and $CaSO_4.2H_2O$. *Soil Science Society of America Journal* **46**, 1201–1207.

Peech, M. (1965). Hydrogen-ion activity. *In* 'Methods of Soil Analysis, Part 2'. (C. A. Black, ed.), pp. 914–926. American Agronomy Society, Madison, Wisconsin.

Ponnamperuma, F. N. (1972). The chemistry of submerged soils. *Advances in Agronomy* **24**, 29–96.

Ponnamperuma, F. N., Loy, T. A. and Tianco, E. M. (1969). Redox equilibria in flooded soils. II. The manganese oxide systems. *Soil Science* **108**, 48–57.

Posner, A. M. (1964). Titration curves of humic acid. Transactions of the 8th Congress of the International Soil Science Society **3**, 161–174.

Pratt, P. F. and Blair, F. L. (1961). A comparison of three reagents for the extraction of aluminium from soils. *Soil Science* **91**, 357–359.

Pyman, M. A. F. (1978). 'Surface Properties of Amorphous Mixed Oxides'. Ph.D. Thesis, University of Western Australia.

Randall, G. W., Schulte, E. E. and Corey, R. B. (1976). Correlation of plant manganese with extractable soil manganese and soil factors. *Soil Science Society of America Journal* **40**, 282–287.

Rashid, M. A. (1974). Absorption of metals on sedimentary and peat humic acids. *Chemical Geology* **13**, 115–123.

Reddy, M. R. and Perkins, H. F. (1976). Fixation of manganese by clay minerals. *Soil Science* **121**, 21–24.

Reeve, N. G. and Sumner, M. E. (1970). Lime requirements of Natal oxisols based on exchangeable aluminium. *Soil Science Society of America Proceedings* **34**, 595–598.

Rengasamy, P. and Oades, J. M. (1977). Interaction of monomeric and polymeric species of metal ions with clay surfaces. III. Aluminium III and chromium III. *Australian Journal of Soil Research* **16**, 53–66.

Rhodes, E. R. and Lindsay, W. L. (1978). Solubility of aluminium in soils of the humid tropics. *Journal of Soil Science* **29**, 324–330.

Richburg, J. S. and Adams, F. (1970). Solubility and hydrolysis of aluminium in soil solutions and saturated-paste extracts. *Soil Science Society of America Proceedings* **34**, 728–734.

Ritchie, G. S. P. and Dolling, P. J. (1985). The role of organic matter in soil acidification. *Australian Journal of Soil Research* **23**, 569–576.

Ritchie, G. S. P. and Posner, A. M. (1982). The effect of pH and metal binding on the transport properties of humic acids. *Journal of Soil Science* **33**, 233–247.

Ritchie, G. S. P., Posner, A. M. and Ritchie, I. M. (1982). The polarographic study of the equilibrium between humic acid and aluminium in solution. *Journal of Soil Science* **33**, 671–677.

Roorda van Eysinga, J. P. N. L., Van Dijk, P. A. and De Bes, S. S. (1978). The available manganese content of soil in the Netherlands determined by various methods. *Communications in Soil Science and Plant Analysis* **9**, 141–151.

Ross, G. J. and Turner, R. C. (1971). Effect of different anions on the crystallization of aluminium hydroxide in partially neutralized aqueous aluminium salt systems. *Soil Science Society of America Proceedings* **35**, 389–392.

Russell, E. W. (1973). 'Soil Conditions and Plant Growth'. Longman, London.

Salcedo, I. H. and Ellis, B. G. (1979). Manganese labile pool and plant uptake. *Soil Science* **127**, 227–234.

Salcedo, I. H., Ellis, B. G. and Lucas, R. E. (1979). Studies in soil manganese. II. Extractable manganese and plant uptake. *Soil Science Society of America Journal* **43**, 138–141.

Sanchez, C. and Kamprath, E. J. (1959). Effect of liming and organic matter content on the availability of native and applied manganese. *Soil Science Society of America Proceedings* **23**, 302–304.

Sanders, J. R. (1983). The effect of pH on the total and free ionic concentrations of manganese, zinc and cobalt in soil solutions. *Journal of Soil Science* **34**, 315–323.

Schnitzer, M. and Hansen, E. H. (1970). Organo-metallic interactions in soils. 8. An evaluation of methods for the determination of stability constants of metal-fulvic acid complexes. *Soil Science* **109**, 333–340.

Schnitzer, M. and Skinner, S. I. M. (1967). Organo-metallic interactions in soils. 7. Stability constants of Pb^{2+}-, Ni^{2+}-, Mn^{2+}-, Co^{2+}-, Ca^{2+}- and Mg^{2+}- fulvic acid complexes. *Soil Science* **103**, 247–252.

Schofield, R. K. (1949). Effect of pH on electric charges carried by clay particles. *Journal of Soil Science* **1**, 1–8.

Schofield, R. K. and Taylor, A. W. (1955). The measurement of soil pH. *Soil Science of America Proceeding* **19**, 164–167.

Schwertmann, U., Fitzpatrick, R. W. and Le Roux, J. (1977). Al substitution and

differential disorder in soil haematites. *Clays and Clay Minerals* **25**, 373–374.
Serna, C. J., White, J. L. and Hem, S. L. (1977). Anion-aluminium hydroxide gel interactions. *Soil Science Society of America Journal* **41**, 1009–1013.
Sheppard, L. J. and Floate, M. J. S. (1984). The effects of soluble-Al on root growth and radicle elongation. *Plant and Soil* **80**, 301–306.
Sherman, G. D., McHargue, J. S. and Hodgkiss, W. S. (1942). Determination of active manganese in soil. *Soil Science* **54**, 253–258.
Shuman, L. M. (1979). Zinc, manganese and copper in soil fractions. *Soil Science* **127**, 10–17.
Shuman, L. M. (1983). Sodium hypochlorite methods for extracting microelements associated with soil organic matter. *Soil Science Society of America Journal* **47**, 656–660.
Shuman, L. M. and Anderson, D. E. (1974). Evaluation of six extractants for their ability to predict manganese concentrations in wheat and soybeans. *Soil Science Society of America Proceedings* **38**, 788–790.
Shuman, L. M., Boswell, F. C., Ohki, K., Parker, M. B. and Wilson, D. E. (1980). Critical soil manganese deficiency levels for four extractants for soybeans grown in sandy soil. *Soil Science Society of America Journal* **44**, 1021–1025.
Sims, J. L. and Patrick, W. H. Jr. (1978). The distribution of micronutrient cations in soil under conditions of varying redox potential and pH. *Soil Science Society of America Journal* **42**, 258–262.
Sims, J. L., Duangpatra, P., Ellis, J. H. and Phillips, R. E. (1979). Distribution of available manganese in Kentucky soils. *Soil Science* **127**, 270–274.
Singh, S. S. (1982a). The formation and coexistence of gibbsite, boehmite, alumina and alunite at room temperature. *Canadian Journal of Soil Science* **62**, 327–332.
Singh, S. S. (1982b). Effect of chloride and sulfate anions on the chemical characteristics of some acid soils. *Canadian Journal of Soil Science* **62**, 549–557.
Singh, S. S. (1982c). Hydrolytic reactions and properties of dilute aluminium salt solutions. *Canadian Journal of Soil Science* **62**, 559–569.
Singh, S. S. and Brydon, J. E. (1969). Solubility of basic aluminium sulfates at equilibrium in solution and in the presence of montmorillonite. *Soil Science* **107**, 12–16.
Smith, R. W. (1971). Relations among equilibrium and nonequilibrium aqueous species of aluminium hydroxy complexes. *In* 'Nonequilibrium Systems in Natural Water Chemistry'. Advances in Chemistry Series 106 (R. F. Gould, ed.), pp. 250–279. American Chemical Society, Washington, D C.
Sparks, D. L. (1984). Ion activities: an historical and theoretical overview. *Soil Science Society of America Journal* **48**, 514–518.
Sparrow, L. A. and Uren, N. C. (1984). 'Manganese Toxicity and Wheat Production'. Special Publication, School of Agriculture, La Trobe University.
Sposito, G. (1981a). The operational definition of the zero point of charge in soils. *Soil Science Society of America Journal* **45**, 292–297.
Sposito, G. (1981b). 'The Thermodynamics of Soil Solutions'. Clarendon Press, Oxford.
Sposito, G. (1984). The future of an illusion: ion activities in soil solutions. *Soil Science Society of America Journal* **48** 531–536.
Stevenson, F. J. (1982). 'Humus Chemistry. Genesis, Composition, Reactions'. John Wiley and Sons, Inc.
Stol, R. J., van Helden, A. K. and de Bruyn, P. L. (1976). Hydrolysis precipitation studies of aluminium (III) solutions. 2. A kinetic study and model. *Journal of Colloid and Interface Science* **57**, 115–131.
Street, J. J. and Elwali, A. M. O. (1983). Fluorite solubility in limed acid sandy soils. *Soil Science Society of America Journal* **47**, 483–485.

Stumm, W. and Morgan, J. J. (1981). 'Aquatic Chemistry. An Introduction Emphasizing Chemical Equilibria in Natural Waters'. Wiley Interscience, New York.

Taylor, R. M., McKenzie, R. M., Fordham, A. W. and Gillman, G. P. (1983). Oxide minerals. *In* 'Soils, An Australian Viewpoint' Division of Soils, CSIRO. pp. 309–334. CSIRO, Melbourne/Academic Press, London.

Theng, B. K. G. (1980). 'Soils with Variable Charge'. New Zealand Society of Soil Science, Lower Hutt.

Thomas, G. W. (1975). The relationship between organic matter content and exchangeable aluminium in acid soil. *Soil Science Society of America Proceedings* **39**, 591.

Tiller, K. G. (1983). Micronutrients. *In* 'Soils, An Australian Viewpoint'. Division of Soils, CSIRO. pp. 365–387. CSIRO, Melbourne/Academic Press, London.

Tsai, P. P. and Hsu, P. H. (1984). Studies of aged OH-Al solutions using kinetics of Al-ferron reactions and sulfate precipitation. *Soil Science Society of America Journal* **48**, 59–65.

Turner, R. C. (1968). Conditions in solution during the formation of gibbsite in dilute Al salt solutions. I. Theoretical treatment of the effect of the formation of mono-and polynuclear hydroxyaluminium ions, precipitation and crystallization on curves representing the titration of $AlCl_3$ with a base. *Soil Science* **106**, 291–296.

Turner, R. C. (1969). Three forms of aluminium in aqueous systems determined by 8-quinolinolate extraction methods. *Canadian Journal of Chemistry* **47**, 2521–2527.

Turner, R. C. (1976). Effect of ageing on properties of polynuclear hydroxyaluminium cations. *Canadian Journal of Chemistry* **54**, 1528–1534.

Turner, R. C. and Sulaiman, W. (1971). Kinetics of reactions of 8-quinolinol and acetate with hydroxyaluminium species in aqueous solutions. I. Polynuclear hydroxyaluminium cations. *Canadian Journal of Chemistry* **49**, 1683–1687.

Uehara, G. and Gillman, G. (1981). 'The Mineralogy, Chemistry and Physics of Tropical Soils with Variable Charge Clays'. Westview Tropical Agriculture Series No. 4, Westview Boulder, Colorado.

Uren, N. C. (1981). Chemical reduction of an insoluble higher oxide of manganese by plant roots. *Journal of Plant Nutrition* **4**, 65–71.

Uren, N. C. (1982). Chemical reduction at the root surface. *Journal of Plant Nutrition* **5**, 515–520.

Uren, N. C. and Leeper, G. W. (1978). Microbial oxidation of divalent manganese. *Soil Biology and Biochemistry* **10**, 85–87.

Uren, N. C. and Sparrow, L. A. (1983). Forms of Manganese in Soils: their Measurement and their Significance to Plants'. Australian Society of Soil Science, Riverina Branch Conference, Yanco, May, 1983.

van Breemen, N. (1976). 'Genesis and Solution Chemistry of Acid Sulfate Soils in Thailand'. Centre for Agricultural Publishing and Documentation, Wageningen.

van Breemen, N. and Wielemaker, W. G. (1974a). Buffer intensities and equilibrium pH of minerals and soils. I. The contribution of minerals and aqueous carbonate to pH buffering. *Soil Science Society of America Proceedings* **38**, 55–60.

van Breemen, N. and Wielemaker, W. G. (1974b). Buffer intensities and equilibrium pH of minerals and soils. II. Theoretical and actual pH of minerals and soils. *Soil Science Society of America Proceedings* **38**, 61–66.

van Breemen, N., Mulder, J. and Driscoll, C.T. (1983). Acidification and alkalinization of soils. *Plant and Soil* **75**, 283–308.

van Raij, B. and Peech, M. (1972). Electrochemical properties of some oxisols and alfisols of the tropics. *Soil Science Society of America Proceedings* **36**, 587–593.

Veith, J. A. (1977). Basicity of exchangeable aluminium, formation of gibbsite, and composition of the exchange acidity in the presence of exchangers. *Soil Science Society of America Journal* **41**, 865–870.

Veith, J. A. (1978). Selectivity and adsorption capacity of smectite and vermiculite for aluminium of varying basicity. *Clays and Clay Minerals* **26**, 45–50.

Veith, J. A. and Sposito, G. (1979). On the average equilibrium OH/Al molar ratio for aluminium adsorbed by a synthetic cation exchanger. *Soil Science* **127**, 161–167.

Viets, F. G. Jr. and Lindsay, W. L. (1973). Testing soils for zinc, copper, manganese and iron. *In* 'Soil Testing and Plant Analysis' (L. M. Walsh and J. D. Beaton, eds.), pp. 153–172. Soil Science Society of America, Madison, Wisconsin.

Vinkler, P., Lakatos, B. and Meisel, J. (1976). Infrared spectroscopic investigations of humic substances and their metal complexes. *Geoderma* **15**, 231–242.

Wada, K. and Okamura, Y. (1983). Net charge characteristics of dystrandept B and theoretical prediction. *Soil Science Society of America Journal* **47**, 902–905.

Wada, S. I. and Wada, K. (1980). Formation, composition and structure of hydroxy-aluminosilicate ions. *Journal of Soil Science* **31**, 457–467.

Walker, J. M. and Barber, S. A. (1960). The availability of chelated Mn to millet and its equilibria with other forms of Mn in the soil. *Soil Science Society of America Proceedings* **24**, 458–488.

Webber, M. D., Hoyt, P. B. and Corneau, D. (1982). Soluble Al, exchangeable Al, base saturation and pH in relation to barley yield on Canadian acid soils. *Canadian Journal of Soil Science* **62**, 397–405.

White, R. E. (1969). On the measurement of soil pH. *Journal of the Institute of Agricultural Science* March, 3–15.

White, R. E. (1980). Retention and release of phosphate by soil and soil constituents. *In* 'Soils and Agriculture'. (P. B. Tinker, ed.), pp. 71–114. Critical Reports on Applied Chemistry, Vol. 2. Blackwell Scientific Publications, Oxford.

White, R. E., Tiffen, L. O. and Taylor, A. W. (1976). The existence of polymeric complexes in dilute solutions of aluminium and orthophosphate. *Plant and Soil* **45**, 521–529.

White, R. P. (1970). Effects of lime upon soil and plant manganese levels in an acid soil. *Soil Science Society of America Proceedings* **34**, 625–629.

Young, S. D. and Bache, B. W. (1985). Aluminium-organic complexation: formation constants and a speciation model for the soil solution. *Journal of Soil Science* **36**, 261–269.

Zende, G. K. (1954). The effect of air-drying on the level of extractable manganese in the soil. *Journal of the Indian Society of Soil Science* **2**, 55–61.

2

Soil Acidification, its Measurement and the Processes Involved

K. R. HELYAR and W. M. PORTER

I. INTRODUCTION

Soil acidification is a natural process which starts when rock surfaces are first colonised by algae and lichens. Acids, largely derived from the carbon and nitrogen cycles, are involved in the dissolution of soil and rock minerals during the process of soil development. In natural ecosystems soils usually become gradually more acidic with time. The older and

SOIL ACIDITY AND PLANT GROWTH
ISBN 0 12 590655 2

more weathered soils are generally more acidic than younger soils. In general, superimposing agricultural production on an ecosystem results in faster rates of soil acidification.

It is important that we understand the *rates* at which agricultural soils acidify for two reasons. First we need to predict how long a soil may be allowed to acidify before production will suffer, and second we need to estimate how much alkali to apply each year to stabilise the soil's acidity. In addition, by understanding the *mechanisms* of acidification, we may be able to devise strategies to reduce rates of acidification other than by applying lime.

In this chapter we develop approaches for analysing the processes and rates of soil acidification. Most of the examples are from Australian agricultural ecosystems, but the principles are the same for all ecosystems.

II. MECHANISMS OF ACIDIFICATION — A THEORETICAL BASIS

A. Concept of an ecosystem H^+ pool

In considering the mechanisms of acidification of soils, we have found it helpful to think in terms of inputs and outputs of H^+ to and from an ecosystem H^+ pool (Fig. 1). The ecosystem can be any volume of the biosphere, though in studies of agricultural soils it would usually include all living plant material above the soil surface and all soil to a depth equal to that of the deepest rooting species grown on that soil. The H^+ pool represents all H^+ in the ecosystem under study. This pool of free H^+ is compartmentalised. Major compartments are H^+ in living organic materials, H^+ in dead organic material on or above the soil surface, H^+ in free water (lakes, rivers), and H^+ in the soil water. The pH usually varies between and within these compartments. The pH of each component of the ecosystem is also buffered to varying degrees, depending upon H^+ association/dissociation reactions, and/or precipitation/dissolution processes that involve H^+ or OH^-. Thus inputs and outputs of H^+ to the ecosystem will not increase the level of H^+ in the pool by an amount equal to the amount of H^+ added or removed. Rather the change in H^+ in the system (represented by the change in pH) is less than the absolute gains and losses because of the pH buffer capacity of the system.

$$\text{pH change} = (\text{moles H}^+ \text{ added.ha}^{-1})\,(\text{pHBC} \times \text{W}) \qquad (1)$$

where pHBC is the pH buffer capacity (mol $H^+.kg^{-1}.pH\ unit^{-1}$) and W is the weight of the component of the ecosystem involved ($kg.ha^{-1}$).

The pH of different parts of the ecosystem responds differently to acid addition or removal. Living plant material is usually maintained at

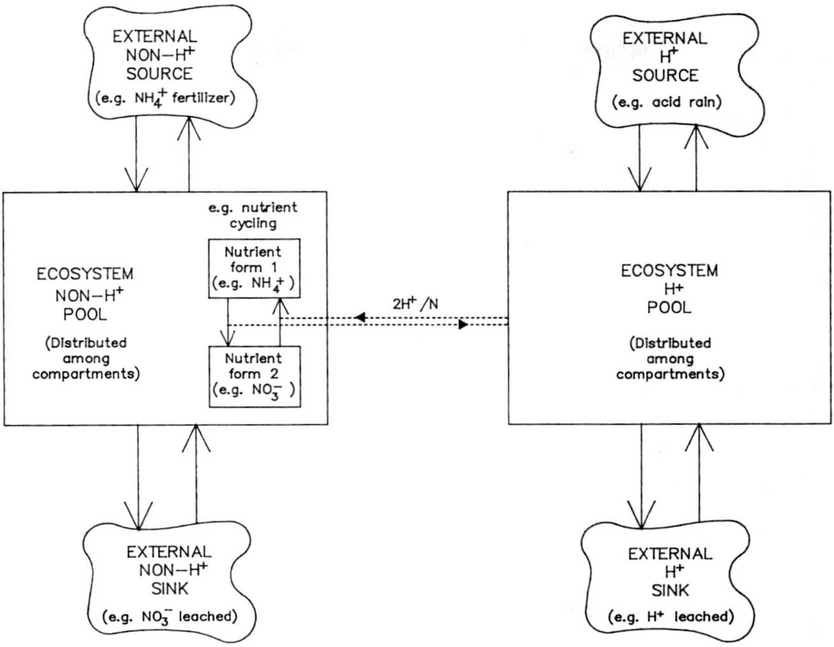

Fig. 1. The size of the ecosystem H^+ pool as affected by inputs and outputs of H^+ and other materials, and by transformations of other materials in the ecosystem.

approximately pH 6.0, by biochemical processes (Raven and Smith, 1976), so acid additions to the ecosystem do not change the pH of living material. This leaves the non-living material (plant litter and soil) and free water to express the pH change of the system. Soil dominates the buffer capacity of the total ecosystem because of its significantly greater mass. For 14 000 t soil.ha^{-1} and 0–50 t plant litter.ha^{-1} the pH buffer capacities are about 100–200 and 0–16 kmol H^+.ha^{-1}.pH unit^{-1}.

B. Sources of H^+

Hydrogen ions may be added or removed from an ecosystem, or they may be produced or consumed in reactions within the ecosystem (Fig. 1). There are few mechanisms of direct input or output of H^+. The most significant inputs are H^+ from the dissociation of water and carbonic acid and from nitric and sulphuric acids in acid rain. In Australian agricultural ecosystems inputs of acids other than water and carbonic acid in rainfall are minor (Hingston and Gailitis, 1976; Blackburn and McLeod,

1983). In contrast there are many processes within the ecosystem which indirectly affect the size of the H^+ pool, and addition or removal of some non-H^+ material can affect the magnitude of these reactions. These processes can be categorised in a number of ways (van Breeman *et al.*, 1983, 1984). We have chosen to discuss them in terms of the biogeochemical cycles. This classification has been adopted because it can be linked to a suite of practical measurements which can be used to determine the causes of acidification in an ecosystem.

1. Carbon cycle

Water in soils contains carbonic acid in equilibrium with CO_2 in the soil air. The partial pressure of CO_2 in the soil air varies between 0.0015 and 0.0065 atm (Russell, 1973), or five to twenty times the usual concentration in the atmosphere (0.0003 atm). As water percolates through the soil, acidification occurs when bicarbonates are leached (Fig. 2). In acid soils

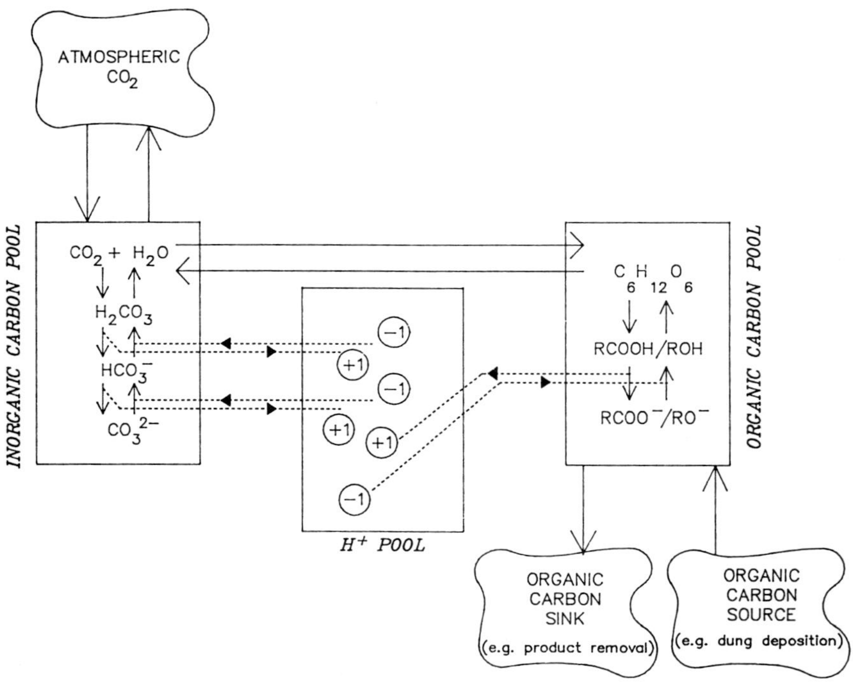

Fig. 2. The carbon cycle and the ecosystem H^+ pool. The numbers in circles represent the numbers of moles of H^+ added to or removed from the H^+ pool for each mole of carbon reacting.

inputs, outputs and the accumulation of HCO_3^- can be significant cause of pH changes.

Within agricultural and natural ecosystems many carboxylic acids are produced, along with other compounds such as phenolic and enolic groups that dissociate H^+ at higher pH values (Coleman and Thomas, 1967). The size of the H^+ pool is increased by an accumulation or export of organic anions which were originally derived from non-dissociated carbon compounds. Organic anions added to the system that subsequently associate with H^+ within the system reduce the size of the H^+ pool (Fig. 2).

2. *Nitrogen cycle*

In many agricultural and probably also many natural ecosystems, the nitrogen cycle plays an important role in causing soil acidification. The fluxes of protons to and from the H^+ pool of an ecosystem, in response to the inputs and outputs of the various forms of nitrogen, and to the

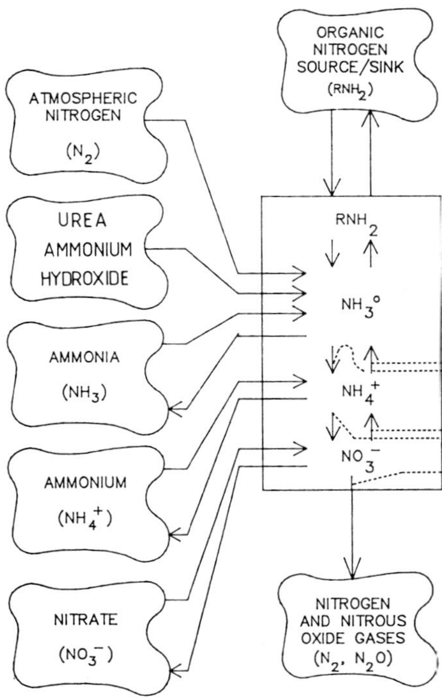

Fig. 3. The nitrogen cycle and the ecosystem H^+ pool. (See Helyar, 1976 for a more comprehensive diagram and analysis).

nitrogen cycle reactions, are summarised in Figure 3. Protons are either produced or consumed in the various redox reactions, and association/dissociation reactions involved. The net effect of nitrogen inputs and outputs depends on the form and quantity of the inputs and the outputs, and on the quantities of the different forms of nitrogen that accumulate in the system (Fig. 3).

3. *Sulphur cycle*

Acid is either produced or consumed during the oxidation of sulphur from the S^{2-} oxidation state in sulphides and organic sulphur forms (e.g. proteins), to sulphate (S^{6+} oxidation state) (Fig. 4). In agricultural ecosystems, acids in the sulphur cycle are sometimes important (acid sulphate soils, acid rain, elemental sulphur or finely divided sulphides in fertilizers). However, in many situations effects of the sulphur cycle are

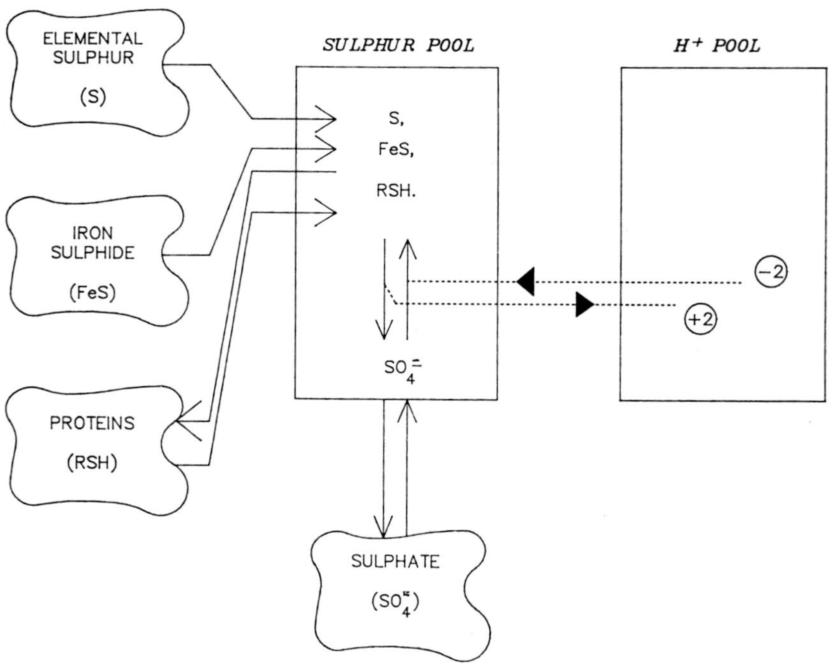

Fig. 4. The sulphur cycle and the H^+ pool.

of minor significance because the fluxes of sulphur are small compared with those of nitrogen and carbon.

4. *Other element cycles*

The same principles as enunciated above for carbon, nitrogen and sulphur apply to the effects of other nutrient cycles on soil acidification. In given situations nutrient cycles can have important effects on the H^+ pool of an ecosystem. Examples are the redox reactions involving the transformations Fe(III) to Fe(II) and Mn(IV) to Mn(II) under alternately flooded (anaerobic) and aerobic conditions in soils. These reactions are important in flooded rice production (Ponnamperuma, 1972), but are largely cyclic and, except over very long periods, are not significantly associated with transport of the reduced or oxidised products from the system. This means that, apart from seasonal pH changes, effects on soil acidification are usually small. Over periods involved in soil formation, however, Fe(III) to Fe(II) and Mn(IV) to Mn(II) transformations have been implicated in leading to acidification (Mann, 1983).

In some conditions, additions of phosphatic fertilizers can affect the soil pH. These include: the dissolution of rock phosphates (e.g. various apatite minerals), which produces alkali; the specific adsorption of phosphate onto aluminium and iron oxides in soil, which produces some alkali below pH 7.0; the addition of $H_2PO_4^-$ to soils of pH > 7.0, which is acidifying; and the addition of $H_2PO_4^{2-}$ to soils of pH > 7.0, which produces alkali. Below pH 7.0 the complete reaction of superphosphate (phosphorus in the form HPO_4^-) with the soil is near neutral, despite the very acidic initial reaction on formation of the 'metastable triple point solution' in the zone of high phosphorus concentration close to a dissolving fertilizer granule (Lindsay, 1979).

5. *Acids added directly*

Acids added directly need to be accounted for if they are not included as one of the terms in a nutrient cycle. The net H^+ addition to the ecosystem through rainfall and leaching can be calculated from rainfall and soil pH values. Furthermore, some substances added to agricultural ecosystems are equivalent to strong acids. Examples are elemental sulphur and finely divided sulphides added to moist aerobic soils, and cations that act as Bronsted-Lowry acids because of strong hydrolysis. Aluminium and iron sulphates are examples of the latter, and the sulphides and sulphur are almost completely oxidised to sulphuric acid in aerobic soils (Weir, 1975).

C. Quantifying the alternative acidification mechanisms

1. An acidification function based on the reference state concept

The amount of acid added to an ecosystem over a period (mol $H^+.ha^{-1}$) can be measured empirically. The techniques are described in Section III of this chapter.

The contribution of different sources of acid and alkali can be assessed by considering each nutrient cycle involved. For each nutrient it is helpful to define a reference compound, or group of compounds, and other 'acid' or 'alkaline' forms of the same element (Fig. 5). Transformations between the compounds in the reference group do not involve

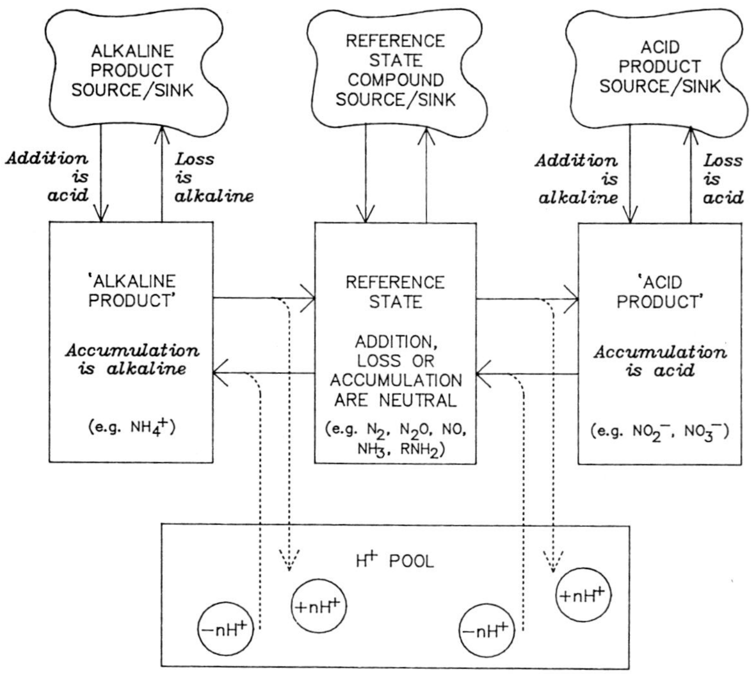

Fig. 5. The reference state convention for estimating the effects of nutrient cycling, losses and gains on the acidity of an ecosystem. 'Acid products' are those associated with acid release when formed from the reference state. 'Alkaline products' are those associated with acid consumption when formed from the reference state compound.
The rates of loss, addition and accumulation (in italics) must be measured to estimate the contribution of the nutrient cycle to acidification of the ecosystem. Nitrogen forms are included as an example.

production or consumption of acid or alkali. For each element two other classes of compounds may exist. These are 'acid products', that are associated with acid production or alkali consumption when formed from reference compounds, and 'alkaline products' associated with alkali production or acid consumption when formed from reference compounds. The reference group usually includes the major form or forms of the nutrient in the ecosystem, but analytical considerations may make it appropriate to define a minor form as the reference state. By convention, losses, gains or accumulation of reference state compounds by the ecosystem are neutral, as are transformations among reference state compounds. Gains of acid products and losses or an accumulation of alkaline products lead to the soil becoming more alkaline (Fig. 5). Similarly, gains of alkaline products, and losses or an accumulation of acid products lead to the soil becoming more acidic. Under this convention fluxes of the reference state compounds can be ignored, because the net acid/alkali effects are fully accounted for by measuring the fluxes and accumulation of the acid and alkaline products.

The system is illustrated for nitrogen in Figure 5. As an example, take the simplest case of adding 1.43 kmol of fertilizer $NH_4^+.ha^-$ (20 kg nitrogen ha^{-1}) and assume it remains as NH_4^+ in the soil, so no acid has been added. The H^+ pool is assumed to increase by 1.43 $kmol.ha^{-1}$ (1 mol $H^+.mol\ NH_4^{+-1}$). However, soil measurements would reveal that the soil NH_4^+ pool has increased by 1.43 $kmol.ha^{-1}$ and so the calculated H^+ pool would be reduced by 1.43 $kmol.ha^{-1}$, leaving no net acidification.

The above conventions have been formalised in Eq.(2), which has been designed with two aims. The first is to define specifically the acidifying effects of the nutrient cycles important in agricultural ecosystems. The second aim is to minimise both the number and difficulty of the analyses needed to define the rate and causes of soil acidification in a given ecosystem.

$$
\begin{aligned}
&\text{Acid addition } (\text{mol.ha}^{-1}\ \text{period}^{-1})\\
&\quad = (\text{carbon cycle effects})\\
&\qquad + (\text{nitrogen cycle effects})\\
&\qquad\quad + (\text{addition of other alkalis})\\
&\qquad\qquad + (\text{addition of acids})\\
&\quad = (OA_{ac} + OA_{ex} - OA_{ad} + HCO^-_{3\,ac} + HCO^-_{3\,ex} - HCO^-_{3\,ad})\\
&\qquad + (NH^+_{4\,ad} - NO^-_{3\,ad} - NH^+_{4\,ac} + NO^-_{3\,ac} + NO^-_{3\,ex} - NH^+_{4\,ex})\\
&\qquad\quad + (-L_{ad})\\
&\qquad\qquad + (H^+_{ad}) - (H^+_{ex})
\end{aligned}
\tag{2}
$$

where OA represents organic anions, L represents liming materials e.g. $CaCO_3$, H^+ represents acids, e.g. strong acids or compounds that almost completely hydrolyse or oxidise to produce acids in most soil conditions. The chemical symbols have their normal meanings. The subscripts ac, ex, and ad represent material accumulated in, exported from, or added to the ecosystem. The units of all terms are mol(+) or mol(−).ha^{-1}. $period^{-1}$. Methods of measuring acid addition (the left hand side of the equation) are described in Section III.

Some potentially acidifying reactions are not accounted for in equation (2). As mentioned in Section II,B, nutrient cycles involving sulphur, iron and manganese also include acid and alkaline reduction and oxidation reactions. These cycles are thought to be of minor importance in causing soil to become progressively acid in the medium term (months to years), but are recognised as being significant causes of seasonal pH fluctuations in periodically flooded soils. It would probably be most practical to define sulphate as the reference state if measurements of acidification from the sulphur cycle were important. This means that accumulations, gains or losses of organic and sulphide sulphur would need to be measured, which is probably an easier task than measuring leaching of sulphate and measuring small changes in the size of the large sulphate pool in the soil. Except for addition of sulphides and elemental sulphur, and sulphuric acid in rainfall, we have ignored the effects of the sulphur cycle in our calculations. Transformations and transport of manganese and iron over very long periods may also be important in processes of soil formation. If they were considered important in a given system, terms for each could be simply added to equation (2).

2. *Measuring nitrogen cycle acidification*

The reference state, and acidic and alkaline forms of nitrogen have been defined in Figure 5. Transformations between the reference state and acid or alkaline products involve production or consumption of one mole of H^+ per mole of nitrogen transformed.

Cursory examination of equation (2) may raise questions such as: does not nitrate loss by denitrification produce alkalinity? If so, why does $NO^-_{3\,ex}$ acidify? Closer examination, however, reveals that nitrate loss by denitrification occurs as one of the gases N_2, N_2O or NO rather than as nitrate itself. The alkali produced on denitrification is equivalent to the reduced size of the acid $NO^-_{3\,ac}$ term — so in an accounting sense the alkali produced in the denitrification reaction is included in the equation.

The quantities of nitrate and ammonium added ($NO^-_{3\,ad}$ and $NH^+_{4\,ad}$) are known in the case of fertilizers of known composition. The addition of nitric acid in rainfall is usually small, and in this scheme can be ignored because the 'alkaline' $NO_{3\,ad}^-$ from the rainfall is equivalent to the acid added. If manures containing significant amounts of ammonium or nitrate are used, then analysis of their content of ammonium and nitrate is required. In organic manures or plant residues, levels of mineral nitrogen are, probably low enough to be ignored.

In aerated, moist soils in the pH range 5–8, and at temperatures above 6°C, accumulations of ammonium and nitrate ($NH_{4\,ac}^+$ and $NO_{3\,ac}^-$) are limited by nitrification and plant uptake. This results in the generalisation that nitrate and readily exchangeable ammonium rarely constitute more than 1 or 2% of the nitrogen present in the soil (Russell, 1973). However, fixed ammonium held within the lattice of montmorillonite and vermiculite-illite type clays can represent up to half of the nitrogen present in subsoils. The absolute levels present are of the order of 0–1 cmol.kg^{-1} of soil or 0–1.4 kmol.ha^{-1}.m^{-1} (Tisdale and Nelson, 1975) which are small in the context of soil acidification rates. Both nitrate and ammonium levels in the soil can be measured using standard techniques (Keeney and Nelson, 1982).

This leaves nitrate and ammonium exports ($NO_{3\,ex}^-$ and $NH_{4\,ex}^+$) to be measured. A major source of nitrogen loss is probably nitrate loss in run-off and leaching beyond the root zone. Run-off and leaching losses of ammonium are usually small because most soils adsorb ammonium. The amounts of free nitrate and ammonium in plant and animal products are also small or negligible. Therefore the $NH_{4\,ex}^+$ term can be neglected or given a small value based on levels of ammonium measured in products and run-off water.

Measurement of nitrate leaching is difficult. It involves continuously intercepting nitrate as it leaves the root zone without disturbing the leaching process. There are many sampling difficulties and large numbers of analyses are required. Our suggestion is that since this is the most difficult term to estimate in (2) (including the acidification rate itself, on the left of the equation), it can be most accurately estimated difference. For the difference calculation to be reasonably reliable in a particular ecosystem, consideration must be given to the validity of the assumptions used to disregard certain terms as negligible (e.g. NH_4^+ leaching, effects of iron, sulphur or manganese cycles).

3. *Measuring carbon cycle acidification*

In the carbon cycle, neutral carbon compounds (e.g. sugar, starch) and undissociated carbonic acid have been defined as the reference state. Organic anions and bicarbonate ions are thus acid products. So OA_{ac} and

OA_{ex} are acidifying and OA_{ad} produces alkali. Similarly $HCO^-_{3\,ex}$ and $HCO^-_{3\,ac}$ are acidifying and $HCO^-_{3\,ad}$ produces alkali (equation (2)).

Organic anion exports and additions (OA_{ex} and OA_{ad}) can be conveniently measured by ashing the material and titrating the ash to a suitable end-point with acid (Pierre *et al.*, 1970; Jarvis and Robson, 1983a). Ash alkalinity values for many plant materials have been measured (Table I). The alkalinity value is equal to the sum of the non-nitrogenous *inorganic* cations absorbed minus the sum of non-nitrogenous *inorganic* anions, plus any sulphate-sulphur that has been reduced in the plant (Jarvis and Robson, 1983a). The excess cations are balanced by *organic* anions.

Alternatively, OA_{ex} and OA_{ad} can be estimated from published data on the organic anion content of various materials. Organic anion content increases from stem to leaf, and decreases with plant age (Table I). Ash alkalinity values are generally very low in grains, and higher in legumes than grasses. Plants grown with nitrate-nitrogen have higher values than those fixing nitrogen, which have higher organic anion contents than plants supplied with ammonium (Jarvis and Robson, 1983b; Table I). For more detailed data refer to Pierre and Banwart (1973).

The measurement of the accumulation of organic anions should take into account the net accumulation of anions during a certain time period and the change in dissociation/association of organic anions due to concomitant changes in pH (see application of this principle in equation 5 Section IV,C).

Accumulation of organic anions in an ecosystem (OA_{ac}) should be measured at the initial soil pH and at the final soil organic matter levels for substitution in equation (2). The initial soil pH is used to take account of organic acids that have accumulated in undissociated forms at the final, lower soil pH, that would dissociate H^+ and organic anions at the initial soil pH (see application of this principle in equation (5), Section IV,C).

Organic anions occur in soil organic matter, litter and standing plant material. In forests, from 60 to 85 per cent of the total organic carbon in the ecosystem has been measured in living plants. This proportion tends to decrease as soil humus levels increase with time (Jenny, 1980). In grassland and agricultural ecosystems however, the biomass of living plants is much lower than in forests, and at times (fallow periods or dead grasslands in dry seasons) is near zero. Thus in these systems most organic anions are in the soil or on the soil surface.

It is important to note that clearing, and the method of clearing, can have very large effects on soil acidity. Oxidation (burning) of the large accumulation of organic anions on site can cause large increases in soil pH

Table I. The ash alkalinity of plants and plant parts.

Plant	Stage of growth plant part	Ash alkalinity ($cmol.kg^{-1}$)	Reference
White clover	pre-flowering	111,209,163	
Subterranean clover	"	80,196,147	Jarvis & Robson (1983b)[a]
Wheat	"	50, 73, –	
Perennial ryegrass	"	61,113, –	
Cereal grains		6 (5–7)	
Cereal hay		44 (26–76)	Pierre & Banwart (1973)[b]
Forage grasses		69 (41–117)	
Legume crops		110 (72–173)	
	Harvest		
Corn	leaves	93	
	stems	44	
	grain	5	
	Pods yellowing		
Soybeans	leaves	254	
	stems	96	
	pods	83	Pierre & Banwart (1973)
	Flowering		
Lucerne	leaves	169	
	stems	91	
	Fruiting		
Tomatoes	leaves	152	
(Fruiting)	stems	107	
	Mature		
Carrots	tops	182	
(Mature)	roots	84	

[a] Values are for plants grown with nitrogen supplied as NH_4^+ fertilizer, NO_3^- fertilizer, or by nitrogen fixation respectively.
[b] Mean values for a number of species and samples with the range in parentheses.

compared with removal of the timber. Similarly, the fate of the soil organic and plant nitrogen accumulated in the system is important. If much more nitrate is mineralised on clearing than can be recovered by the new plant system, then strong acidification may result from nitrate leaching. Alternatively, if a proportion of the nitrate is denitrified, proportionately less acidification will occur. An extreme example of the contrasting effects of slash and burn as compared with timber removal has been given by Sanchez and Salinas (1981) for the clearing of Brazilian rainforests. The slash and burn technique caused a rise in soil pH before a steady decrease, while removal of the timber led to a rapid and permanent drop in soil pH. There are also clear implications for contrasting methods of forest management such as clear felling, selective logging with partial canopy maintenance, different degrees of biomass removal, and different methods of post-harvest management of the litter.

The organic anion content of soil organic matter is not easily measured, because of problems of extracting all the soil organic matter and changes during extraction. An ashing and titration procedure, comparable with that used for above-ground organic matter, could be used after adjusting the organic matter extract to the soil pH value and drying it. Alternatively, the cation exchange capacity (CEC) of the organic matter can be determined at various pH levels but the results depend on the method used (Allison, 1973).

Another approach to evaluating OA_{ac} in soil organic matter is to calculate it from the soil organic matter content and soil pH. Helling *et al.* (1964) measured the whole soil CEC of 60 soils at several soil pH values. Then a multiple regression was used to allocate CEC values to the clay and organic matter components. Their results and the soil humic acid titration curve data reviewed by Kononova (1966) are expressed by equation (3), with the assumption that the titration curve for soil organic matter is close to linear over the pH range 3.5–8.0.

Thus, for pH 3.5–8.0,

$$\mathrm{OM} - \mathrm{CEC}\ (\mathrm{cmol}\ (+).\mathrm{kg}^{-1}\ \mathrm{OM}) = a(\mathrm{pH} - 1.5) \tag{3}$$

The slope coefficient, *a*, for humic acids varies slightly between soils but is near 83 cmol(+).kg^{-1} OM (Kononova, 1966). In contrast, calculations from the data of Helling *et al.* (1964) averaged over 60 soils, give an *a* value of 32 cmol(+).kg^{-1} for 'soil OM'. For both sets of data, linear extrapolation gives a pH for zero net charge of the organic matter of 1.5. This pH is similar to the pK_a values for the stronger, naturally occurring organic acids such as oxalic acid ($pK_a = 1.2$). The difference between the soil and humic acid *a* values probably reflects the fact that increasing humification leads to increased OM − CEC values (Allison, 1973). Furthermore, assuming that ash alkalinity values provide a close estimate of the CEC of plant material (Jarvis and Robson, 1983a), and assuming that the pH of plant tissue is 6.0, ash alkalinity values can be substituted into equation (5) to calculate *a* values for fresh plant residues. Substitution of ash alkalinity values from Table I yields *a* values varying from 7 cmol(+).kg^{-1} for cereal straw to 44 cmol(+).kg^{-1} for young legume material grown with a nitrate-nitrogen source.

These calculations show that the mean slope of the titration curve for soil organic matter plus surface litter (i.e. all the material required to calculate the OA_{ac} value for a soil) will lie between 7 cmol(+).kg^{-1}, the probable lower value for plant litter, and 83 cmol(+).kg^{-1} for humic acids. The mean value of 32 cmol(+).kg^{-1} calculated from the data of Helling *et al.* (1964) is probably representative of many soils. Higher or lower values are representative of soils with very high or very low proportions of their organic matter in a humified form.

Thus the OA_{ac} level for substitution in equation (2) in soil can be estimated from knowledge of the soil organic matter content (or organic C or N measurements and estimated C/N and OM/C ratios), the initial soil pH, and the *a* value in equation (3). A C/N ratio of 155/10 (Williams and Donald, 1957) and an OM/C ratio between 1.7/1 and 1.9/1 (Allison, 1973; Jenny, 1980) can be used to estimate soil OM from C or N measurements. Table II has been calculated using equation (3), and an *a* value of 32 cmol(+).kg^{-1} OM. The variation OA_{ac} values with soil organic matter and soil pH is shown in Table II along with the amount of $CaCO_3$ equivalent to the acidity dissociated from these organic anions.

Acidification resulting from bicarbonate fluxes to and from the system (the $HCO^-_{3\,ac}$, $HCO^-_{3\,ac}$ and $HCO^-_{3\,ex}$ terms) can be estimated from measurements of the amount and pH of water contained in, entering, and leaving the system, and from estimates of the partial pressure of CO_2 (G) with which the water is equilibrated. Standard CO_2 (G) solubility and H_2CO_3 dissociation constants (Lindsay, 1979) are then used to calculate the three bicarbonate terms.

Net bicarbonate balance calculations are given in Table III for a range of rainfall and leaching conditions. The assumption of zero run-off used in these calculations could be easily changed where loss of bicarbonate in run-off is significant. The extra values needed are the amount of water lost by run-off and the CO_2(G) concentration with

Table II. The cation exchange capacity of organic matter, and its equivalent in t $CaCO_3$.ha^{-1} over 0–10 cm for different soil organic matter and pH levels (soil bulk density assumed to be 1.37 g cm^{-3}). Values calculated from OM − CEC = a(pH − 1.5).

Soil OM (%)	OM − CEC (cmol(+).kg soil^{-1})			$CaCO_3$ equivalent (t.ha^{-1} 10 cm)		
	pH 4	pH 6	pH 8	pH 4	pH 6	pH 8
1	0.79	1.36	1.92	0.54	0.93	1.32
2	1.58	2.72	3.84	1.08	1.86	2.63
5	3.95	6.80	9.60	2.71	4.66	6.58
10	7.90	13.60	19.20	5.41	9.32	13.20
20	15.80	27.20	38.40	10.80	18.60	26.30

Table III. Net moles HCO_3^- balance per hectare ($HCO^-_{3\,ad} - HCO^-_{3\,ex}$) for different amounts of rainfall, rainwater pH values, soil pH values and leaching rates[a] (Positive values indicate increased acidity, negative values reduced acidity).

Rainfall[b]		Soil and leached water[c]		Net HCO_3^- lost ($mol.ha^{-1}$)			
pH (Nature of rain)	HCO_3^- concentration ($mol.100\ mm^{-1}.ha^{-1}$)	pH	HCO_3^- concentration ($mol.100\ mm^{-1}.ha^{-1}$)	Rainfall ($mm.year^{-1}$)			
				200	400	800	1600
				Proportion rainfall leached below root zone			
				0	0.05	0.20	0.50
5	0.454	4	0.454	−0.908	−1.72	−2.91	−3.63
(mildly acid		5	4.540	−0.908	−0.90	3.63	−29.0
rain)		6	45.40	−0.908	7.27	69.0	356
5.67	2.14	4	0.454	−4.28	−8.47	−16.4	−30.7
(water/atmospheric		5	4.540	−4.28	−7.65	−9.36	2.08
CO_2 equilibrium)		6	45.40	−4.28	0.52	55.5	329

[a] The following relationship (Lindsay, 1979) has been assumed: $pHCO_3^- = pCO_2 - pH + 7.82$
Where $pHCO_3^-$ is the negative logarithm of the HCO_3^- concentration and pCO_2 is the negative logarithm of the partial pressure of CO_2 in atmospheres. A further assumption is that run-off is zero.
[b] Atmospheric CO_2 partial pressure assumed to be 0.0003 atm.
[c] Soil air CO_2 partial pressure assumed to be 0.003 atm.

which it is equilibrated. Atmospheric $CO_2(G)$ would be a good initial estimate.

The most interesting feature of these results is that the net effect of the bicarbonate fluxes: (1) is always slightly alkaline (negative values) at a soil pH of 4.0, and at all pH values where there is no leaching; (2) is usually alkaline or very slightly acid at soil pH 5.0; and (3) at pH 6.0 reaches an acidification rate of about 0.3 kmol $H^+.ha^{-1}.year^{-1}$, but only at high rainfall and leaching rates (Table III). Including losses of bicarbonate in run-off water would make all the values slightly more positive (i.e. more acidic).

4. *Measuring direct application of acid and alkali (H^+_{ad}, H^+_{ex}, L_{ad})*

The quantity of H^+ ions gained (H^+_{ad}) and lost (H^+_{ex}) in water by an ecosystem, can be calculated from the pH values of the rainfall, leachate, and run-off. This accounts for these forms of H^+ addition and loss whether derived from weak acids (e.g. H_2O, H_2CO_3) or strong acids (e.g. H_2SO_4, HNO_3). In many cases these calculations will account for the total H^+_{ad} and H^+_{ex} terms. In the case of HNO_3 in rainfall, accounting in this section has to be consistent with calculations of the $NO^-_{3\,ad}$ term discussed in Section II,C,*2* (i.e. if nitrate in rainfall is ignored in the $NO^-_{3\,ad}$ term, the equivalent H^+ must be ignored here).

The net H^+ term ($H^+_{ad} - H^+_{ex}$) under various rainfall and leaching conditions, for rainfall in equilibrium with atmospheric CO_2 (G) and for slightly acid rain, is shown in Table IV. In the absence of leaching, H^+ addition in rainfall is obviously acidifying. Secondly, the soil pH at which the term ($H^+_{ad} - H^+_{ex}$) becomes negative or alkaline decreases as the rainfall acidity increases, and as the proportion of the rainfall leached decreased (Table IV). The net effect of HCO_3^- and H^+ fluxes in water is discussed in Section V,D which considers acidification due to acid and normal rainfall.

The contribution to the H^+_{ad} term of human additions of acids and materials that rapidly hydrolyse or oxidise to produce acid (Fe^{3+}, Al^{3+}), can be calculated from the quantities added per hectare and the compound composition.

The L_{ad} term represents the addition of alkaline inorganic compounds other than bicarbonate ions to the ecosystem. Liming materials such as $CaCO_3$, CaO, $Ca(OH)_2$, $MgCO_3$ and mixtures of these, are the main compounds encountered in agricultural ecosystems. The accounting system assumes these compounds have completely reacted, and the OH^- in water fluxes is subtracted from the H^+ before determining H^+_{ad} and H^+_{ex} (see Table IV).

Table IV. Net moles H^+ gained or lost ($H^+_{ad} - H^+_{ex}$) due to H^+ entering in rainfall and leaving as leached soil water, as a function of rainfall amount and pH, and leachate amount and pH (mol $H^+.ha^{-1}.year^{-1}$). Run-off is assumed to be zero.

Rainfall		Soil and leached water		Net $[H^+]$ gained[a] ($mol.ha^{-1}.yr^{-1}$)			
				Rainfall ($mm.yr^{-1}$)			
				200	400	800	1600
				Proportion rainfall leached			
pH	$[H^+] - [OH^-]$ ($mol.100\ mm^{-1}.ha^{-1}$) or $moles.L^{-1}$)	pH	$[H^+] - [OH^-]$ ($mol.100\ mm^{-1}.ha^{-1}$) or $moles.L^{-1}$)	0	0.05	0.20	0.50
5.0	10.0	4	100	20.0	20.0	−80.0	−640
(mildly acid		5	10.0	20.0	38.0	64.0	80.0
rain)		6	0.99	20.0	39.8	78.4	152
5.67	2.13	4	100	4.26	−11.5	−143.0	−766
(water/atmospheric		5	10.0	4.26	6.52	1.04	−45.9
CO_2 equilibrium)		6	0.99	4.26	8.32	15.5	26.1

[a] $\underbrace{([10^{-pH}] - [10^{-pOH}]) \times \text{Volume}}_{\text{Rainfall}} - \underbrace{[10^{-pH}] - [10^{-pOH}] \times \text{Volume}}_{\text{Leachate}}$

D. Carbonate soils

The rate of dissolution of carbonates is clearly affected by the addition of acid but is also strongly affected by the water flux through the soil, and dissolution rates increase with lowered temperature (Jenny, 1980). The principles discussed above do apply to carbonate soils, but the measure of 'acidification' is the rate of disappearance of $CaCO_3$ rather than the product of the change in pH and the buffer capacity. Secondly, the effect of the water flux on $CaCO_3$ dissolution would need to be considered. Arkley (1963) estimated that at 13°C, 100 mg $CaCO_3$ would be dissolved per litre of water passing through a carbonate horizon. No further consideration will be given to acidification processes in carbonate soils.

III. MEASURING RATES OF ACIDIFICATION

The rate of acidification of a soil can be expressed in two ways that are of interest in agricultural systems: (1) the rate of change in soil pH (pH units.period^{-1}); and (2) the rate of addition of acids which must be neutralised to maintain the soil pH constant (mol.ha^{-1}.period^{-1}). The latter measure is calculated from the former, using other soil parameters:

$$\text{Acidification rate (mol H}^+\text{.ha}^{-1}\text{.period}^{-1}) = \sum_{i=1}^{n} \Delta pH_i \times pHBC_i \times BD_i \times V_i \qquad (4)$$

where, for each soil horizon or litter layer (i), ΔpH_i is the change in pH over the period, $pHBC_i$ is the pH buffer capacity at the end of the period (mol H$^+$.kg^{-1}. pH unit^{-1}), BD_i is the bulk density (kg.m^{-3}) and V_i is the volume (m^3.ha^{-1}).

A. Absolute and relative estimates of acidification

Studies which measure components of equation (4) can be divided into two groups: those which measure *absolute* changes in acidity; and those which estimate changes in acidity *relative* to some control soil. An example of the former is the study of Watson (1969) who measured soil pH at a site before and after a 10-year experiment. Examples of studies into relative rates of acidification are those of Williams and Donald (1957), who compared the acidity of developed pasture soils with that of nearby undeveloped pasture soils, and of Russell (1960), who compared trial plots which had received superphosphate over the previous 30–37 years with plots which had received no superphosphate.

B. Sampling techniques

Whether a study measures absolute or relative acidification, three sources of error must be accounted for: (1) horizontal, spatial variability; (2) temporal variability; and (3) variability in a control soil which is different to variability in the treated soil.

1. Horizontal, spatial variability

Horizontal, spatial variability in pH, buffer capacity, bulk density and horizon depth can be analysed using conventional statistical methods (e.g. Snedecor and Cochran, 1967; Sokal and Rohlf, 1969) and so an adequate sampling plan can be determined. Raupach (1951) could draw no robust conclusions in a study on the magnitude of seasonal variability in pH because his sampling plan was inadequate to account for horizontal, spatial variability.

2. Temporal variability

In general, the soil pH fluctuates within any period over which a net change in pH is being estimated. So, for example, soil samples collected before and after a 10-year period should at least be taken at the same time of year to allow for any annual cycle in pH (Kelly, 1923; Swanback and Morgan, 1930, both cited by Kanehiro *et al.*, 1951). Preferably, monthly samples should be collected to characterise the annual cycles at the start and finish of the period.

Because acidification studies usually involve long time periods, there is potential for significant changes in soil properties which, in shorter-term studies, are usually assumed to be stable.

Changes in *buffer capacity* over time are most likely to result from changes in the organic matter content. A change in the content of clay, the other major contributor to soil buffer capacity, is less likely. The *bulk density* of a horizon can decrease, and so its depth increases through organic matter accumulation (Skene, 1966), or through the loosening of the horizon during cultivation. The reverse, an increase in bulk density and reduction in the horizon depth, results from soil compaction. Finally, erosion can reduce the *volume* of a surface horizon without any effect on its bulk density. The implication of any change in the volume of a soil horizon is that a depth of sampling which is appropriate at the beginning of a study period may not be appropriate at the end of that period. Any change in the depth of a horizon will have most significance on the accuracy of a study in which adjacent horizons have markedly different pH values, buffer capacities or bulk densities.

Table V. Variation in soil pH (1:1 soil:water) of six cultivated and six uncultivated Hawaiian top soils between 7 January 1949 and 22 March 1950 (from Kanehiro *et al.*, 1951).

Soil	Cultivated area		Uncultivated area	
	Average	Variation[a]	Average	Variation
Lualualei	7.9	1.2	7.8	1.0
Pearl City	6.4	1.2	6.4	1.1
University Farm	5.9	1.5	6.8	0.7
Kaneohe	5.8	1.5	6.3	0.4
Waimanalo	6.6	1.2	7.1	0.9
Koko Head	7.1	1.0	7.2	1.1
Average	6.6	1.3	6.9	0.9

[a] Variation = highest pH obtained minus lowest pH obtained.

3. *Variability in control soil*

The third potential source of error in studies of relative acidification rates results from the variability in a control soil being different to variability in the treated soil. Short-term fluctuations in pH of the two soils may be different. Control soils are commonly uncultivated or undisturbed soils adjacent to the soil of interest. Kanehiro *et al.* (1951) found that the pH of six cultivated soils varied much more within a season than six uncultivated soils nearby (Table V). Taking their most extreme example, the difference in pH between the cultivated and uncultivated Kaneohe soil ranged from − 0.2 units to + 0.9 units depending on the time of year at which paired samples were taken.

C. Measurement techniques

1. *Measuring soil pH*

In acidification studies, the pH of a soil should be measured in a suspension made using a dilute salt solution (e.g. 1:5 soil:0.01 M $CaCl_2$). Many techniques for measuring soil pH have been used in acidification studies. They vary in the ionic strength of the suspending solution and the soil:solution ratio. In some studies (e.g. Pierre *et al.*, 1971) the soil has been leached before the pH was measured.

Water is not an appropriate liquid to suspend soil in for measuring pH in soil acidification studies. The salt content of a soil, and so the ionic strength of the soil solution, is likely to change over the period of a soil acidification study. Changes in the ionic strengths of suspending solutions used in pH measurement can have marked effects on the measured pH. Dolling and Ritchie (1985) found that the ionic strength of soil solutions increased from 3.9×10^{-3} to 9.6×10^{-3} as the period of cultivation increased from 3 to 15 years. They also found that by increasing the ionic strength of the suspending solutions from near zero (distilled water) to 5×10^{-3}, the measured pH fell by 0.3–1.0 units, with an average fall of 0.59 pH units.

2. Measuring pH buffer capacity of soil

Outside the pH range 4.5–7.5 the soil pH is often very strongly buffered by precipitation-dissolution reactions involving carbonate minerals at high pH (Russell, 1973; Jenny, 1980) and aluminium hydrous oxides at low pH (e.g. Hartikainen, 1986). In these soils, acid addition causes little pH change because mineral dissolution buffers the soil pH.

In the pH range 4.5–7.5, soil pH decreases with acid addition. The pH is mainly buffered by association/dissociation reactions of H^+ ions. Dissolution/precipitation processes are relatively slow in this pH range.

The soil pH buffer capacity is the measure of the rate of acid or alkali addition per unit change in soil pH. It is determined by titrating soil with acid or base. Often the buffer capacity is approximately constant within the pH range 4.5–6.0 (Magdoff and Bartlett, 1985) and so a single value is used. In any acidification study the titration curve should be characterised.

In titrating a soil there are three variables: the reagent used, the conditions of equilibration, and the method of measuring pH. Although many techniques have been used to measure soil buffer capacity (Table VI), there have been no studies on the suitability of alternatives for these variables. However, the following comments apply:

(a) Reagent
- can be either acid or base.
- concentrations used should give enough points on the titration curve to define its shape adequately over the pH range of interest.
- the associated cation or anion should preferably be one that is common in the soil solution.

(b) Conditions of equilibration
- soil:liquid ratio should probably be high to allow gas exchange readily. It may be appropriate to periodically dry then rewet the soil.
- the equilibration time should be as long as possible. There is little information on the role of slow reactions in buffering soil pH.

Table VI. Some techniques used to estimate pH buffer capacity of soils.

Author	Reagent	Dilution (soil:water)	Period	Temperature	pH method	comments
Abruna *et al.* (1958)	$Ca(OH)_2$	1:5	1 d	?	1:5 H_2O	
Abruna *et al.* (1958)	fine ground dolomitic limestone 50% neutral value.	'moist'	7 d	warm	1:1 H_2O	
Pierre *et al.* (1971)	pptd. $CaCO_3$	1/3 bar	400 d	8°C	0.01 M $CaCl_2$	
Williams (1980)	0.04 N $Ca(OH)_2$	1:5	20 h	?	1:5 H_2O	
Webber *et al.* (1977)	$CaCO_3$	In field	30 d	lab.	0.02 M $CaCl_2$	dried before pH measured
Fox (1980)	pptd. $CaCO_3$	Field capacity: dry = 1 mo:1 mo	6 mo (3 cycles)	Room temp.		
Magdoff and Bartlett (1985)	$CaCO_3$ and H_2SO_4	Field capacity	30 d	Glass house	2:1 0.01 M $CaCl_2$	

- the temperature should restrict nitrification and other biologically mediated transformations, if the equilibration times are long enough for those to be significant. High temperatures will accelerate any important slow reactions but very high temperatures make it difficult to maintain a constant moisture content.

(c) Method of measuring pH

- dilute salt solution to suspend the soil, though not so dilute that the variable ionic strength of the different acid/base reagent treatments will affect the measured pH.

IV. RATES AND MECHANISMS OF ACIDIFICATION OBSERVED IN ONE ECOSYSTEM

Understanding the mechanisms of acidification operating in an ecosystem is the starting point for designing the best acidity management practices. In this section, an example of one agricultural system is given to illustrate how equation (2) can be used to estimate the relative importance of different acidification processes in an ecosystem.

Acidification under grazed subterranean clover/annual grass pastures on the Southern Tablelands of New South Wales has been studied extensively (Williams and Donald, 1957; Siman and Flemons, 1970; Williams, 1980; Bromfield *et al.*, 1983; Cumming and Williams, 1983). These studies provide much of the information required to define the sources of acidification in the soil using equation (2). The total acidification during periods of pasture production in two groups of yellow duplex soils developed from granitic and sedimentary rocks is shown in Table VII.

The pHBC, soil organic matter and bulk density data have been assembled from various sources and some assumptions have been made. Bulk densities were estimated from the data given by Stace *et al.* (1968) for similar yellow podsolic soils. The original level of soil organic matter was assumed to be 1.9%. in the 0–10 cm layer (Williams and Donald, 1957; Williams, 1980) and 1.0% in the 10–20 cm layer (Stace *et al.*, 1968). The increment in organic matter with time was calculated using the relationships: total soil N = 0.0223 × (years pasture) + 0.0664 (Williams, 1980), and assuming a C:N ratio of 15.5:1.0 (Williams and Donald, 1957) and OM:C ratio of 1.8:1.0 (Jenny, 1980). The absolute increase in organic matter in the 10–20 cm zone was assumed to be 40% of the increase in the 0–10 cm layer. No increase was allowed for in deeper layers and the values used were estimated from yellow podzolic profiles described by Stace *et al.* (1968). The pHBC at time 2 (t_2) and the soil pH data were taken from Bromfield *et al.* (1983). The reduced pHBC for the surface horizons at time 1 (t_1) was calculated assuming that the

Table VII. Estimates of total profile acidification in two groups of yellow duplex soils (Dy 3.42 and Dy 3.41; Northcote, 1971) during different periods (from t_1 to t_2) under grazed pastures of subterranean clover/grass.

Horizon depth (cm)	Bulk density ($g.cm^{-3}$)	Organic matter (%)		Soil pH buffer capacity ($cmol.kg^{-1}.pH\ unit^{-1}$)		Soil pH 1:5 soil:water		Soil acidification ($kmol\ H^+.ha^{-1}$)
		t_1	t_2	t_1	t_2	t_1	t_2	
Soil: Granite origin, Dy 3.42, $t_2 - t_1 = 55$ years.								
0–10	1.4	1.90	5.31	1.56	2.65	6.15	4.75	49.3
10–20	1.5	1.0	1.7	0.88	1.1	6.50	4.50	33.0
20–30	1.5	0.4	0.4	0.6	0.6	6.80	4.7	18.9
30–40	1.5	0.3	0.3	0.4	0.4	7.00	4.7	13.8
40–50	1.6	0.3	0.3	0.6	0.6	7.00	4.8	21.1
50–60	1.6	0.3	0.3	0.7	1.7	7.00	5.0	54.4
0–60								190.5 ($3.46.year^{-1}$)
Soil: Sedimentary rock origin, Dy 3.41, $t_2 - t_1 = 25$ years.								
0–10	1.4	1.70	3.45	2.65	3.15	5.7	4.95	30.2
10–20	1.5	1.0	1.3	1.71	1.8	5.7	5.0	18.9
20–30	1.5	0.4	0.4	1.20	1.2	5.7	5.0	12.6
30–40	1.5	0.3	0.3	4.10	4.1	5.6	5.2	24.6
40–50	1.6	0.3	0.3	6.00	6.0	5.6	5.4	19.2
50–60	1.6	0.3	0.3	11.6	11.6	5.5	5.5	0.0
0–60								105.5 ($4.22.year^{-1}$)

Note: 10 $kmol\ H^+.ha^{-1}$ is equivalent to 0.5 t $CaCO_3.ha^{-1}$.

contribution of organic matter to buffer capacity was 32 cmol(+).kg^{-1} OM.pH unit^{-1} (see Section II,C,*3*).

A. The acidification rate

The total acid input for each soil was determined using equation (4) and the data in Table VII. The acidification rate was slightly greater in the soil developed from sediments than in that from granite (4.22 compared with 3.46 kmol H^+.ha^{-1}.year^{-1} respectively (Table VII)). The difference is partly due to an inadequate depth of sampling from the granitic soil in which some acidification had occurred below the 60 cm sampling depth (Table VII).

In these cases the soil pHBC increased during the acidification period due to substantial increases in the organic matter content of the soils. The total acidification has been calculated using the final pH buffer capacity.

B. Nitrogen cycle acidification

These soils have not received nitrogenous fertilizer so the $NO^-_{3\,ad}$ and $NH^+_{4\,ad}$ terms are zero (Table VIII). The data of Simpson (1962) and Williams (1968) show nitrate and ammonium levels are about 5 μg NO_3^--N and 2 μg NH_4^+-N.kg soil^{-1}. These values convert to 3.24 and − 1.26 kmol.ha^{-1} respectively for terms $NO^-_{3\,ac}$ and $NH^+_{4\,ac}$.

Of the nitrogen loss terms, $NH^+_{4\,ex}$ has been estimated as zero on the assumption of little or no NH_4^+ leaching. However, the $NO^-_{3\,ex}$ term has been estimated by difference as 1.40–2.2 kmol NO_3^-.ha^{-1}.year^{-1} lost by leaching and run-off. Since this term has been determined by difference, some direct evidence is required.

Complete leaching from the root zone of 20–27 μg NO_3^--N. soil^{-1} g in the surface 10 cm layer, would account for all the acidification attributed to $NO^-_{3\,ex}$. Simpson (1962) and Williams (1968) have demonstrated that NO_3^--N accumulated to levels as high as 60 μg N.g soil^{-1} in the surface soil, over the summer period when pasture growth rates are minimal. These levels are reduced to less than 5 μg.g. soil^{-1}, partly due to plant uptake during periods of active plant growth. However, Simpson (1962) also demonstrated the leaching of NO_3^--N to at least 80-90 cm, and NO_3^--N accumulated at that depth to 19 μg N.g soil^{-1} following a period of rainfall sufficient to cause leaching before, or early in the pasture growing season. Thus one or two leaching events to beyond the rooting depth per year, in late summer or autumn, are sufficient to account for the calculated $NO^-_{3\,ex}$ rate.

Table VIII. Total soil acidification and sources of acid/alkali for two yellow duplex soils under grazed pastures of subterranean clover/annual grass for long periods.

Acidification component (from Eq. (2))	Acidification ($kmol\ H^+.ha^{-1}$)[a] Granitic soils			Sedimentary soils		
	per 55 years	per year	% of total	per 25 years	per year	% of total
N cycle terms						
$NH^+_{4\,ad}$	0	0	0	0	0	0
$NO^-_{3\,ad}$	0	0	0	0	0	0
$NH^+_{4\,ac}$	−1.26	−0.02	−0.6	−1.26	−0.05	−1.2
$NO^-_{3\,ac}$	3.24	0.06	1.7	3.24	0.13	3.0
$NH^+_{4\,ex}$	0	0	0	0	0	0
$NO^-_{3\,ex}$ by difference	76.9	1.40	40	53.7	2.15	51
C cycle terms						
OA_{ac} (pH = t_1)	81.8	1.49	43	36.5	1.46	34
OA_{ac} (pH = t_2)[b]	(26.3)	(0.48)	(–)	(20.6)	(0.86)	(–)
OA_{ex}	28.3	0.52	15	12.9	0.52	12
OA_{ad}	0	0	0	0	0	0
Mass flow (H^+, OH^-, HCO_3^-)	2.0	0.036	1.0	0.9	0.036	0.8
Other Acids						
H^+_{ad}	0	0	0	0	0	0
Other Alkalis						
L_{ad}	0	0	0	0	0	0
Total acidification	191	3.46	100	106	4.22	100

[a] 10 $kmol\ H^+.ha^{-1}$ is equivalent to 0.5 t $CaCO_3.ha^{-1}$.

[b] This term is the actual organic anion content at the end of the acidification period. The term above it (OA_{ac} (pH = t_1) is the correct estimate of acidification caused by accumulation of weak acids such as organic acids. The organic anion accumulation, calculated at the initial pH (see text), includes the final organic anion content plus undissociated weak acids that are dissociated at the initial pH.

C. Carbon cycle acidification

The OA_{ex} term comprises organic anions removed in hay, in animal products (wool and lambs), and in transport of waste products to sheep camp sites (Hilder, 1964). Based on the assumption that organic anion removal and addition in hay were equal, the hay components of OA_{ex} (acid) and OA_{ad} (alkaline) were ignored. The waste product and animal product components of OA_{ex} were based on an 'animal loss factor' of 0.7 kg $P.ewe^{-1}.ha^{-1}.year^{-1}$ developed by Cornforth and Sinclair (1982). At five breeding ewes.ha^{-1}, 0.75 per cent P in the dry products plus waste products, and an organic anion content of 110 cmol.kg^{-1} organic matter in these products (mean value for subterranean clover and grasses from Table I), the OA_{ex} value less the hay component is 0.52 kmol.ha^{-1}.year^{-1} (Table VIII).

The OA_{ac} term was estimated by substituting the initial soil pH and the initial and final organic matter data from Table VII into equation (5):

$$OA_{ac}\ (\text{kmol.ha}^{-1}) = (BD \times VOL \times 0.32(OM\%_{t_2}(pH_{t_1} - 1.5) - (OM\%_{t_1}(pH_{t_1} - 1.5)))i \quad (5)$$

In these examples the soil organic matter level increased as the soils acidified. This resulted in a greater pHBC after acidification (Table VII), and greater total acidification when this was taken into account in equation (2) (Table VIII). This effect on acidification, due to accumulation of a weak acid, is accounted for by substituting the initial pH (pH_{t_1}) into both terms of equation (5). Thus organic anion accumulation is measured with respect to the initial soil pH, rather than determining the actual accumulation of organic anions. The large effect of this correction is shown in Table VIII.

D. Acids and alkalis added directly

The terms L_{ad} and H^+_{ad} are zero because no addition of these materials had been made other than the mass flow of H^+, OH^- and HCO_3^- in rainfall and leachate accounted for by the valves in Tables III and IV for 800 mm of pH 5.67 rainfall, and a soil pH of 6.0.

E. Importance of the various sources of acid

The acidification of these soils has been caused mainly by nitrate losses (40–51 per cent), organic matter accumulation (34–43 per cent) and organic anion exports in products and waste products (12–15 per cent) (Table VIII). Other sources of acid and alkaline effects ($NH^+_{4\,ac}$, $NO^-_{3\,ac}$, $HCO^-_{3\,ad}$ and $HCO^-_{3\,ex}$) have accounted for less than 4 per cent of the total effect.

The lime required to neutralise the total acidification on the granitic and sedimentary soils respectively, is 173 and 211 kg $CaCO_3$.ha^{-1}.year^{-1}. This provides a good estimate of the maintenance lime requirements of these pasture ecosystems. This maintenance lime rate could be reduced by more efficient use of nitrate, less export of waste products, and by minimising the increase in soil organic matter (e.g. by use of crop rotation).

V. VARIATION IN IMPORTANCE OF ACIDIFYING FACTORS IN DIFFERENT SYSTEMS

The contribution of the various terms in equation (2) to acidification varies between different agricultural ecosystems. Systems clearly differ in exports of organic anions. The different products (e.g. wool, milk, grain, hay) vary widely in their organic anion content (Table I), and the quantity of product removed (10–1000 $kg.ha^{-1}.year^{-1}$ for wool and meat; 1–30 $t.ha^{-1}.year^{-1}$ for hay and grain). Similarly, nitrate leaching and soil organic matter accumulation rates vary. Examples are given below to illustrate these differences.

A. Rainfall seasonality and nitrate leaching

The rate of acidification of the surface 10 cm soil on the Southern Tablelands of New South Wales is compared with surface acidification rates at two other experimental sites in the intermediate and high rainfall coastal zone of New South Wales (Fig. 6). The trends in soil pH at the coastal sites were from treatments adequately fertilized with all nutrients except nitrogen. The grass/legume pastures were grazed at set stocking rates by young steers or heifers. The intermediate moisture site supported a pasture dominated by white clover, carpet grass (*Axonopus affinis*), and paspalum (*Paspalum dilatatum*) at Grafton, New South Wales (Mears and Havilah, unpublished data). The highest rainfall site at Wollongbar, New South Wales, supported a pasture dominated by kikuyu (*Pennisetum clandestinum*), white clover, paspalum and carpet grass (Mears and Humphreys, 1974; Colman and Mears, unpublished data).

The coastal sites have higher and more evenly distributed rainfall patterns than the Southern Tablelands, where warm temperature and low soil moisture conditions during summer/autumn allow nitrification to proceed faster than nitrate uptake, so soil nitrate accumulates (e.g. Simpson, 1962; Williams, 1968). The acidification rate of the surface soil increases with decrease in the mean moisture index (mean MI) (Fig. 6A), and the number of dry weeks in the year (Fig. 6C).

Under perennial grass/legume pastures in non-seasonal humid conditions, soil nitrate evidently does not accumulate to levels sufficient to allow significant nitrate leaching. The more seasonal (Southern Tablelands) environment provides more opportunity for nitrate leaching

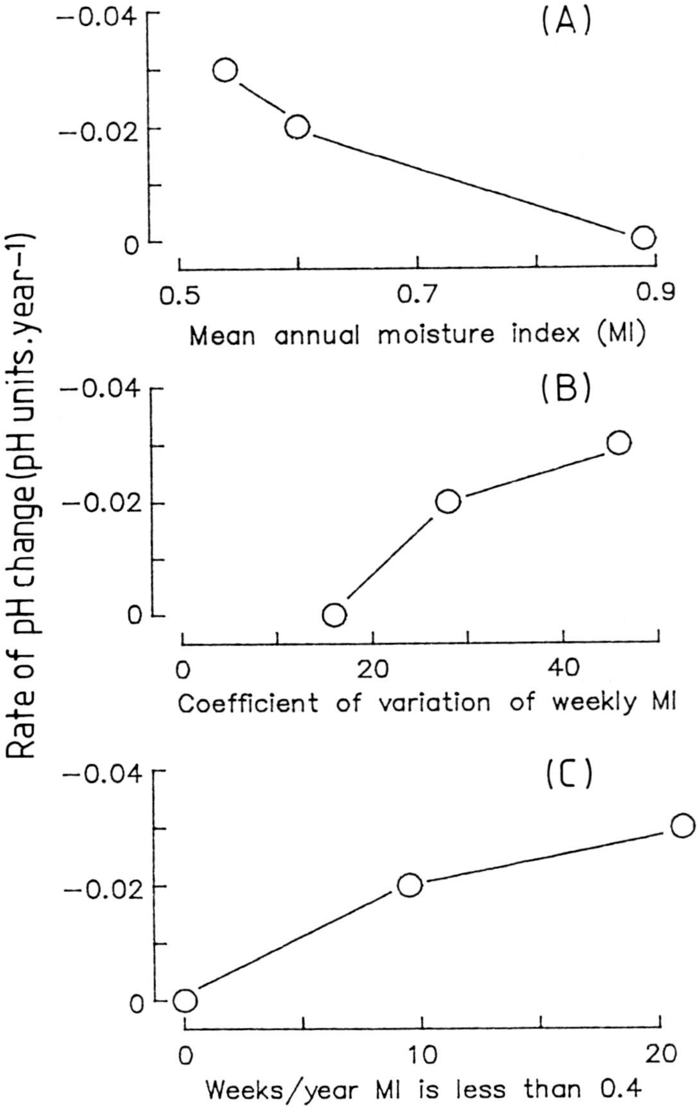

Fig. 6. Soil acidification rates under legume/grass pastures as a function of various measures of the moisture regime: A, the mean annual moisture index (MI). (The moisture status is calculated from climatic records and a soil moisture budget according to Fitzpatrick and Nix (1970). It varies from 0 to 1 representing the range from no plant growth to no moisture restriction on plant growth); B, the coefficient of variation of the weekly MI throughout the year; and C, the number of weeks the MI value is less than 0.4.

Table IX. Soil pH and pH changes under kikuyu pastures and effects of the form and rate of nitrogen and the stocking rate in two adjacent experiments.

Nitrogen form and rate (kg $N.ha^{-1}.year^{-1}$)	Stocking rate (weaner $steers.ha^{-1}$)	0–10 cm pH (0.01 M $CaCl_2$)	Annual change in pH (Difference from N_o[a])
Experiment 1			
Ammonium nitrate			
0	3.3	4.92	(0)
	4.9	4.88	−0.0036
125	4.9	4.89	−0.0027
336	4.9	4.63	−0.026
	7.4	4.47	−0.041
690	11.1	4.00	−0.084
	($LSD_{0.05}$ = 0.12)		
Experiment 2			
Ammonium sulphate			
0	2.5–5.0 $cows.ha^{-1}$	4.95	(0)
336		4.32	−0.158

Experiment 1 reported by Mears and Humphreys (1974).
Experiment 2 conducted before experiment 1 on an adjacent area, and reported initially by Awad and Edwards (1977).
[a] N_o = pH when no fertilizer applied.

following periods of nitrate accumulation. These trends occurred despite similar or greater organic matter accumulation (an acidifying process) in the wetter environments. In a highly seasonal, tropical environment at Katherine, Northern Territory (Mean MI = 0.4; CV of MI = 102; Weeks MI < 0.4 = 31), Wetselaar and Norman (1960) found soil nitrate accumulated during periods of poor growing conditions and was leached to depths of 120–150 cm by heavy rainfall during the wet season. Thus the relationships shown in Figure 6 may also apply to tropical areas. They need to be quantitatively modified wherever dry seasons are so dry that nitrification as well as plant growth is zero (Wetselaar and Norman, 1960).

B. Nitrogen fertilizer and grazing pressure in a perennial grass pasture system

Mears and Humphreys' (1974) data allow more detailed observations on the high rainfall point in Figure 6. Ten years after the establishment of a nitrogen rate by stocking rate experiment grazed by weaner steers, the surface soil was sampled and soil pH values were measured in

0.01 M $CaCl_2$ (Table IX). Nitrogenous fertilizer at 125 kg $N.ha^{-1}$ had no effect on acidification but higher rates increased it. The pasture varied from a lightly grazed, vigorous, kikuyu-dominant pasture at the lowest stocking rate to a very short, heavily grazed pasture at the highest stocking rate (Mears and Humphreys, 1974). Increasing the stocking rate when fertilizing at 336 kg $N.ha^{-1}$ increased the acidification rate (Table IX).

These observations are consistent with expected effects of the treatments on the nitrogen cycle. Increasing additions of ammonium nitrate should result in higher rates of nitrate leaching, and hence acidification once the ability of the plant system to absorb the applied nitrogen is exceeded. Ammonium sulphate should at least double the soil acidification rate caused by ammonium nitrate, so long as nitrification is complete and other conditions are similar. The statistically significant effect of increasing the stocking rate when fertilizing at 336 kg $N.ha^{-1}.year^{-1}$ (Table IX) in increasing soil acidification probably resulted from greater nitrate leaching as the heavy stocking diminished the plants' capacity to absorb nitrogen.

C. Nitrogen form, and grain or hay production — North Dakota, USA

Power *et al.* (1972) compared soil acidification and nitrate leaching under a corn/fallow rotation with that under a bromegrass pasture managed for hay production. In the corn production system with 110 kg N applied.$ha^{-1}.year^{-1}$, up to 168 kg NO_3^--$N.ha^{-1}$ accumulated below the soil zone from which roots extracted moisture (0–120 cm depth), while no accumulation occurred under bromegrass (Fig. 7). Another major difference was the level of export of organic anions. The rate of 110 kg $N.ha^{-1}$ increased yields of corn by 1.07 $t.ha^{-1}.year^{-1}$ and yields of bromegrass hay by 2.65 $t.ha^{-1}.year^{-1}$. At organic anion contents of 5 and 67 cmol(−).kg^{-1} for corn grain and bromegrass hay respectively (Pierre and Banwart, 1973), the increase in OA_{ex} values due to nitroen fertilization of the two systems was very different: 0.054 and 1.78 kmol(−).$ha^{-1}.year^{-1}$.

Thus, although the potential for acidification from effects of nitrate leaching was much greater in the periodically cultivated corn system than in the bromegrass sod system, exports of organic anion caused much greater acidification in the bromegrass system. The result was a similar effect of nitrogen fertilization on the rate of soil acidification in each system (Table X).

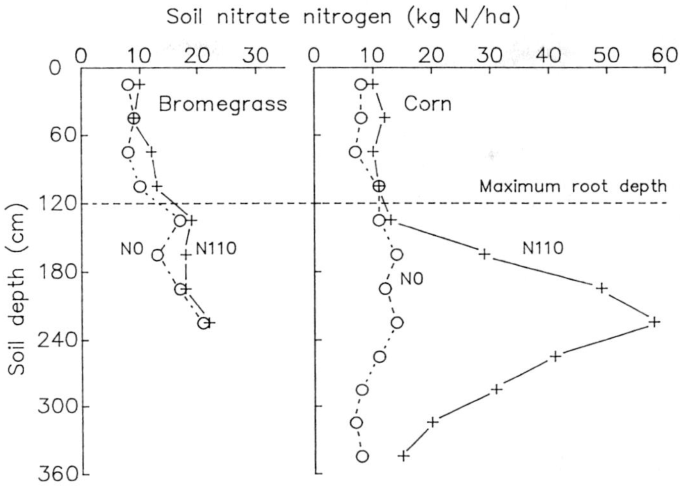

Fig. 7. Soil nitrate distribution with depth after 4 years fertilization of corn and bromegrass with either no nitrogen or 110 kg N.ha^{-1}.year^{-1}. (Redrawn from Power *et al.*, 1972).

Table X. Soil acidification rates in nitrogen fertilized systems of production of corn/fallow and bromegrass/hay in North Dakota (data from Power *et al.* 1972).

System	Nitrogen fertilizer form and rate (kg.ha^{-1}.year^{-1})		Soil pH$_{(paste)}$ 1962	1966	Annual change in pH due to N fertilizer[a]
Bromegrass	Nil	0	6.4	7.0	0.0
	$Ca(NO_3)_2$	110	6.5	6.9	−0.05
	NH_4NO_3	110	6.8	6.7	−0.18
	Urea	110	6.6	6.8	−0.10
	$(NH_4)_2SO_4$	110	6.6	6.4	−0.20
Corn	Nil	0	6.5	7.0	0.0
	$Ca(NO_3)_2$	110	6.4	6.9	0.0
	NH_3NO_3	110	6.6	6.5	−0.10
	Urea	110	6.4	6.6	−0.08
	$(NH_4)_2SO_4$	110	6.6	6.2	−0.23

[a] Change in pH = $pH_{1966} - pH_{1962}$. Annual change in pH due to fertilizer N = ((Change in pH_{N+}) − (Change in pH_{N_0}))/4. Soil pH was measured in spring of both years.

The effect of the form of nitrogen fertilizer is also illustrated in this example where the order of acidifying effect was ammonium sulphate > ammonium nitrate > urea > calcium nitrate. Such a pattern could be expected if nitrification was complete in the soil, and losses of nitrogen were similar for each fertilizer.

D. Natural and acidified rainfall

Theoretical calculations of the net acid additions (moles $H^+.ha^{-1}.year^{-1}$) due to HCO_3^-, H^+ and OH^- fluxes in water movement to and from an ecosystem are shown in Figures 8 and 9. Several important observations can be made from this data.

First, for rainfall unpolluted by acids from industrial pollution (unpolluted rain in Figure 8 at pH 5.67 where the only acids present are H_2O and H_2CO_3), there is no net acidification if leaching is zero. The net input of HCO_3^- equals the net input of H^+ (Table III,IV). When leaching occurs, soil pH rises at soil pH values < 5.2 and falls at soil pH values > 5.2. A pH of 5.2 is the pH at which HCO_3^- and H^+ levels are equivalent in leachate in equilibrium with soil air containing 0.003 atm $CO_2(G)$. The null point pH on Figure 8 would be higher or lower for lower and higher soil air $CO_2(G)$ partial pressures respectively. Lastly, the quantity of alkali or acid added to the ecosystem due to the water fluxes is proportional to the amount of water leached from the system (Fig. 8).

This means that the soil pH will approach 5.2 with time in natural ecosystems with little or no export or import of organic anions, which have stable levels of organic matter accumulated, and that use nitrogen very efficiently (i.e. little or no nitrate leaching). The rate at which the pH approaches 5.2 will depend largely on the soil pH buffering capacity, on the initial soil pH, and on the amount of water leached from the soil per year. At leaching rates of 960 mm per year and at soil pH values of 4.0 and 6.O, the alkali and acid addition rates respectively are equivalent to 48 and 21 kg $CaCO_3.ha^{-1}.year^{-1}$.

If dissociated acids in addition to H_2O and H_2CO_3 are contained in the rainfall, the pH will be < 5.67. The extra acidity in the rainfall results in net addition of acid when there is no leaching and reduces the soil pH value at which net addition of acid is zero (Fig. 9). Furthermore, the null point soil pH for a given volume of leachate ($mm.year^{-1}$), decreases as rainfall increases.

These calculations demonstrate the serious consequences of acid rain for stability of soil pH. Rainfall of pH 4.0, a frequently recorded level of atmospheric pollution (Likens *et al.*, 1979), moves the null point soil pH from 5.2, at which aluminium levles in solution are usually non-toxic, to below 4.0 where aluminium toxicity and other acid soil problems (H^+ and manganese toxicity and calcium deficiency) are highly likely.

In a review of 'acid rain' in North America and Europe, Likens *et al.*, (1979) summarised data showing that the pH of rainfall varied from 5.8–2.4. Unpolluted rain had pH values of 5.4–5.8, near the theoretical atmospheric CO_2(G)/water equilibrium of 5.67. In areas with varying degrees of pollution of rainfall with sulphuric and nitric acids, values as low as 3.9, and in extremely acid rainshowers as low as 2.4, were recorded. Over large regions of North America and Europe, average annual pH values of 4.0 have been recorded. The long-term effects on the null point soil pH of these levels of rainfall acidity can be calculated using the methods outlined in this chapter. Furthermore, the relatively more serious consequences for the environment of sulphuric compared with nitric acid pollution is apparent, because a greater proportion of the nitrate than sulphate ions are likely to be chemically reduced (a process producing alkali) in the soil/plant system. This reduction reaction has the potential of balancing the pH effects of the acid rain.

F. Other systems and acidity management onions

Many more instances of agricultural systems in which various sources of acid are more or less important can be found in the literature. For example, the oxidation of sulphides to sulphuric acid is important in acid sulphate soils (Dost and van Breemen, 1982).

Identification of the important sources of acid in a given system enables consideration of management options to minimise acidification. Obvious candidates are: minimising soil nitrate accumulation (minimum tillage; continuous, deep-rooted plant cover); using nitrogenous fertilizers that are less acidifying; avoiding over-fertilizing with nitrogen; returning more organic matter to the soil (e.g. removing grain and returning straw); feeding hay in the paddock rather than in concentrated feeding areas; minimising the 'camping behaviour' of grazing animals; and avoiding grazing pressures high enough to reduce the capacity of the plant to absorb nitrate. Chapter 7 discusses the amelioration of soil acidity in more detail.

VI. THE DISTRIBUTION OF PROFILE-ACIDITY

Within an ecosystem, acidification often proceeds faster in some soil layers than in others. Diffusion and mass flow of H^+ and OH^- are strongly limited by the soil pH buffer capacity and by the low concen-

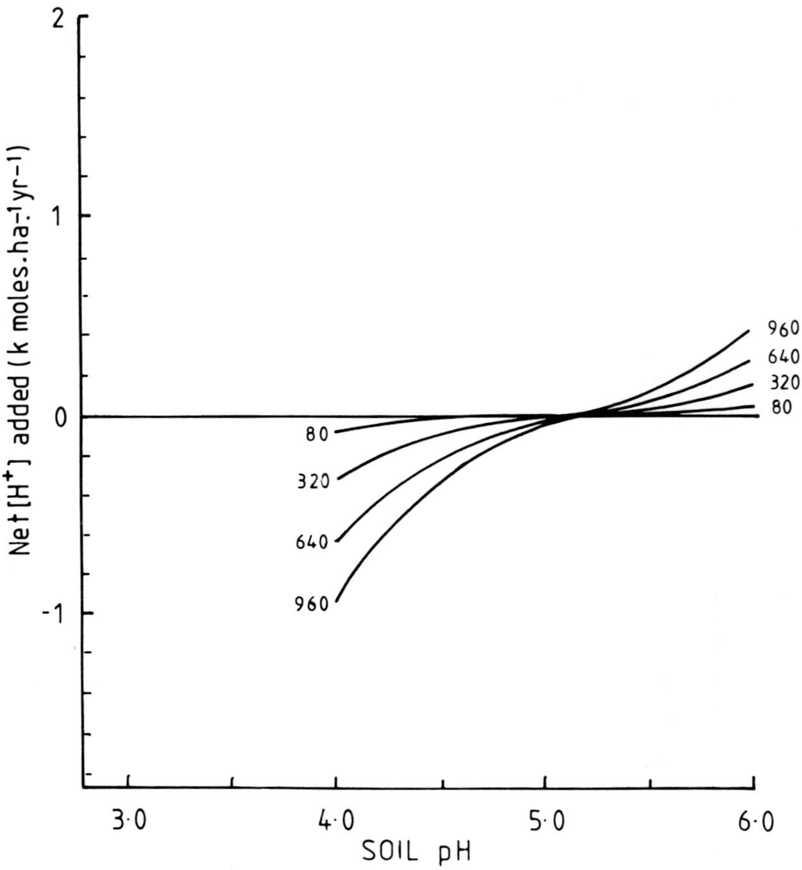

Fig. 8. Acid addition or loss rates for any amount of unpolluted rainfall of pH 5.67 (the only sources of acid being H_2O and H_2CO_3 at atmospheric CO_2 (G) of 0.0003 atm., acid soil air CO_2 (G) of 0.003 atm.) and the amount of leachate (mm. yr^{-1}) indicated on the curves.

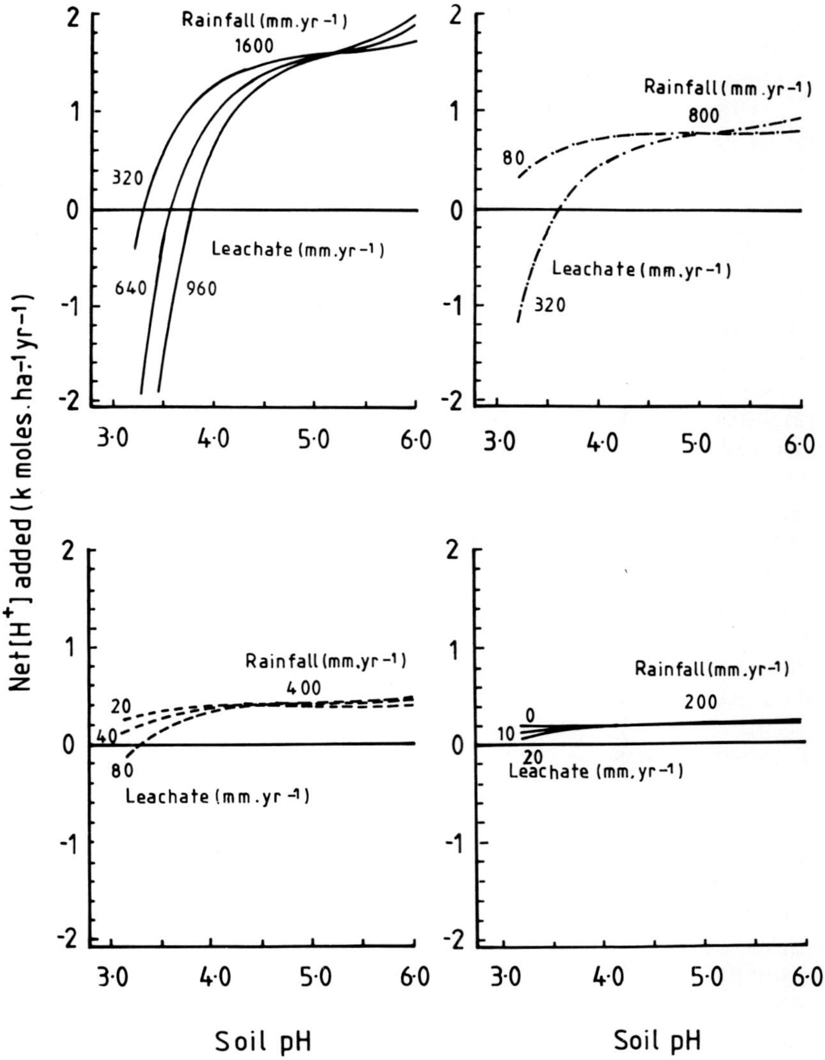

Fig. 9. Net acid addition or loss rates for various quantities of polluted rainfall at pH 4.0, and for the amounts of leachate mm.yr^{-1} indicated on the curves.

trations of these ions in the soil solution. Thus, pH differences, once established, are persistent.

Many soils in natural ecosystems and under permanent pastures have a higher surface than sub-surface pH, and the pH of the deeper subsoil again rises towards that of the weathering parent material (see Stace *et al.*, 1968 for a wide range of Australian soils, and Williams, 1980; Bromfield *et al.*, 1983 for specific examples). This trend probably results from an amelioration of the acidifying processes in the surface soil by the deposition and oxidation of organic anions at or near the surface ($RCOO^- + H^+ \rightarrow H_2O + CO_2$). Furthermore, if plant uptake of nitrate occurs predominantly from the topsoil, acid excretion by roots will be minimised in this zone and maximised in deeper soil layers. This reflects the fact that nitrate absorption is usually most of the total anion absorption, and thus usually controls the balance of cation to anion absorption by roots, and of $H^+/(OH^-, HCO_3^-)$ excretion.

Manipulation of the pH of the surface soil in agriculture by liming is straightforward. Increasing the subsoil pH is more difficult. Liming subsoil is costly, and the effect of lime moves down the profile slowly (e.g. Adams and Pearson, 1969). However, nitrate is easily leached into the subsoil where it is available for absorption by deep roots. If these roots absorb sufficient nitrate to achieve a net excess of anion absorption, then they excrete OH^- or HCO_3^- into the subsoil. Adams and Pearson (1969) provide a clear example in which nitrate fertillizers ($NaNO_3$ and $Ca(NO_3)_2$ were used to correct subsoil acidity to depths of 60–80 cm within 4 years. In the same period, lime application to the surface had only a marginal effect on the pH 15–30 cm below the limed soil.

VII. CONCLUSION

The topic of soil acidification is of fundamental importance when considering the longterm stability of agricultural production systems. Rates of acid addition to agricultural ecosystems vary with different forms and methods of production, and the rate of pH decline also varies with the evenness of distribution of the acid in the soil profile, and with the soil pH buffer capacity. In all acidifying systems, however, corrective additions of liming materials or other residually alkaline material (e.g. organic matter), is necessary to maintain highly productive soils.

This chapter has outlined methods for measuring acidification, and for identifying the forms and quantities of acids causing acidification in a given ecosystem. This knowledge is of central importance to the development of efficient and stable systems of agricultural production. These systems will be characterised by acidification rates held as low as possible by more effective management of the carbon and nitrogen cycles, and by commensurately low rates of lime (or other alkaline amendment) rates required to maintain stable, productive soils.

REFERENCES

Abruna, F., Pearson, R. W. and Elkins, C. B. (1958). Quantitative evaluation of soil reaction and base status changes resulting from field application of residually acid-forming nitrogen fertilizers. *Soil Science Society of America Proceedings* **22**, 539–542.

Adams, F. and Pearson, R. W. (1969). Neutralizing soil acidity under bermudagrass sod. *Soil Science Society of America Proceedings* **33**, 737–742.

Allison, F. E. (1973). 'Soil organic matter and its role in crop production'. Elsevier, London.

Arkley, R. J. (1963). Calculation of carbonate and water movement in soils from climatic data. *Soil Science* **96**, 239–248.

Awad, A. S. and Edwards, D. G. (1977). Reversal of adverse effects of heavy ammonium sulphate application on growth and nutrient status of a kikuyu pasture. *Plant and Soil* **48**, 169–185.

Blackburn, G. and McLeod, S. (1983) Salinity of atmospheric precipitation in the Murray-Darling drainage division, Australia. *Australian Journal of Soil Research* **21**, 411–434.

Bromfield, S. M., Cumming, R. W., David, D. J. and Williams, C. H. (1983). Change in soil pH, manganese and aluminium under subterranean clover pasture. *Australian Journal of Experimental Agriculture and Animal Husbandry* **23**, 181–191.

Coleman, N. T. and Thomas, G. W. (1967). The basic chemistry of soil acidity. *In* 'Soil Acidity and Liming'. (R. W. Pearson and F. Adams, eds.), pp. 1–41. American Society of Agronomy, Madison, Wisconsin.

Conforth, I. S. and Sinclair, A. G. (1982). Model for calculating maintenance phosphate requirements for grazed pastures. *New Zealand Journal of Experimental Agricultura* **10**, 53–61.

Cumming, R. W. and Williams, C. H. (1983). Soil pH in the Pejar catchment area, Goulburn. *Journal of the Soil Conservation Service of New South Wales* **39**, 106–112.

Dolling, P. J. and Ritchie, G. S. P. (1985). Estimates of soil solution ionic strength and the determination of pH in West Australian soils. *Australian Journal of Soil Science* **23**, 309–314.

Dost, D. and van Breemen, N. (1982). Proceedings of the Bangkok symposium on acid sulphate soils. International Institute for Land Reclamation and Improvement, Wageningen, Publication 31.

Fitzpatrick, E. A. and Nix, H. J. (1970). The Climatic Factor in Australian Grassland Ecology. *In* 'Australian Grasslands'. (R. M. Moore, ed.) pp. 3–26. Australian National University Press, Canberra.

Fox, R. L. (1980). Comparison of several lime requirement methods for agricultural soils in Pennsylvania. *Communications in Soil Science and Plant Analysis* **11**, 57–69.

Hartikainen, H. (1986). Acid- and base-titration behaviour of Finnish mineral soils. *Zeitschrift für Pflanzenernaerhrung und Bodenkunde* **149**, 522–532.

Helling, C. S., Chesters, G. and Corey, R. B. (1964). Contribution of organic matter and clay to soil cation-exchange capacity as affected by the pH of the saturating solution. *Soil Science Society of America Proceedings* **28**, 517–520.

Helyar, K. R. (1976). Nitrogen cycling and soil acidification. *Journal of the Australian Institute of Agricultural Science* **42**, 217–221.

Hilder, E. J (1964). The distribution of plant nutrients by sheep at pasture. *Proceedings of the Australian Society of Animal Production* **5**, 241–248.

Hingston, F. J. and Gailitis, V. (1976). The geographic variation of salt precipitated over Western Australia. *Australian Journal of Soil Research* **14**, 319–335.

Jarvis, S. C. and Robson, A. D. (1983a). The effects of nitrogen nutrition of plants on the development of acidity in Western Australian soils. I. Effects with subterranean clover grown under leaching conditions. *Australian Journal of Agricultural Research* **34**, 341–353.

Jarvis, S. C. and Robson, A. D. (1983b). The effects of nitrogen nutrition of plants on the development of acidity in Western Australian soils. II. Effects of differences in cation/anion balance between plant species grown under non-leaching conditions. *Australian Journal of Agricultural Research* **34**, 355–365.

Jenny, H. (1980). 'The soil resource. Origin and behaviour'. Springer-Verlag, New York.

Kanehiro, Y., Matsusaka, Y. and Sherman, G. D. (1951). The seasonal variation in pH of Hawaiian soils. *Hawaii Agricultural Experiment Station Technical Bulletin* No. 14.

Keeney, D. R. and Nelson, D. W. (1982). Nitrogen-inorganic forms. *In* 'Agronomy Monograph No. 9'. (A. L. Page, ed.) American Society of Agronomy, Madison, Wisconsin.

Kononova, M. M. (1966). 'Soil organic matter, it's nature, it's role in soil formation and in soil fertility'. 2nd English edition, Pergamon Press, London.

Likens, G. E., Wright, R. F. and Galloway, J. N. (1979). Acid rain. *Scientific American* **241**, 39–48.

Lindsay, W. L. (1979). 'Chemical equilibria in soils'. John Wiley and Sons, New York.

Magdoff, F. R. and Bartlett, R. J. (1985). Soil pH buffering revisited. *Soil Science Society of America Journal* **49**, 145–149.

Mann, A. W. (1983). Hydrogeochemistry and weathering on the Yilgarn block, Western Australia — ferrolysis and heavy metals in continental brines. *Geochimica et Cosmochimica Acta* **47**, 181–190.

Mears, P. T. and Humphreys, L. R. (1974). Nitrogen response and stocking rate of *Pennisetum clandestinum* pastures. I. Pasture nitrogen requirement and concentration, distribution of dry matter, and botanical composition. *Journal of Agricultural Science, (Camb)* **83**, 451–467.

Northcote, K. H. (1971). 'A Factual Key for the Recognition of Australian Soils'. Rellim Technical Publication, Adelaide.

Pierre, W. H. and Banwart, W. L. (1973). Excess-base and excess-base/nitrogen ratios of various crop species and plant parts. *Agronomy Journal* **65**, 91–96.

Pierre, W. H., Meisinger, J. and Birchett, J. R. (1970). Cation-anion balance in crops as a factor in determining the effect of nitrogen fertilizers on soil acidity. *Agronomy Journal* **62**, 106–112.

Pierre, W. H. , Webb, J. R. and Shrader, W. D. (1971). Quantitative effects of nitrogen fertilizer on the development and downward movement of soil acidity in relation to level of fertilization and crop removal in a continuous corn cropping system. *Agronomy Journal* **63**, 291–297.

Ponnamperuma, F. N. (1972). The chemistry of submerged soils. *Advances in Agronomy* **24**, 29–96.

Raupach, M. (1951). Studies in the variation of soil reaction. II. Seasonal variations at Barooga, N.S.W. *Australian Journal of Agricultural Research* **2**, 73–82.

Raven, J. A. and Smith, F. A. (1976). Nitrogen assimilation and transport in vascular land plants in relation to intracellular pH regulation. *The New Phytologist* **76**, 415–431.

Russell, E. W. (1973). 'Soil conditions and plant growth'. 10th edition, Longman, London.

Russell, J. S. (1960). Soil fertility changes in the long term experimental plots at Kybybolite,

South Australia. I. Changes in pH, total N, organic carbon and bulk density. *Australian Journal of Agricultural Research* **11**, 902–926.

Sanchez, P. A. and Salinas, J. G. (1981). Low input technology for managing oxisols and ultisols in tropical America. *Advances in Agronomy* **34**, 279–406.

Siman, A. and Flemons, K. (1970). Goulburn lucerne failures linked with induced manganese toxicity. *Agricultural Gazette of New South Wales* **81**, 662–663.

Simpson, J. R. (1962). Mineral nitrogen fluctuations in soils under improved pasture in southern New South Wales. *Australian Journal of Agricultural Research* **13**, 1059–1072.

Skene, J. K. M. (1966). Errors in accretion studies of soil organic matter. (letter) *Journal of the Australian Institute of Agricultural Science* **32**, 208–209.

Snedecor, G. W. and Cochran, W. G. (1967). 'Statistical Methods'. Iowa State University Press, Iowa.

Sokal, R. R. and Rohlf, F. J. (1969). 'Biometry, the principles and practice of statistics in biological research'. W. H. Freeman and Co., San Francisco.

Stace, H. C. T., Hubble, G. D., Brewer, R., Northcote, K. H., Sleeman, J. R., Mulcahy, M. J. and Hallsworth, E. G. (1968). 'A handbook of Australian soils'. Rellim Technical Publication, Adelaide.

Tisdale, S. L. and Nelson, W. L. (1975). 'Soil fertility and fertilizers'. 3rd edition. Macmillan Publishing Co. Inc., New York.

van Breemen, N., Mulder, J. and Driscoll, C. T. (1983). Acidification and alkalinisation of soils. *Plant and Soil* **75**, 283–308.

van Breemen, N., Driscoll, C. T. and Mulder, J. (1984). Acidic deposition and internal proton sources in acidification of soils and waters. *Nature* **307**, 599–604.

Watson, E. R. (1969). The influence of subterranean clover pastures on soil fertility. III. The effect of applied phosphorus and sulphur. *Australian Journal of Agricultural Research* **20**, 447–456.

Webber, M. D., Hoyt, P. B. and Nyborg, M. (1977). A comparison of lime requirement methods for acid Canadian soils. *Canadian Journal of Soil Science* **57**, 361–370.

Weir, R. G. (1975). The oxidation of elemental sulphur and sulphides in soil. *In* 'Sulphur in Australasian Agriculture' (K. D. McLachlan, ed.), pp. 40–49. Sydney University Press, Sydney.

Wetselaar, W. and Norman, M. T. J. (1960). Recovery of available soil nitrogen by annual fodder crops at Katherine, Northern Territory. *Australian Journal of Agricultural Research* **11**, 694–704.

Williams, C. H. (1968). Seasonal fluctuations in mineral sulphur under subterranean clover pasture in southern New South Wales. *Australian Journal of Soil Research* **6**, 131–139.

Williams, C. H. (1980). Soil acidification under clover pasture. *Australian Journal of Experimental Agriculture and Animal Husbandry* **20**, 561–567.

Williams, C. H. and Donald, C. M. (1957). Changes in organic matter and pH in a podzolic soil as influenced by subterranean clover and superphosphate. *Australian Journal of Agricultural Research* **8**, 179–189.

3

Symbiotic Nitrogen Fixation and Soil Acidity

D. R. COVENTRY and J. EVANS

I. INTRODUCTION

The nitrogen fixing symbioses between legumes and rhizobia are used widely in agricultural systems to supply leaf and grain protein and replenish soil nitrogen. Their importance has led to extensive study and reviews of the topic (Quispel, 1974; Hardy and Gibson, 1977; Vincent *et al.*, 1977; Andrew and Kamprath, 1978; Gibson and Newton, 1981; Evans and Herridge, 1987). Soil acidity has long been recognised as harmful to nitrogen fixation (Bryan, 1923; Dollas, 1930; Jensen, 1943), but the problem is current in both tropical and temperate regions of the world. This chapter considers the impact of soil acidity on the legume-rhizobia symbiosis.

Vincent (1980) has highlighted the many phases in establishment of a functional symbiosis. The breadth of the potential role of acidity in undermining the effectiveness of this association is indicated in Figure 1. Following discussion of the factors highlighted in Fig. 1, we consider practical aspects, namely problem recognition, consequences and solutions. An attempt is made throughout to illuminate deficiencies in basic knowledge and its applicability to the field. We use the term rhizobia as a general term for bacteria within both *Rhizobium* and *Bradyrhizobium*.

SOIL ACIDITY AND PLANT GROWTH
ISBN 0 12 590655 2

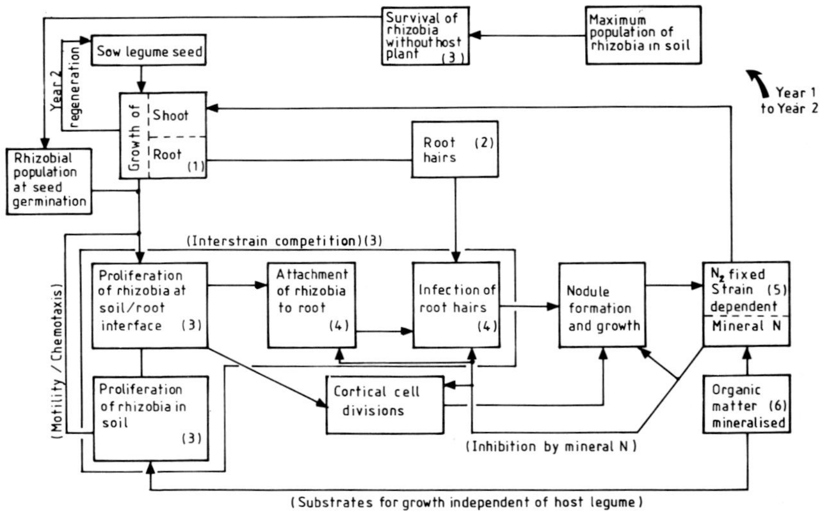

Fig. 1. Flow diagram of interactions leading to the formation of nitrogen fixing nodules and of the steps sensitive to acid soil factors. Numbers with-in boxes represent involvement of particular components of soil acidity as follows:
(1) Al, P, Mn, Ca, Mg, Zn, B, Mo; (2) Al; (3) H, Al, P; (4) Ca, H, Al; (5) Mo; (6) H

II. LEGUME SYMBIOSIS

A. Biology of rhizobia

1. Nutrients and toxicities

The growth and survival of rhizobia are essential for sustaining nodule formation on roots. In acid soils the populations of some *Rhizobium* spp. can be low and certain species or strains may be absent (Bryan, 1923; Petersen and Gooding, 1941; Vincent, 1958; Jensen, 1969; Robson and Loneragan, 1970a; Rice *et al.*, 1977). For example, *R. meliloti* is particularly difficult to detect in soils of pH < 5.5 (1/2.5, water) Petersen and Gooding, 1941), and *Bradyrhizobium japonicum* serotype 135 is scarce, at least in nodules of soybean, in soils of pH < 5.9 (Damirgi *et al.*, 1967) where other serotypes predominate. Acidity was identified as a cause of low numbers of rhizobia in soil following frequent observations of larger populations when acidity was reduced (Spencer, 1950; Vincent, 1958). This response continues to be shown (Loos and Louw, 1965; Bromfield and Jones, 1980; Lowendorf *et al.*, 1981; Mahler, 1983; Evans and Hochman, 1984; Coventry *et al.*, 1985a). In addition, controlled laboratory experiments have demonstrated the sensitivity of rhizobia to H^+ ions (Fred and Davenport, 1918; Graham and Parker, 1964; Keyser and Munns, 1979a).

Recent experiments using defined media have also assessed the role of other chemical components of acid soils, i.e. those increasing in availability with acidification (aluminium, manganese) and those usually less available (phosphorus, calcium). Aluminium was shown to be particularly toxic to rhizobia. With tropical cowpea *Bradyrhizobium*, adding 50 μM aluminium increased the time before division of rhizobial cells commenced, increased the time taken for cells to divide, and, in some situations, led to incomplete cell division (Keyser and Munns, 1979a,b; 1981). Adding 50 μM aluminium also inhibited multiplication of strains of *R. trifolii* (Wood *et al.*, 1984b). High levels of manganese (200 μM) had little direct effect on rhizobial growth (Wilson and Reisenauer, 1970; Holding and Lowe, 1971; Keyser and Munns, 1979b). Low levels of phosphate ($< 10\ \mu$M) significantly decreased the growth of strains of cowpea rhizobia and *Br. japonicum* (Keyser and Munns, 1979b; Cassman *et al.*, 1981). The specific requirement for calcium as a nutrient for *Rhizobium* is low (10–25 μM) (Norris, 1959; Bergersen, 1961; Vincent, 1962) and calcium concentrations are usually believed to be sufficient in soil solutions for rhizobial growth. However, certain strains of rhizobia were unable to grow at low levels of calcium (50 μM) in a defined medium (Keyser and Munns, 1979b). It has been suggested that the calcium concentration in the rhizosphere could be lowered enough to limit the growth of *R. trifolii* (Lowther and Loneragan, 1970). Rerkasem (1977) has indicated that calcium can ameliorate the effect of acidity on the survival of fast-growing *Rhizobium*. This effect was strain-specific and biologically insignificant for the slow-growing, cowpea *Bradyrhizobium* (Keyser and Munns, 1979b).

These defined media experiments indicate that aluminium and phosphate (both directly and indirectly), as well as H^+, have the greatest effect on the growth of rhizobia. The solution concentrations of aluminium and phosphate were chosen to be realistic in terms of acid soil solutions and rhizospheres, indicating that aluminium toxicity and phosphate deficiency may limit nodulation by limiting rhizobial numbers in the soil. The experiments also suggest that, since the effects of phosphate and aluminium are additive to those of H^+, they may need to be considered together with the soil pH to diagnose the likely stress on rhizobial growth in different soils. However, since rhizobial strains differ in sensitivity to the chemical composition of soil solutions, it would be imprecise to define critical chemical values at the species level. It may be more productive to consider the requirements of specific, beneficial strains, such as those selected as commercial legume inocula.

Less is known of the reactions of rhizobia to the other major nutrients and trace elements. Norris (1959) and Vincent (1962) reported an essential requirement for magnesium of 100 μM and a non-specific need for divalent ions of 400–600 μM (Vincent, 1962). A response to an increase in soil pH by both fast- and slow-growing rhizobia was only achieved when sufficient divalent cations were supplied (Norris, 1959). However, more data on nutrient concentrations in soil and rhizosphere solutions in acid soils are needed to predict the likelihood of deficiencies for rhizobial growth.

Rhizobia also require specific vitamins, amino acids and energy substrates (reviewed by Parker *et al.*, 1977). These are normally supplied in ample quantities by decaying organic matter and root exudates. In acid soils, the flow of natural energy sources and growth factors to rhizobia may be diminished due to reduced organic matter decomposition (Chapter 4) and plant growth, or the composition of substrates may be altered. Because of diversity between strains in nutritional requirements, the latter may result in changes in the relative abundance of different strains (Parker *et al.*, 1977; Chakrabaiti *et al.*, 1981). However, there is little definition and quantification of the organic nutrition of rhizobia in soil and the rhizosphere, let alone the possible effects due to acidification.

The presence of organic matter may reduce the availability of aluminium (Chapter 1; Thomas, 1975). Thus, with long-term retention of organic matter in agricultural systems, conditions for growth and survival of rhizobia may improve with time and allow nodulation without the addition of lime (Anderson and Moye, 1952). However, the improvement may be localised within the soil since accumulation of organic matter has also been linked with increases in soil acidity.

2. *Survival and colonisation of rhizobia in acid soils*

Whilst nutritional deficiencies and toxicities may directly limit the survival and colonisation of acid soil by rhizobia, such effects may be more pronounced due to the activities of other microflora and changes in physical conditions.

In soils and rhizospheres, rhizobia reside with numerous and varied microorganisms (Rovira, 1961; Rovira and Stern, 1961). In the absence of their host legume, numbers of a particular rhizobial species can be substantially less than 0.1 per cent of the total number of aerobic bacteria, though this may increase at particular times of the year to 8–9 per cent when legumes are cultivated (Mahler and Wollum, 1981; 1982). Not surprisingly, interactions between the various microorganisms occur and may involve competition for substrates, predation, parasitism, lysis and abiotic excretions. Various studies have illustrated the vulnerability of rhizobia to these interactions (Hely *et al.*, 1957; Danso and Alexander, 1974; Habte and Alexander, 1977; Trinick and Parker, 1982). In this environment, the survival and abundance of rhizobia will depend on their ability to locate and colonise sites of suitable nutrition under pressure from other competing microorganisms. Suitable sites need to be located because they are not ubiquitous, as illustrated by the fact that the bulk of bacteria are distributed unevenly, both in the soil and at the root surface. In soil, bacterial colonies appear to be small and widely spaced and may occupy less than 0.04 per cent of the exposed surface area of mineral and organic matter (Gray *et al.*, 1967). At the root surface, bacteria occur in non-random patterns (Gray *et al.*, 1967; Newman and Bowen, 1974). In this situation, where organisms compete for similar nutrients available within confined areas, Bowen and Rovira (1976) have suggested that a capacity for high growth rate is likely to be particularly important in determining the extent to which specific microorganisms will be successful colonisers, and in determining their survival and abundance.

Therefore, where the growth rate of a species or strain of rhizobia is impaired by soil acidity or its related factors, more so than the rate of its competitors, then the rhizobia may be more easily displaced in the competitive struggle to colonise suitable niches. Even mild sensitivity to acidity may be important in non-sterile soil as compared with sterile soil. Although there are few data on the significance of microbial competition in aggravating effects of soil acidity, Lowendorf *et al.* (1981) reported that the abundance of cowpea *Bradyrhizobium* and *R. phaseoli* declined more with increasing acidity in non-sterile soil compared with gamma-irradiated soil. In the same study, two strains of *R. meliloti*, differing in tolerance to pH, showed similar patterns of survival in sterile soils above pH 4.8, but the more acid-tolerant strain survived better in non-sterile soil. Rovira (1961) has also presented data indicating that the proportion of *R. trifolii* to other bacteria increased in rhizospheres of paspalum and red clover with alleviation of acidity.

Rhizobia are also sensitive to physical conditions in the soil, and a decline in abundance often occurs with dessication and high temperature (Marshall, 1964; Pena-Cabriales and Alexander, 1983). Dessication and high temperature may have greater effects on nodulation if soil acidity limits rhizobial growth when physical conditions are favourable. The interaction of physical conditions and acidity on rhizobial survival has been examined briefly by Chao and Alexander (1982). The observation by Spencer (1950), that better nodulation occurred in an acid, yellow podsolic soil if the soil were maintained at field capacity, may have its origins in the interactive effect of moisture stress and acidity on rhizobial growth and distribution.

It has been suggested that motility is important in aiding the colonisation of new, nutrient-rich, material (Bowen and Rovira, 1976). Although little is known about the role of motility in the survival and abundance of rhizobia in soil, the observation of Bowra and Dilworth (1981) that motility of *R. leguminosarum* is dependent upon pH, may prove significant.

3. Consequences of soil pH for rhizobia as symbionts

Numbers of rhizobia influence nodulation and so effects of acidity on the abundance of these bacteria can be expressed in their symbiotic associations. Spencer (1950) was able to improve nodulation of subterranean clover in an acid soil by increasing the amount of inoculum of rhizobia added with the seed; similar observations have been made by others (Munns, 1965a; 1968; Jensen, 1969; Anderson, 1970; Robson and Loneragan, 1970b). Lime may be used in acid soils to increase rhizobial numbers (Rovira, 1961; Loos and Louw, 1965; Coventry *et al.*, 1985a) and increase nodulation (Anderson and Spencer, 1948; Andrew and Norris, 1961; Munns, 1965a; Lowther and Loneragan, 1970). This effect of liming may not arise only from increasing rhizobial abundance in the rhizosphere; there are several accounts of a requirement for both calcium and high pH for nodulation associated with root hair infection (Spencer, 1950; Loneragan and Dowling, 1958; Lowther and Loneragan, 1968; Munns, 1968). However, it may be necessary to increase the numbers of rhizobia before a response to calcium can be achieved. Wood *et al.* (1984a) showed that at pH 4.7 and below, no nodules formed on *Trifolium repens* grown in hydroponic culture irrespective of the calcium concentration (50–1000 μM). At these low pH levels, both root hairs and rhizobia were present but the rhizobia did not increase in number. At higher pH (5.5), rhizobial numbers and nodulation responded to calcium. In the field, Evans and O'Connor (1985) have shown that despite a plentiful addition of calcium, an increase in the numbers of *R. trifolii* under *Trifolium subterraneum* was essential to increasing nodulation in an acid soil.

The combined effects of acidity, microbial competition and physical extremes in suppressing rhizobial growth may decrease the ability to occupy potentially available available niches in the soil habitat, thus reducing the frequency of rhizobia in a soil matrix. It seems reasonable that the more frequent the occurrence of rhizobia, the greater the probability of developing roots contacting microcolonies, and the greater the probability of early and sustained nodulation. Even after seed inoculation there may be a need for subsequent contact of roots with rhizobia. Graham *et al.* (1982) created discrete sites abundant in rhizobia by introducing inoculum in granular form. Compared with slurry inoculation of seed, granular inoculation significantly increased root nodulation. In the acid soils used in their study, seed-applied rhizobia may have failed to migrate adequately with the enlarging root system. The need for repeated inoculation to achieve adequate nodulation would depend on the significance and occurrence of water movement and motility as dynamic mechanisms enabling continuing contact of rhizobia with new root tissue.

Raising the pH may increase both the rate of rhizobial growth on root surfaces and nodulation in nutrient solutions (Munns, 1968; Lie, 1969). In all of these studies, the importance of growth rate of rhizobia for nodulation has perhaps been under-emphasised. Although Lie (1969) suggested that numbers of *R. leguminosarum* were not related to poor nodulation of pea at pH 4.5, this conclusion was deduced indirectly. In fact no data were obtained on the activity or rate of survival of the rhizobia on the roots of plants grown at pH 4.5 and inoculated with different levels of rhizobia. Subsequent nodulation of plants at pH 6.5, inoculated with rhizobia from cultures at pH 4.5, could have occurred following rapid growth of an otherwise small, residual population of bacteria surviving at pH 4.5. By recovering rhizosphere populations of *R. meliloti*, Munns (1968) demonstrated the need to attain a sufficient population of organisms before nodulation occurred in solution. Thus, delayed growth of rhizobia and low levels of inoculum meant that nodulation of *Medicago sativa* was not maximal at pH 4.6, even though a specific pH requirement for root hair curling had been met. At higher pH levels, faster growth rates were obtained and the level and rate of nodulation were improved. Low levels of inoculum and retarded rhizobial division are likely to resemble conditions in the field for self-generating pasture legumes in acid soils, following a dry summer.

Rhizobial growth rate is potentially important because the root hairs are only transiently susceptible to infection by rhizobia. Some investigations (Purchase and Nutman, 1957; Bhuvaneswari *et al.*, 1980) show that regions of tissue just behind the tips of the growing roots are most susceptible to nodule initiation. These zones transform relatively quickly into a non-infectible condition. The probability of rhizobia initiating nodules in this region is increased when the population of rhizobia there is large; as the infectible area ages the requirement for more rhizobia increases (Purchase and Nutman, 1957). Thus, nodule initiation can be considered a function of rhizobial numbers and time. The rate of rhizobial proliferation has been associated with nodule abundance (Evans *et al.*, unpublished data) and the relative nodulating success of serotypes of *R. trifolii* (Bottomley *et al.*, 1986).

From the foregoing discussion, we postulate that effects of acidity resulting in increased generation time and impaired distribution (through displacement by other more tolerant microorganisms in a competitive environment) will establish conditions less conducive to rhizobia as a symbiont. Thus, an ability of rhizobia to maintain an abundant presence in the vicinity of new root tissue, either through motility or frequent occurrence in the soil matrix, or to increase rapidly in number, should be advantageous to nodule formation.

4. Variation in rhizobial strain tolerance to acidity

The soil population of rhizobia is usually a mixture of strains competing for nutrients and occupancy of nodules. Many strains in the soil population may not be highly effective in nitrogen fixation (Bergersen *et al.*, 1971). However, as strains also differ in their tolerance to acidity (Graham and Parker, 1964), there is a possibility that the more acid-tolerant are competitive and persistent but are not necessarily the most effective in nitrogen fixation. In one study (Holding and King, 1963), the effectiveness of indigenous populations of *R. trifolii* on a cultivar of white clover was positively correlated with base saturation of the soil exchange capacity.

Relatively few studies are concerned with the impact of acidity on strain competitiveness and persistence, and on subsequent nodulation and nitrogen fixation. Although the host legume influences the composition of strains occupying nodules (Vincent and Waters, 1953), environmental factors have the potential to alter composition in favour of otherwise less successful strains (Evans *et al.*, 1979). In particular, Jones and Morley (1981) showed than an acid-tolerant strain of *R. trifolii* could improve its occupancy of nodules on white clover relative to a competing, less acid-tolerant strain following a reduction in pH. Similarly, Dughri and Bottomley (1983) observed that pH influenced the relative representation of serogroups of *R. trifolii* forming nodules with *T. subterraneum*. Such changes may involve differences in strain effectiveness and may, therefore, be agronomically significant (Jones *et al.*, 1964; Russell and Jones, 1975; Jones and Morley, 1981).

B. Nodule initiation and formation

The initiation, establishment and control of the association between the host legume and rhizobia is complex and sensitive to acidity. The sequence of events following proliferation of rhizobia in the rhizosphere is: attachment to the host root; invasion of the root; development of the infection towards, and ultimately into, the root cortex; and finally nodule development (Goodchild and Bergersen, 1966; Dart, 1977; Broughton, 1978; Schmidt, 1979; Bauer, 1981). The acidity factors influencing these events are summarised in Figure 1.

The physiological and morphological events leading to infection require recognition between rhizobia and the legume, and in many cases this recognition is very selective (Vincent and Waters, 1953; Marques-Pinto *et al.*, 1974; Bromfield and Jones, 1980). Recognition involves specific biochemical reactions, probably involving lectins and rhizobial carbohydrates (Bohlool and Schmidt, 1973; Dazzo and Hubbell, 1975; Broughton, 1978; Dazzo *et al.*, 1978). Attachment (pre-infection) of rhizobia to the host root is quite rapid, occurring within two hours of inoculation in *Pisum sativum* (Broughton *et al.*, 1980).

Legumes vary in the development of the infection from recognition of rhizobia to initiation of the nodule. The best description of infection is demonstrated in the temperate legumes. Strong attachment and invasion appear to be preceded by root hair curling (Yao and Vincent, 1976), and the rhizobia pass via an infection thread into the root cortex (Dart, 1974). For some legumes infection threads have not been found (Dart, 1974). With *Arachis hypogaea* and *Stylosanthes* spp., different modes of invasion have been observed (Ranga Rao, 1977); Chandler, 1978; Chandler *et al.*, 1982). Rhizobia enter through spaces between epidermal cells at the junctions of lateral roots. The subsequent manner of infection varies though; with *A. hypogaea*, intercellular zones of infection are formed (Chandler, 1978), but infection of *Stylosanthes* causes the progressive collapse of invaded cells (Chandler *et al.*, 1982).

1. Events in the infection process sensitive to acidity

Step-wise transfer of legumes between nutrient solutions with differing concentrations of H^+, calcium and aluminium have been used to define events in the infection process sensitive to acidity and related factors. Results obtained with different legumes consistently show that the early infection stages (attachment, invasion and infection) are most sensitive to acidity (Munns, 1968; Lie, 1969; de Carvalho *et al.*, 1982). Attachment was shown recently to be sensitive to pH and divalent cations in solution (Caetano-Annolles *et al.*, 1985; Kijne *et al.*, 1985), and rhizobial attachment can have a pH-dependence similar to that of nodulation (Dazzo and Hubbell, 1975). Sensitivity would appear to involve a rhizobial factor because alteration in the proportional representation of strains of *R. trifolii* in nodules can occur with pH changes (Jones and Morley, 1981). Also, the evidence from many reports of acidity affecting root morphology and nodulation (Munns, 1968; de Carvalho *et al.*, 1982; Franco and Munns, 1982; Wood *et al.*, 1984b) is consistent with rhizobial invasion and root infection being the most acid-sensitive stages.

a. Calcium: Low levels of soil calcium have long been known to reduce nodulation and the growth of legumes (Albrecht and Davis, 1929). Using *T. subterraneum* in culture solutions, Loneragan and Dowling (1958) showed that nodulation had a higher requirement for calcium than did host plant growth. Lowther and Loneragan (1968) confirmed this result using a flowing culture technique and identified infection and/or nodule initiation as the stage most sensitive to calcium deficiency. Munns (1970) showed that in *M. sativa* this calcium-demanding stage lasted for about one day after inoculation. Root growth and root hair development were not restricted by calcium concentrations (246 μM) at which nodulation was reduced (Lowther and Loneragan, 1968). The earliness of this calcium-sensitive stage suggests that the recognition and association stage may be involved.

With many legumes, calcium and H^+ have interdependent effects on nodulation in solution (Loneragan and Dowling, 1958; Munns, 1970; Andrew, 1976), suggesting that they may operate together in a single process (Munns, 1978). However this interaction does not occur with all legumes (Munns, 1977a).

b. Aluminium: Legume growth is known to be affected by aluminium, and within legumes there is a range of tolerance to high concentrations of aluminium (Andrew *et al.*, 1973; de Carvalho *et al.*, 1980; Kim *et al.*, 1985). The effect of aluminium on nodulation may also vary considerably amongst legumes; with some legumes nodulation may be more sensitive to aluminium than growth of the host plant (de Carvalho *et al.*, 1981; Murphy *et al.*, 1984). The interpretation of effects of aluminium on legume growth and nodulation can be difficult because the form and activity of aluminium in solution are affected by the concentration of other ions in solution (Asher, 1981; Blamey *et al.*, 1983; Wood *et al.*, 1983).

de Carvalho *et al.* (1981; 1982) studied the effect of aluminium concentrations on nodulation of *Stylosanthes* spp. known to be adapted to acid soil conditions (Fig. 2). Aluminium delayed nodulation in five of six species studied, with *S. hamata* the exception. The number of plants nodulated was reduced in the case of *S. guianensis*, *S. scabra* and *S. viscosa.* Aluminium also reduced the number and dry weight of nodules

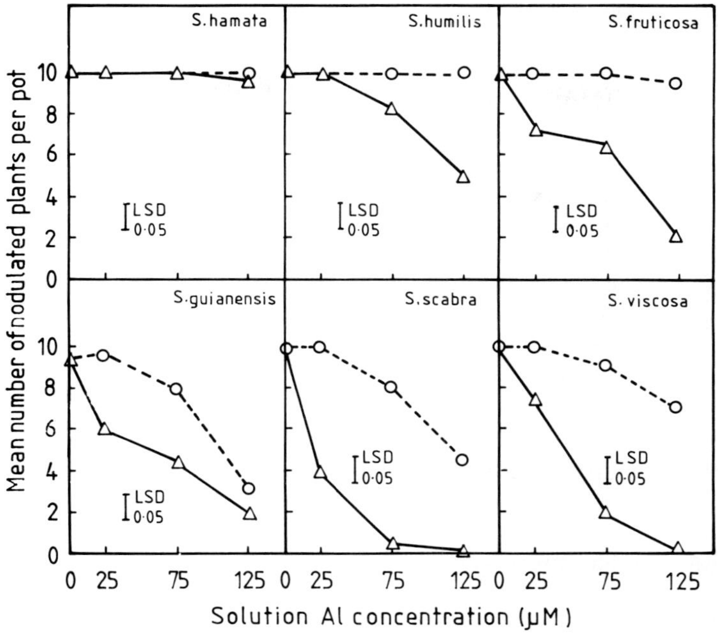

Fig. 2. Effect of solution aluminium concentration on the number of nodulated plants per plot of six *Stylosanthes* species, harvested at 4 and 8 weeks after germination. Bars show LSD at 5% level. △ 4 weeks; ⊙ 8 weeks. After de Carvalho *et al.* (1981).

in all species including *S. hamata.* Reduction of root growth and lateral root production in these tropical legumes does not explain fully the more severe reduction in nodulation (de Carvalho *et al.*, 1981; 1982; Murphy *et al.*, 1984). Further investigation using solution transfer and the hosts *S. hamata* and *S. scabra* found that delay in nodulation due to aluminium was much reduced (in *S. scabra*) by an initial short exposure to solution free of aluminium. This result suggests that an early stage of infection is sensitive to aluminium (de Carvalho *et al.*, 1982). Nodule appearance on *T. subterraneum* was also delayed by high concentrations of aluminium in solution and nodule number and weight were reduced on plants which formed some nodules (Kim, 1985). The effects of aluminium on root growth and nodulation can be alleviated by increasing the concentration of calcium in solution (Kamprath, 1978; Franco and Munns, 1982; Wood *et al.*, 1984b); however it is difficult to distinguish whether the effect of calcium is on the activity of aluminium in solution or due to calcium itself.

c. Hydrogen ion: Acidity was the major factor limiting early nodulation in *Phaseolus vulgaris* (Franco and Munns, 1982) and restricted nodulation of *T. repens*. However, with aluminium in solution (Wood *et al.*, 1984b) the effect of pH on nodulation appears to worsen.

d. Manganese: Few studies have examined the effect of manganese on nodulation of legumes. Whilst there are reports of high concentrations of manganese reducing nodule number (Vose and Jones, 1963; Dobereiner, 1966; Souto and Dobereiner, 1969), it is difficult to separate this response from the intolerance of the host plant itself to manganese. Vose and Jones (1963) obtained reduction in both nodule number and volume in *T. repens* with increasing manganese, but the plants also showed obvious symptoms of manganese toxicity. Supplying calcium reduced the adverse effect of manganese on both the growth of the plant and nodulation. Evans *et al.* (1987) observed that the major effect of manganese toxicity in reducing nodulation in *T. subterraneum* is due to host sensitivity to manganese.

C. Nodule development and function

In the field, acidity can reduce nodule number but there is only limited evidence that acidity affects nodule development independently of direct effects on host growth. Compensation in nodule size when the number is reduced (Nutman, 1952) has been reported for a range of legumes grown on acid soils (Dobereiner, 1966; Munns *et al.*, 1977; Keyser *et al.*, 1979; Hartel and Alexander, 1983; Coventry *et al.*, 1985a). However, where nodule numbers are reduced, legume growth may also suffer (Keyser *et al.*, 1979; Evans and Hochmann, 1984). Coventry *et al.* (1985b) showed that nitrogen fixation (C_2H_2 reduction) per g nodule was less with plants of *T. subterraneum* having large but few nodules compared with plants with more nodules formed in response to inoculation or liming. It is thus possible that the volume of nitrogen-fixing tissue per plant per season is smaller where legumes have fewer but larger nodules. The effect of failing to compensate nodule activity fully for dry matter production may vary between environments in response to plant demand for nitrogen.

1. Nutrients and toxicities

a. Aluminium: The transfer of legumes between solutions of varying pH and aluminium concentrations shortly after inoculation gives some insight into the effect of these factors on nodule development. Transfer of two-day old seedlings of *M. sativa* from solutions at pH 5.7 or with adequate calcium to solutions at pH 4.7 or low calcium respectively, produced very little disruption to the development of nodules that had already been initiated or formed (Munns, 1968; 1970). However nodule formation was retarded when *S. scabra* was transferred to 100 μM aluminium after an initial short exposure to 0 or 25 μM aluminium (de Carvalho *et al.*, 1982). The possibility that aluminium may interfere with multiplication of rhizobia in the invaded cells has been raised by Murphy *et al.* (1984) to explain this delay in the speed of early nodule development. In addition, where aluminium reduced the number of nodules, the dry weight of the nodules was also reduced (de Carvalho *et al.*, 1981; 1982). The effect was greater on nodule weight than on root weight (de Carvalho *et al.*, 1982), suggesting perhaps that with *Stylosanthes* the reduction in nodule weight involves some aspect of the symbiosis rather than an indirect effect on nodule development via an inhibition of plant growth. However, with *P. vulgaris*, nodule growth was less sensitive to aluminium than either root or shoot growth (Franco and Munns, 1982).

b. Molybdenum: Nodule function is more sensitive than host metabolism to molybdenum deficiency (Anderson, 1942; Anderson and Spencer, 1950; Gupta and Lipsett, 1981). The requirements of nitrogenase for molybdenum far exceed the requirements for molybdenum elsewhere in the plant. Spectacular yield increases with legumes have been obtained through the use of trace additions of molybdenum on acid soils (Anderson, 1956; Dawson and Bhella, 1972). Although liming may release molybdenum (Anderson and Moye, 1952), in some soils molybdenum levels are so low that additional yield responses can be obtained with the application of both molybdenum and lime (Gupta, 1979; Franco and Munns, 1982; Coventry, 1985).

c. Calcium and acidity: Increases in plant growth in response to liming were obtained by Banath *et al.* (1966) with *T. subterraneum* and by Andrew (1976) with a range of tropical and temperate legumes in situations where levels of calcium or pH were considered adequate for nodulation. This suggested enhancement of nitrogen fixation. Evidence that acidity affects the rate of nitrogen fixation was also obtained by Munns *et al.* (1977) in a field experiment with eighteen legume species (including *T. subterraneum*). In this study, growth responses due to liming were associated with increases in C_2H_2 reduction per g nodule. Only with one species, *Coronilla varia*, could the response to liming be explained by an increase in nodule number (Munns *et al.*, 1977). Keyser *et al.* (1979), using the dry matter of *Vigna unguiculata* as an indicator of its nitrogen fixation, also suggested that acidity was inhibiting the functioning of the nodules. Thus, with some rhizobial strains dry matter production was less at pH 4.6, even though there was little difference in nodule number. Use of solution cultures to study the effects of acidity on nodule function has been scarce. An exception is that of Franco and Munns (1982), who showed solution aluminium had little effect on C_2H_2 reduction by nodules of *P. vulgaris.*

III. PRACTICAL CONSIDERATIONS

A. Problem recognition

Accurate diagnosis of the nature of field problems associated with soil acidity is important agronomically. It can influence the choice of symbiont, strategies for liming, fertilizer use and soil management. Recognition of the problem can be deduced from field observations, though the scope for this is limited. Observations by farmers of poor legume and pasture growth may be due to factors other than acidity, for example, changes in fertilizer practice, seasonal differences, soil compaction, waterlogging and changes in rotation. Thus there is a need for caution in problem interpretation based on field observations. The shortcomings of field observations may be overcome by relevant pot and field experiments. A more thorough diagnosis may inevitably depend on chemical analyses of soil and plant material.

1. Field observations

a. Plant symptoms: Nodule appearance and number can vary considerably with legumes grown on acid soils (Diatloff, 1972). The position and size of nodules can indicate whether inoculation techniques were successful, and the colour of nodule tissue is useful as an indicator of nodule activity. If a legume is not nodulated, or the nodule is ineffective, then the leaves may be chlorotic depending on soil reserves of nitrogen. However in general, little can be gained from observations of leaf symptoms. With the possible exception of manganese toxicity (Siman *et al.*, 1974; Cregan, 1980), symptoms of the various nutritional disorders associated with soil acidity are usually non-specific in the field.

Other plant observations which may associate soil acidity with poor legume growth are seasonality of growth, changes in botanical composition and better growth on ash and urine patches. Legume growth in acid soils is often very seasonal, possibly due to inhibition of root growth by aluminium limiting the access of roots to moisture. Thus during periods of moisture stress, growth may be enhanced by liming; for example, in summer and autumn in New Zealand (During *et al.*, 1984; Shannon *et al.*, 1984) and Australia (Rowe, 1982).

Increasing soil acidity may change the botanical composition of pasture over a number of years or may dictate which crops can be grown successfully (Cregan, 1980). Frequently there is better growth of legumes on ash patches (Loneragan *et al.*, 1955) and on animal camps (Hilder, 1964) where the pH is higher compared with the soil in the bulk of the field.

b. Field surveys: A more rigorous approach to problem definition is to survey affected areas, detailing soil chemistry and factors related to nitrogen fixation by legumes. For example, Jones and Curnow (1985) surveyed populations of *R. trifolii* and nodulation of *T. subterraneum* at 44 sites over two seasons in central Victoria. Nodulation was negatively correlated with percentage exchangeable aluminium and positively with percentage exchangeable calcium in the soil, but they considered populations of *R. trifolii* were adequate. Because only 49 and 62 per cent of plants were effectively nodulated eight to ten weeks after germination in 1982 and 1983 respectively, they suggested that nodule initiation was affected by the acidity. On the other hand, in a survey of cultivated soils in north-east Victoria, it was found that numbers of *R. trifolii* were less than 1000 per g soil for 89 per cent of the 28 sites sampled and numbers were positively correlated with soil pH (Coventry *et al.*, 1985a). Acid soils in the USA have also been surveyed for rhizobial populations to evaluate persistence and effectiveness of inoculant rhizobial strains (Hagedorn, 1978; Barber, 1980; Dughri and Bottomley, 1984).

2. *Diagnosis*

a. Fertilizer additions: Of particular use, diagnostically, is a comparison of the growth of legumes supplied with or without fertilizer nitrogen (Sewell and Ganey, 1930; Munns, 1965a; Helyar and Anderson, 1970; Robson, 1978; Coventry *et al.*, 1987). Presumptive evidence of an acidity-impaired symbiosis is obtained if a response to fertilizer nitrogen occurs without lime but is reduced in a similar comparison at elevated pH levels (Fig. 3). The diagnosis relies on the additional mineral nitrogen reducing dependency of the legume on symbiotic nitrogen. The improved growth of the legume following this treatment suggests poor nodulation or inefficient nitrogen fixation. Experiments of this nature have generally

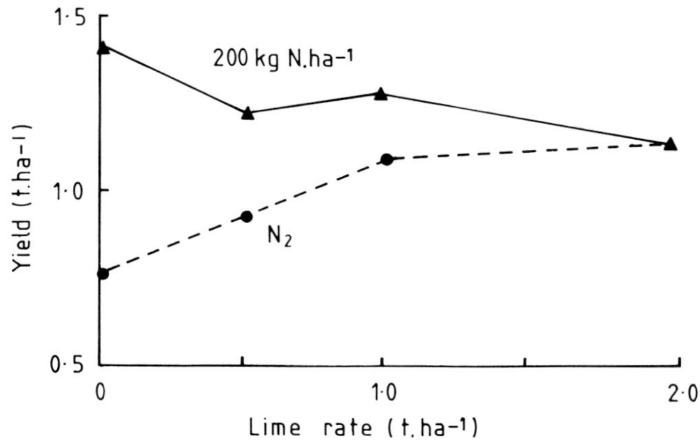

Fig. 3. Effect of lime application on the production of dry matter by subterranean clover on an acid soil at Yerong Creek and either supplied with mineral nitrogen (●) or reliant on symbiotic nitrogen fixation (○) (Evans *et al.* unpublished data).

highlighted the greater sensitivity of symbiotic plants to acidity, although this is not always so (Munns *et al.*, 1981).

Other fertilizer additions may be used in conjunction with the nitrogen treatments to improve diagnosis. In theory, high levels of phosphate may be used to precipitate ionic aluminium (Blamey *et al.*, 1983) and resolve the relative contribution of H^+ and aluminium to symbiotic failure; and $CaSO_4$ and $MgCO_3$ or NaOH may be manipulated to identify effects due to calcium deficiency and excess H^+ (Loneragan and Dowling, 1958; Munns, 1965b). Convergence in the dry matter yields of symbiotic and nitrogen-fertilized plants, as acidity is reduced, may also be due to release of molybdenum (Anderson and Moye, 1952). Consequently addition of this element could be used to assess the degree to which lack of molybdenum contributes to a disorder in nitrogen fixation.

b Nitrogen concentrations in shoots: Where acidity has impaired nitrogen fixation, the nitrogen concentration of the plant tissue should be low, provided that the legume does not have access to mineral nitrogen. Weir (pers. comm.) has suggested the following critical concentrations of nitrogen: 4.5 per cent for *T. repens* and *T. subterraneum* (actively growing up to commencement of flowering), 4.5 per cent for *M. sativa* (young mature leaves 5–20 cm from top of plant at early flowering), and 3.0 per cent for *Glycine max* (uppermost fully developed leaves at commencement of flowering). In each case the critical value for molybdenum was suggested as 0.4 $\mu g.g^{-1}$ provided it is associated with low total soil nitrogen.

c. Soil analysis: Soil pH has been the primary measurement used to predict the likelihood of difficulty in establishing an optimal symbiosis. For example, Munns (1977b) has collated published information and produced a table of critical pH values for nodulation. The table provides an upper pH value above which nodulation is not limited and a lower pH limit below which nodulation does not occur. Where comparisons have been made, the values obtained in solution culture agree reasonably well with observations in soil systems.

As a guide, the table is particularly valuable. It is difficult to be precise about critical values of this type, however, because they depend significantly upon the rhizobial strain (Munns, 1965a; Robson and Loneragan, 1970a) and legume cultivar. The effect of pH on nodulation below the 'non-limiting' value may also vary quantitatively between legumes. For example, the percentage of plants of *M. sativa* nodulated fell by 20 per cent with strain U45 when pH was decreased by 0.3 units (from 5.7), and by nearly 60 per cent with strain SU47 (Munns, 1965a).

d. Nodulation: Visual assessments of nodulation are often made, though interpretation may be difficult. We consider that the ability of legumes to compensate for a loss in nodule formation by an increase in nodule volume may depend on the otherwise balanced nutrition of the plant, the extent of inhibition of nodule formation, and longevity of nodule tissue for nitrogen fixation. To rely on this compensation may be hazardous. Evans *et al.* (unpublished data) have obtained significant correlations between nodule number per g root and the dry matter production of subterranean clover in pastures, indicating that too few nodules cannnot meet demand for nitrogen under all field conditions. Sparse nodulation may also indicate poor survival of rhizobia, which may produce nodulation failure in subsequent years, particularly after soil disturbance (Coventry *et al.*, 1985c). An additional danger is the possibility that, as acidity increases, the nodules may be formed by less effective rhizobial strains, and/or the efficiency of nitrogen fixation *per se* may be reduced.

e. Improving diagnosis: Although pH, as such, is perhaps the most useful diagnostic measurement, it is clear from defined culture work that both rhizobial growth and nodulation can be affected by other factors: soil calcium, aluminium and phosphate levels, rhizobial strain, and legume (see Sections II,A,*1* and II,B,*3*). For example, the inhibition of nodulation of *T. repens* by aluminium (50 μM) was reduced as phosphate was increased from 10 μM, and the formation of nodules between pH 5.0 and pH 6.0 was only markedly affected by acidity either if calcium was low (50 μM) or if aluminium was present (Wood *et al.*, 1984b). Strains of cowpea rhizobia of similar acid tolerance at pH 4.6 showed different rates of growth in the presence of high aluminium (50 μM) and low phosphate (10 μM) (Keyser and Munns, 1979a; Cassman *et al.*, 1981). These studies indicate the imprecision that may be encountered when using pH as the sole diagnostic measurement. They also emphasise the need to interpret the results of soil analysis in terms of both the species and cultivar of legume to be grown, as well as the likely fertilizer practice, particularly in regard to phosphate, gypsum (calcium) and organic matter additions. Furthermore, prediction of symbiotic performance will depend on the nature of the strains of rhizobia inhabiting or introduced into the soil.

B. Consequences of acidity for nitrogen fertility

The legume-*Rhizobium* symbiosis cycles gaseous N_2 back into soils to sustain nitrogen fertility. Thus the sensitivity of symbiotic nitrogen fixation to soil acidity can reduce soil fertility, the productivity of the legume and that of other plants grown as part of a rotation.

Loss in nitrogen fixation capacity must carry economic penalties. However, in the absence of published data on the quantitative effect of acidity on nitrogen fixation, only tentative comments on the economic consequences of acidity for nitrogen fertility are given.

Several estimates of the nitrogen benefits from clovers and medics in southern Australia have been made (Cook, 1939; Donald and Williams, 1954; Russell, 1960; Ford, 1968); these generally range between 47 and 115 $kg.ha^{-1}.year^{-1}$ (Greenland, 1971). Whilst none of these estimates involved quantification of nitrogen fixed *per se*, Bergersen and Turner (1983), using ^{15}N isotope techniques, indicated that in grass-clover swards the contribution of nitrogen fixation to total nitrogen can reach 90 per cent. Given an average clover nitrogen yield of 150 kg $N.ha^{-1}$ for southern Australia, and a conservative contribution from nitrogen fixation of 50 per cent, then the mean benefit of 70 kg $N.ha^{-1}$ suggested by Ford (1968) seems reasonable.

The potential loss of dry matter in clover pasture affected by acidity has been observed to be as high as 50 per cent (Evans and Hochman, 1984) but we have assumed a loss of 20 per cent in the calculation below. Further, if we assume for simplicity that dry matter reduction incurs no penalty in terms of nitrogen concentrations in tissue, and that there is a consistent, proportional contribution (50 per cent) of symbiotic nitrogen to total nitrogen, then for each 10^6 ha of acidity affected land in southern

Australia the loss in nitrogen benefit is 14 kg N.ha^{-1}. This loss would be worth about \$6 million per 10^6 ha if valued at the current price of nitrogen fertilizer. The cost of preventing this loss, if broadcast lime is the appropriate method to overcome the harmful effects of acidity, would amount to about \$12 million.year^{-1}.10^6 ha^{-1} (assuming a required application rate of 2000 kg lime.ha^{-1}, a cost of application of \$30.ha^{-1} and a five-year residual effect of the lime).

C. Correction of acid soil disorders

1. Plant and rhizobial tolerance to acidity

a. Objectives of selection: Increasing nitrogen fixation in legumes grown in acid soils may be profitable through increasing the tolerance to acidity of the rhizobia, the legume, or their ability to form a symbiosis. The objective will depend on which factor is the most sensitive to the particular acidity stress. Commonly, the host legume is less sensitive than the rhizobia and the stages involved in initiating nodules. Efficient selection must counter the most stressful factors limiting effective symbiosis and must produce sufficient tolerance to be of practical value. The factors identified as the most likely causes of an impaired symbiosis are excessive amounts of H^+ and aluminium and deficiencies of calcium and phosphorus.

A crucial plant factor for achieving tolerance is maintenance of root growth, and particularly formation of infectible root cells in the presence of aluminium. In addition, the specific requirements for high calcium and low H^+ in the infection process of some legumes need to be mitigated, though it is not clear whether these requirements occur at the plant or microbial level, or both. Since acidity and low soil calcium may adversely influence the effectiveness of nodules for nitrogen fixation, the established symbiosis also needs to be screened for efficiency for nitrogen assimilation under these stresses (Munns *et al.*, 1979).

A valuable criterion for selecting rhizobia is to attain a capacity for rapid rhizobial growth, particularly in situations of high H^+ and aluminium and, perhaps, low phosphate availability.

The effects of individual acidity-related factors on both rhizobial growth (Keyser and Munns, 1979a,b: Thornton and Davey, 1983a) and root nodulation (Hohenberg and Munns, 1984), appear to be determined independently. Tolerance to one acidity factor does not necessarily carry tolerance to the other factors. Thus, to obtain symbionts which tolerate a range of acid environments, a systematic step-wise approach to screening may be most effective.

b. Existence of variability: It is possible to select legume species with symbioses tolerant of aluminium. de Carvalho *et al.* (1981; 1982) studied the symbiotic tolerance to aluminium (up to 125 μM) of *Stylosanthes* species. They showed marked differences between species in the ability to nodulate under this stress (Fig. 2). Cultivars of species also differ in the sensitivity of symbiosis to acidity. Variation between cultivars in the time to nodule initiation or in nodule number, given stresses involving aluminium, H^+ and calcium, have been observed for *Vigna unguiculata* (Hohenberg and Munns, 1984), and similarly in the temperate species *T. subterraneum* (Kim *et al.*, 1985), *T. repens* (Wood *et al.*, 1984b) and *Medicago* spp. (Howieson and Ewing, 1986).

For any cultivar, variability between rhizobial strains in ability to nodulate in acid media is also well-established for both temperate and tropical species. Rhizobial strain influenced the nodulation of *T. repens* and *T. subterraneum* inoculated with *R. trifolii* in soil adjusted in pH (Blair, 1967; Thornton and Davey, 1983b), the nodulation of *Medicago* spp. inoculated with *R. meliloti* (Munns, 1965a; Robson and Loneragan, 1970a; Howieson and Ewing, 1986), the nodulation of *P. vulgaris* (Spain *et al.*, 1974 cited by Munns, 1977a; Graham *et al.*, 1982), and the nodulation of *V. unguiculata* (Keyser *et al.*, 1979; Hartel *et al.*, 1983).

Variation in the survival and growth of different strains of various rhizobial spp. under the stress of high H^+ levels has been known for a long time. Jensen (1943) observed a change in the proportion of strains of *R. trifolii* and *R. meliloti* able to grow at pH 4.9–5.5 and 5.3–5.7 respectively; and subsequently, Graham and Parker (1964) established strain differences in acid tolerance for each of the seven major rhizobial species. Strain variability in tolerance to aluminium, low calcium and low phosphate, at low pH, has been demonstrated more recently for tropical rhizobia (Keyser and Munns, 1979a; Cassman *et al.*, 1981; Ayanaba *et al.*, 1983) and temperate rhizobia (Thornton and Davey, 1983a; Howieson and Ewing, 1986).

c. Screening procedures: Comparison of legumes for ability to nodulate under acidity stresses has been carried out in soil (e.g. Thornton and Davey, 1983b), sand (e.g Andrew, 1976) and liquid media (e.g. Wood *et al.*, 1983). The mechanisms causing acid stress to legumes and the ability to select against single factor stresses are greatly facilitated by studies in flowing nutrient culture. It is necessary to use realistic solution concentrations in flowing culture techniques. These should be similar to soil solution concentrations, particularly in regard to aluminium, phosphate, calcium, pH and total ionic strength. The following concentrations, as described by Kim (1985), are likely to be appropriate: calcium

(500 μM), phosphate (1 μM); sulphate (850 μM), ionic strength (2276 μM), pH control ± 0.1 (at pH 4.3), and aluminium up to 200 μM. Flow rates and volumes need to be sufficient to avoid nutrient depletion. Solutions should not contain amounts of inorganic nitrogen that inhibit rapid nodulation, and should initially be aseptic if different rhizobial strains are to be compared.

It is of course difficult to simulate rhizosphere gradients in flowing nutrient culture, and the system is also likely to facilitate interaction between rhizobial cells and roots compared with soil systems. Nonetheless, it is possible that with more stringent conditions than have previously applied, flowing nutrient culture should enable selection of more acid-tolerant plants and symbioses.

A short-term, non-flow system has been described by Wood *et al.* (1983). They emphasise the need for a large volume of rooting solution in order to maintain buffering of pH at low phosphate concentrations. In their studies (Wood *et al.*, 1984a,b), the solution concentrations used were in the following ranges: aluminium (0–50 μM), phosphate (10–100 μM), calcium (50–1000 μM) and pH (4.3–5.5).

Large numbers of rhizobial strains can be screened in broth culture or on nutrient agar surfaces. The primary concern with these methods is the shift in pH due to metabolic activity of the rhizobia (Norris, 1965; Tan and Broughton, 1981). This shift can be mitigated by the use of pH buffers and by adjusting the composition of the growth medium (Date and Halliday, 1978; Tan and Broughton, 1981; Howieson, 1985). A particularly useful buffer for selecting more acid-tolerant *R. meliloti* is 2-(N-morpholino) ethane sulphonic acid (Howieson, 1985).

Suitable broth cultures have been defined by Keyser and Munns (1979a,b); their criterion for tolerance is the ability of a strain to grow to visible turbidity (which occurs at approximately 10^7 cells.mL^{-1}) in 25 days from an initial population of 10^3–10^4 cells.mL^{-1}). Up to this point in growth, solution pH changes very little, although Howieson (1985) detected a pH change at 10^6 cells.mL^{-1} with *R. meliloti*. Cassman *et al.* (1981) have described a technique for maintaining low levels of orthophosphate in broth culture, by using phosphated iron oxide powder in a dialysis membrane.

The use of solid media to screen rhizobia allows many isolates to be observed by spotting cells of individual strains onto nutrient agar plates. Screening is based on visible growth without a pH change, and an appropriate pH-sensitive indicator is required in the medium. This technique is complicated by the quantity of cells in the initial inoculum (Bromfield and Jones, 1980; Ayanaba *et al.*, 1983). Bromfield and Jones (1980) also indicated that using citrate-phosphate to buffer pH for the screening of *R. trifolii* may be inefficient due to metabolism of the citrate by rhizobia. Ayanaba *et al.* (1983) used small loopfuls of culture (0.04 L) containing 400–4000 cells; the method was considered appropriate providing observations were frequent enough to measure growth before pH changes were detected.

d. Effectiveness of selection: The variability in the ability both of rhizobia to proliferate and of legumes to nodulate under acid stress establishes a basis for correcting acid soil disorders through a genetic approach.

The effectiveness of this approach depends first on whether there is a sufficient level of tolerance to be beneficial practically. Second, the capacity for effective symbiotic nitrogen fixation, i.e. the competence of rhizobia to colonise soils and rhizospheres and to initiate nodules, must be maintained. Symbiotic tolerance of soil acidity and effectiveness in nitrogen fixation are inherited independently (e.g. Keyser *et al.*, 1979). Thus, some level of screening combinations of potential symbionts under field conditions is necessary (Munns, 1984).

Screening rhizobia for survival and growth in acidic laboratory media has had some success in improving the nodulation of host legumes in acid soils (Keyser *et al.*, 1979; Graham *et al.* ,1982; Hartel *et al.*, 1983; Thornton and Davey, 1983a), no doubt because growth is a prerequisite for nodule formation. Among strains of cowpea rhizobia screened by Keyser *et al.* (1979), a few strains tolerant of both H^+ and aluminium (by their criteria) produced abundant nodulation and efficient nitrogen fixation with *V. unguiculata* in acid (pH 4.6), aluminium-rich soils. Strains with a combined tolerance of both H^+ and aluminium were generally better under laboratory conditions than strains tolerant of H^+ only, but many with the combined tolerance did not perform optimally in establishing symbioses in the field. Similarly, Thornton and Davey (1983b) classified strains of *R. trifolii* for acid-tolerance in the laboratory in the absence of plants, and then compared them for symbiotic performance with *T. subterraneum* in acid/aluminium adjusted soil. Over 70 per cent of strains classified as tolerant ranked in the top half of symbiotic performance at pH 4.4 (aluminium, 57 per cent of saturation exchange capacity), indicating that tolerance in nodulation could be combined with effectiveness in nitrogen fixation. However, in no instance did nodulation by 'tolerant' strains approach that achievable through raising soil pH with lime. As a group, the 'tolerant' strains produced dry matter and total nitrogen quantities to 68 per cent and 75 per cent that of the limed treatments.

The inefficiency of laboratory screening of rhizobial cultures (based on growth) for the purpose of optimising nodulation and nitrogen fixation in the field may have several origins. In the studies of both Keyser *et al.* (1979) and Thornton and Davey (1983b), instances of the effects of acidity on efficiency of nitrogen fixation *per se* were cited. This phenomenon is determined by genetics of both the rhizobia and legume, and thus a further round of selection, involving both host plant and bacteria in, for example, acidified flowing culture, could reduce the number of strains for field testing. Secondly, the infection of roots is sensitive to H^+, calcium and aluminium and this may inhibit nodule formation, even if rhizobia have sufficient tolerance to colonise roots. Thus, Hartel *et al.* (1983) suggested that the contrasting abilities of two strains of *Bradyrhizobium* spp. to nodulate *V. unguiculata* in acid soil could not be due to differences in growth of the strains but, rather, to one strain being able to overcome the effect of aluminium in inhibiting nodule initiation. Close examination of this paper, however, indicates differences between the strains in the size of surviving populations consistent with the trends in nodulation. Also, in the complementary paper (Hartel and Alexander, 1983), the aluminium-sensitive strain grew less well in sterile soil and rarely achieved or maintained population levels as high as those of the aluminium-tolerant strain in non-sterile soil. Thirdly, small differences in the rates of growth of strains in acidic media and in the length of lag phase may be critical given the short period during which nodules can be initiated and the competitive environment in natural soils (Section II,A,*3*). It is noteworthy in this regard that, even with so-called 'tolerant' strains, published growth rates in acidic media are well below potential rates indicated in less acidic broths. For example, the mean generation time of rhizobial species in favourable media was 8.5 h compared to generation times of 'tolerant' strains of 10–23 h in low phosphate media and 13–99 h with 50 μM aluminium present (Keyser and Munns, 1979a). It is conceivable that rates of growth achieved so far in the screening of strains on acidic media represent threshold levels of adequate symbiotic performances, making small differences between strains significant. A more stringent screening procedure should consider a rapid determination of generation times. Acquisition of sufficient tolerance may additionally require genetic manipulations to increase the ability of rhizobia to grow more rapidly, particularly in the presence of aluminium.

Besides the ability to nodulate effectively in acid soils, it is important agronomically that rhizobia persist through seasons subsequent to their introduction. Soil acidity may adversely affect the persistence of rhizobia (Jones, 1966; Lowendorf *et al.*, 1981; Thornton and Davey, 1983a,b). However, Thornton and Davey (1983a,b) indicated that laboratory selection for acid-tolerant growth, at least with *R. trifolii*, could be useful for promoting longer-term survival in acid, non-sterile soil. Greater tolerance to acidity has also been associated with improved persistence of *R. meliloti* (Lowendorf *et al.*, 1981; Howieson and Ewing, 1986). The association of acid-tolerance with survival is consistent with observations of better survival of the more acid-tolerant rhizobial species, such as those of the cowpea miscellany, compared with *R. phaseoli* and *R. meliloti* (Lowendorf *et al.*, 1981). Faster growth rates under acid stresses may assist in reducing the displacement of these rhizobial strains from limited niches, as discussed earlier (Section II,A,*2*).

Little use has yet been made of mutation and genetic engineering to improved rhizobial performance in acid soils. Rai and Prasad (1983) used N-methyl-N-nitro-nitrosoguanadine (NTG) to mutate *R. leguminosarum*, and two mutants were isolated with improved ability to nodulate *Lens esculenta* at low pH. Rakotoarisoa *et al.* (1981) also reported the use of NTG to improve the growth of *R. meliloti* in acidic broths.

e. Management by genera and species: Legume genera and species vary markedly in their tolerance to acidity stresses (e.g. Bryan, 1923; Munns, 1977a) and, broadly, the rhizobia of these species also show a similar order of tolerance. Thus, Fred (1918) observed that *R. lupini* was more tolerant than *R. meliloti* and that *R. trifolii*, *R. phaseoli*, *R. leguminosarum* and *R. japonicum* were intermediate. Rhizobia of the cowpea (*Vigna*) group are also particularly tolerant (Rerkasem, 1977; Lowendorf *et al.*, 1981). The genea *Lupinus* and *Vigna* tolerate acid soils better than *Trifolium* and *Pisum*, and the latter better than *Medicago* (Munns, 1977a). Thus, limited opportunity exists for the deliberate choice of legumes possessing more tolerance to acidity. The temperate crop legume, *Lupinus angustifolius*, for example, may be used in wheat rotations in acid soils in preference to *Pisum sativum.*

2. *Inoculation and neutralisation of acidity*

Inoculation of legume seed with selected rhizobial strains is often beneficial to yield. However, this single process, particularly for some of the temperate pasture legumes, is insufficient for the initial establishment of seedlings in some acid soils (Anderson and Moye, 1952). Seedlings fail to establish due to lack of nodulation. This is believed to be due to the difficulty that acid-sensitive rhizobia experience, having left the seed coat. The broadcasting or drilling of lime was adopted as a general solution. Liming materials can be used to neutralise acidity within the soil profile, or in a spatially defined region, when pasture legumes are being established. Grain and forage legumes can also benefit from applications of lime (Sartain and Kamprath, 1975; Elkins *et al.*, 1976; Zaroug and Munns, 1980; Ssali, 1981; Blamey, 1983). The methods used are: broadcast lime plus incorporation; lime drilled with the fertilizer within bands in the soil; and lime pelleted with the inoculum on the surface of the legume seed (see Chapter 6). The quantity of lime required for pelleting the seed is considerably less than that needed when lime is drilled in close proximity to the seed, which again is much less than that required when lime is broadcast and incorporated into the soil (Loneragan *et al.*, 1955).

Lime-pelleting seed is a standard recommendation for the establishment of many temperate legumes in soils of pH less than 6.0 (1:5 soil:water) (Hastings and Drake, 1960; Cass-Smith and Goss, 1964; Roughley *et al.*, 1966). The lime coat results in improved nodulation and growth of these legumes. This stimulatory effect can persist in subsequent years (Drake and Kehoe, 1954; Loneragan *et al.*, 1955; Paton, 1960; Simmons, 1962; Cordero and Blair, 1978). In some situations, the best result in the year after establishment was obtained with a lime and superphosphate mixture drilled with the seed (Paton, 1960; Simmons, 1962). However, lime-pelleting seed of temperate legumes does not always stimulate nodulation (Shipton and Parker, 1967) and with many tropical legumes it can be harmful (Norris, 1971). Many of the rhizobia required for tropical legumes are the slower-growing 'cowpea type' which are better adapted to acid soils. This grouping has been used as the basis for a guide distinguishing the need for lime-pelleting seed of tropical legumes (Diatloff, 1971; Date, 1977).

There have been few reports on the efficacy of methods of inoculation as an alternative to introduction with seed. Graham *et al.* (1982) used a granular, soil-applied, inoculum to introduce *R. phaseoli* under *P. vulgaris* sown into acid soil ranging in pH from 3.8 to 4.6. This procedure resulted in approximately 80 per cent of plants being nodulated regardless of the pH. With seed-applied (lime-free) inoculation, nodulation was significantly reduced. Since soil disturbance may aggravate the survival of acid-sensitive rhizobia (Coventry *et al.*, 1985c), perhaps by disturbing fragile niches protected from bulk soil pH, the incorporation of granular inoculum may permit the survival of rhizobia in larger numbers, albeit at discrete foci.

In addition to seed inoculation, a heavier application of lime than on a seed pellet is required in some soils for satisfactory legume growth and nitrogen fixation (Fitzpatrick, 1958; Paton, 1960; Diatloff and Luck, 1972; Ssali, 1981). These acid soils usually have substantial exchangeable aluminium and/or manganese. However, not all strongly acid soils are responsive to liming, and there are reports of nil responses or negative responses with liming. The effective rates of lime, broadcast on lime-responsive soils, can vary from 1.0 $t.ha^{-1}$ to 5.0 $t.ha^{-1}$ (Munns, 1977a). Consequently, the extent to which lime is used on such soils will depend on the economics of the agricultural enterprise, the magnitude of the legume's response, and the cost of applying and incorporating the lime.

REFERENCES

Albrecht, W. A. and Davis, F. L. (1929). Relation of calcium to the nodulation of soybeans on acid and neutral soils. *Soil Science* **28**, 261–279.

Anderson, A. J. (1942). Molybdenum deficiency on a South Australian ironstone soil. *Journal of the Australian Institute of Agricultural Science* **8**, 73–75.

Anderson, A. J. (1956). Molybdenum as a fertilizer. *Advances in Agronomy* **8**, 163–202.

Anderson, A. J. (1970). Trace elements for sheep pastures and fodder crops in Australia. *Journal of the Australian Institute of Agricultural Science* **36**, 15–29.

Anderson, A. J. and Moye, D. V. (1952). Lime and molybdenum in clover development on acid soils. *Australian Journal of Agricultural Research* **3**, 95–110.

Anderson, A. J. and Spencer, D. (1948). Lime in relation to clover nodulation at sites on the southern tablelands of New South Wales. *Journal of the Australian Institute of Agricultural Science* **14**, 39–41.

Anderson, A. J. and Spencer, D. (1950). Molybdenum in nitrogen metabolism of legumes and non-legumes. *Australian Journal of Scientific Research* (Series B) **3**, 414–430.

Andrew, C. S. (1976) Effect of calcium, pH and nitrogen on the growth and chemical composition of some tropical and temperate pasture legumes. I. Nodulation and growth. *Australian Journal of Agricultural Research* **27**, 611–623.

Andrew, C. S. and Kamprath, E. J. (1978). 'Mineral Nutrition of Legumes in Tropical and Subtropical Soils.' CSIR0, Melbourne.

Andrew, C. S. and Norris, D. O. (1961). Comparative responses to calcium of five tropical and four temperate pasture legume species. *Australian Journal of Agricultural Research* **12**, 40–55.

Andrew, C. S., Johnson, A. D. and Sandland, R. L. (1973). Effect of aluminium on the growth and chemical composition of some tropical and temperate pasture legumes. *Australian Journal of Agricultural Research* **24**, 325–339.

Asher, C. J. (1981). Limiting external concentrations of trace elements for plant growth: use of flowing solution culture techniques. *Journal of Plant Nutrition* **3**, 163–180.

Ayanaba, A., Asanuma, S. and Munns, D. N. (1983). An agar plate method for rapid screening of *Rhizobium* for tolerance to acid-aluminium stress. *Soil Science Society of America Journal* **47**, 256–258.

Banath, C. L., Greenwood, E. A. N. and Loneragan, J. F. (1966). Effect of calcium deficiency on symbiotic nitrogen fixation. *Plant Physiology* **41**, 760–763.

Barber, L. E. (1980). Enumeration, effectiveness, and pH resistance of *Rhizobium meliloti* populations in Oregon soils. *Soil Science Society of America Journal* **44**, 537–539.

Bauer, W. C. (1981). Infection of legumes by rhizobia. *Annual Review of Plant Physiology* **32**, 407–449.

Bergersen, F. J. (1961). The growth of *Rhizobium* in synthetic media. *Australian Journal of Biological Sciences* **14**, 349–360.

Bergersen, F. J. and Turner, G. L. (1983). An evaluation of ^{15}N methods for estimating nitrogen fixation in a subterranean clover-perennial ryegrass sward. *Australian Journal of Agricultural Research* **34**, 391–402.

Bergersen, F. J., Brockwell, J., Gibson, A. H. and Schwinghamer, E. A. (1971). Studies of natural populations and mutants of *Rhizobium* in the improvement of legume inoculants. *Plant and Soil*, Special Volume, 3–16.

Bhuvaneswari, T. V., Turgeon, B. G. and Bauer, W. D. (1980). Early events in the infection of soybean (*Glycine max* L. Merr) by *Rhizobium japonicum*. I. Localization of infectible root cells. *Plant Physiology* **66**, 1027–1031.

Blair, I. D. (1967). Studies on *Rhizobium* strains. II. *Rhizobium trifolii. New Zealand Journal of Agricultural Research* **10**, 237–252.

Blamey, F. P. C. (1983). Acid soil infertility effects on peanut yields and yield components. *Communications in Soil Science and Plant Analysis* **14**, 373–386.

Blamey, F. P. C., Edwards, D. G. and Asher, C. J. (1983). Effects of aluminium, OH : Al, and P : Al molar ratios, and ionic strength on soybean root elongation in solution culture. *Soil Science* **136**, 197–207.

Bohlool, B. B. and Schmidt, E. L. (1973). Persistance and competition aspects of *Rhizobium japonicum* observed in soil by immunofluorescence microscopy. *Soil Science Society of America Journal* **37**, 561–564.

Bottomley, P. J., Almendvas, A. S., Leung, K. T., Valvidia, B., Demezas, D. H., Dughri, M. H. and Busse, M. D. (1986). Significance of *Rhizobium trifolii* in the introduction of subterranean clover. Proceedings of the 8th Australian Nitrogen Fixation Conference. *AustralianInstitute of Agricultural Science Occasional Publication* **25**, 185–188.

Bowen, G. D. and Rovira, A. D. (1976). Microbial colonization of plant roots. *Annual Review of Phytopathology* **14**, 121–144.

Bowra, B. J. and Dilworth, M. J. (1981). Motility and chemotaxis towards sugars in *Rhizobium leguminosarum. Journal of General Microbiology* **126**, 231–235.

Bromfield, E. S. P. and Jones, D. G. (1980). Studies on acid tolerance of *Rhizobium trifolii* in culture and soil. *Journal of Applied Bacteriology* **48**, 253–264.

Broughton, W. J. (1978). A review: Control of specificity in legume-*Rhizobium* associations. *Journal of Applied Bacteriology* **45**, 165–194.

Broughton, W. J., van Egeraat, A. W. S. M. and Lie, T. A. (1980). Dynamics of *Rhizobium* competition for nodulation of *Pisum sativum* cv. Afghanistan. *Canadian Journal of Microbiology* **26**, 562–565.

Bryan, O. C. (1923). Effect of acid soils on nodule-forming bacteria. *Soil Science* **15**, 37–40.

Caetano-Annolles, G., Lagares, A., Martinez, E. N. and Favelukes, G. (1985). Rhizobial attachment to roots: Dependence on divalent cations and pH. Proceedings of 10th North American *Rhizobium* Conference, NifTAL. 34.

Cass-Smith, W. P. and Goss, O. H. (1964). Inoculation and lime pelleting leguminous seeds. *Journal of Agriculture, Western Australia* (Series 4) **5**, 225–231.

Cassman, K. G., Munns, D. N. and Beck, D. P. (1981). Growth of *Rhizobium* strains at low concentrations of phosphate. *Soil Science Society of America Journal* **45**, 520–523.

Chakrabaiti, S., Lee, M. S. and Gibson, A. H. (1981). Diversity in the nutritional requirement of strains of various *Rhizobium* species. *Soil Biology and Biochemistry* **13**, 349–354.

Chandler, M. R. (1978). Some observations on the infection of *Arachis hypogaea* L. by *Rhizobium. Journal of Experimental Botany* **219**, 749–755.

Chandler, M. R., Date, R. A. and Roughley, R. J. (1982). Infection and root-nodule development in *Stylosanthes* species by *Rhizobium. Journal of Experimental Botany* **33**, 47–57.

Chao, W. L. and Alexander, M. (1982). Influence of soil characteristics on the survival of *Rhizobium* in soils undergoing drying. *Soil Science Society of America Journal* **46**, 949–952.

Cook, L. J. (1939). Further results secured in 'top-dressing' poor south-eastern pasture lands with phosphate fertilizers. *Journal of Agriculture, South Australia* **42**, 791–808.

Cordero, S. and Blair, G. J. (1978). The effects of lime-pelleting and lime-superphosphate fertilizer on the growth of three annual legumes in an acid sandy soil. *Plant and Soil* **50**, 257–268.

Coventry, D. R. (1985). Changes in agricultural systems on acid soils in southern Australia. Proceedings of the 2nd Australian Agronomy Conference pp. 126–145.

Coventry, D. R., Hirth, J. R., Reeves, T. G. and Jones, H. R. (1985a). Development of *Rhizobium trifolii* and nodulation of subterranean clover following the cropping phase in crop-pasture rotations in south-eastern Australia. *Soil Biology and Biochemistry* **17**, 17–22.

Coventry, D. R., Hirth, J. R., Reeves, T. G. and Burnett, V. F. (1985b). Growth and nitrogen fixation by subterranean clover in response to inoculation, molybdenum application and soil amendment with lime. *Soil Biology and Biochemistry* **17**, 791–796.

Coventry, D. R., Hirth, J. R., Reeves, T. G. and Jones, H. R. (1985c). Changes in soil populations of *Rhizobium trifolii* following crop establishment in a clover-ley rotation in S. E. Australia. XV International Grasslands Congress 1985, pp. 541–542.

Coventry, D. R., Hirth, J. R. and Fung, K. K. H. (1987). Nutritional restraints on subterranean clover grown on acid soils used for crop-pasture rotation. *Australian Journal of Agricultural Research* **38**, 163–178.

Cregan, P. D. (1980). Soil acidity and associated problems — guidelines for farmer recommendations. AG Bulletin 7, 1980. Department of Agriculture, New South Wales.

Damirgi, S. M., Frederick, L. R. and Anderson, I. C. (1967). Sero-groups of *Rhizobium japonicum* in soybean nodules as affected by soil types. *Agronomy Journal* **59**, 10–12.

Danso, S. K. A. and Alexander, M. (1974). Survival of two strains of *Rhizobium* in soil. *Soil Science Society of America Journal* **38**, 86–89.

Dart, P. J. (1974). The infection process. *In* 'The Biology of Nitrogen Fixation' (A. Quispel, ed.), pp. 381–472. North Holland Publishing Company, Amsterdam.

Dart, P. J. (1977). Infection and development of leguminous nodules. *In* 'A Treatise on Dinitrogen Fixation, Section III, Biology'. (R. W. F. Hardy and W. S. Silver, eds.), pp. 377–472. Wiley, New York.

Date, R. A. (1977). Inoculation of tropical pasture legumes. *In* 'Exploiting the Legume-*Rhizobium* Symbiosis in Tropical Agriculture'. (J. M. Vincent, A. S. Whitney, J. Bose, eds.), pp. 293–312. Miscellaneous Publication 145, University of Hawaii.

Date, R. A. and Halliday, J. (1978). Selecting *Rhizobium* for acid infertile soils of the tropics. *Nature* **277**, 62–64.

Dawson, M. D. and Bhella, H. S. (1972). Subterranean clover (*Trifolium subterraneum* L.) yield and nutrient content as influenced by soil molybdenum status. *Agronomy Journal* **64**, 308–311.

Dazzo, F. B. and Hubbell, D. H. (1975) Cross-reactive antigens and lectins as determinants of symbiotic specificity in the *Rhizobium*-clover association. *Applied Microbiology* **30**, 1017–1033.

Dazzo, F. B., Yanke, W. E. and Brill, W. J. (1978). Trifolii: a *Rhizobium* recognition protein from white clover. *Biochimica et Biophysica Acta* **539**, 276–286.

de Carvalho, M. M., Andrew, C. S., Edwards, D. G. and Asher, C. J. (1980). Comparative performance of six *Stylosanthes* species in three acid soils. *Australian Journal of Agricultural Research* **31**, 61–76.

de Carvalho, M. M., Edwards, D. G., Andrew, C. S. and Asher, C. J. (1981). Aluminium toxicity, nodulation and growth of *Stylosanthes* species. *Agronomy Journal* **73**, 261–265.

de Carvalho, M. M., Edwards, D. G., Asher, C. J. and Andrew, C. S. (1982). Effects of aluminium on nodulation of two Stylosanthes species grown in nutrient solution. *Plant and Soil* **64**, 141–152.

Diatloff, A. (1971). Pelleting tropical legume seed. *Queensland Agriculture Journal* **97**, 363–366.

Diatloff, A. (1972). Recognizing a nodulation problem. *Tropical Grasslands* **6**, 70–72.

Diatloff, A. and Luck, P. E. (1972). The effects of the interactions between seed inoculation, pelleting and fertilizer on growth and nodulation of desmodium and glycine on two soils in S. E. Queensland. *Tropical Grasslands* **6**, 33–38.

Dobereiner, J. (1966). Manganese toxicity effects on nodulation and nitrogen fixation of beans (*Phaseolus vulgaris* L.), in acid soils. *Plant and Soil* **24**, 153–166.

Donald, D. M. and Williams, C. H. (1954). Fertility and productivity of a podzolic soil as influenced by subterranean clover (*Trifolium subterraneum* L.) and superphosphate. *Australian Journal of Agricultural Research* **5**, 664–687.

Dollas, G. Z. (1930). Local variation of soil acidity in relation to soybean inoculation. *Soil Science* **30**, 273–282.

Drake, F. R. and Kehoe, J. K. (1954). Subterranean clover-establishment on acid sandy soil in East Gippsland. *Journal of Agriculture (Victoria)* **52**, 385–388.

Dughri, M. H. and Bottomley, P. J. (1983). effect of acidity on the composition of an indigenous soil population of *Rhizobium trifolii* found in nodules of *Trifolium subterraneum* L. *Applied and Environmental Microbiology* **46**, 1207–1213.

Dughri, M. H. and Bottomley, P. J. (1984). Soil acidity and the composition of an indigenous population of *Rhizobium trifolii* in nodules of different cultivars of *Trifolium subterraneum* L. *Soil Biology and Biochemistry* **16**, 405–411.

During, C., Jackson, B. L. J. and Dyson, C. B. (1984). Lime effects on hill country. 2. Relation of soil moisture, pH and aluminium values with pasture yield. *New Zealand Journal of Agricultural Research* **27**, 531–536.

Elkins, D. M., Olsen, F. J. and Gower, E. (1976). Effects of lime and lime-pelleted seed on legume establishment and growth in south Brazil. *Experimental Agriculture* **12**, 201–206.

Evans, J. and Herridge, D. F. (1987). Nitrogen inputs and utilisation in crop legumes. *In* 'Nitrogen Cycling in Agricultural Systems of Temperate Australasia'. Proceedings Soil Science Society of Australia (in press).

Evans, J. and Hochman, Z. (1984). Rhizobium and subterranean clover performance in acid soils. *Australian Institute of Agricultural Science Occasional Publication* **12**, 129–130.

Evans, J. and O'Connor, G. E. (1985). Population dynamics of *Rhizobium trifolii* influenced by pH: significance to nodulation. Proceedings of the 10th North American *Rhizobium* Conference, Wailea, Mani. HI., p. 54.

Evans, J., Barnet, Y. M. and Vincent, J. M. (1979). Effect of bacteriophage on the colonization and nodulation of clover roots by paired strains of *Rhiobium trifolii. Canadian Journal of Microbiology* **27**, 974–978.

Evans, J., Scott, B. J. and LilI, W. J. (1987). Manganese tolerance in subterranean clover (*Trifolium subterraneum* L.) genotypes grown with ammonium nitrate or symbiotic nitrogen. *Plant and Soil* **97**, 207–215.

Fitzpatrick, E. N. (1958). Pasture development on peaty sands. *Journal of Agricultural, Western Australian* (Series 3) **7**, 505–510.

Ford, G. W. (1968). Partly humidified organic matter in soils. Ph.D. Thesis, University of Adelaide.

Franco, A. A. and Munns, D. N (1982). Acidity and aluminium restraints on nodulation, nitrogen fixation and growth of *Phaseolus vulgaris* in solution culture. *Soil Science Society of America Journal* **46**, 296–301.

Fred, E. B. and Davenport, A. (1918). Influence of reaction on nitrogen-assimilating bacteria. *Journal of Agricultural Research* **14**, 317–336.

Gibson, A. H. and Newton, W. E. (1981). 'Current Perspectives in Nitrogen Fixation'. Australian Academy of Sciences, Canberra.

Goodchild, D. J. and Bergersen, F. J. (1966). Electron microscopy of the infection and subsequent development of soybean nodule cells. *Journal of Bacteriology* **92**, 204–213.

Graham, P. H. and Parker, C. A. (1964). Diagnostic features in the characterisation of root-nodule bacteria of legumes. *Plant and Soil* **20**, 383–396.

Graham, P. H., Viteri, S. E., Mackie, F., Vargas, A. T. and Palacios, A. (1982). Variation in acid soil tolerance among strains of *Rhizobium phaseoli*. *Field Crops Research* **5**, 121–128.

Gray, T. R. G., Baxbi, T., Hill, I. R. and Goodfellow, M. (1967). Direct observation of bacteria in soil. *In* 'The Ecology of Soil Bacteria' (T. R. G. Gray and D. Parkinson, eds.), pp. 171–197. Liverpool University Press, Liverpool.

Greenland, D. J. (1971). Changes in the nitrogen status and physical condition of soil under pastures, with special reference to the maintenance of the fertility of Australian soils used for growing wheat. *Soils and Fertilizer* **34**, 237–258.

Gupta, U. C. (1979). Effect of methods of application and residual effect of molybdenum on the molybdenum concentration and yield of forages on podzol soils. *Canadian Journal of Soil Science* **59**, 183–189.

Gupta, U. C. and Lipsett, J. (1981). Molybdenum in soils, plants and animals. *Advances in Agronomy* **34**, 73–117.

Habte, M. and Alexander, M. (1977). Further evidence for the regulation of bacterial populations in soils by Protozoa. *Archives of Microbiology* **113**, 181–183.

Hagedorn, C. (1978). Effectiveness of *Rhizobium trifolii* populations associated with *Trifolium subterraneum* L. in south west Oregon soils. *Soil Science Society Journal* **42**, 447–451.

Hardy, R. W. F. and Gibson, A. H. (1977). 'A Treatise on Dinitrogen Fixation, Section IV, Agronomy and Ecology'. Wiley, New York.

Hartel, P. C. and Alexander, M. (1983). Growth and survival of cowpea rhizobia in acid, aluminium-rich soils. *Soil Science Society of America Journal* **47**, 502–506.

Hartel, P. G., Whelan, A. M. and Alexander, M. (1983). Nodulation of cowpeas and the survival of cowpea rhizobia in acid, aluminium-rich soils. *Soil Science Society of America Journal* **47**, 514–517.

Hastings, A. and Drake, A. D. (1960). Inoculation and pelleting of clover seed. *New Zealand Journal of Agriculture* **101**, 619–621.

Hely, F. W., Bergersen, F. J. and Brockwell, J. (1957). Microbial antagonism in the rhizosphere as a factor in the failure of inoculation of subterranean clover. *Australian Journal of Agricultural Research* **8**, 24–44.

Helyar, K. R. and Anderson, A. J. (1970). Responses of five pasture species to phosphorus, lime and nitrogen on an infertile acid soil with a high phosphate sorption capacity. *Australian Journal of Agricultural Research* **21**, 677–692.

Hilder, E. J. (1964). The distribution of plant nutrients by sheep at pasture. *Proceedings of the Australian Society of Animal Production* **5**, 241–248.

Hohenberg, J. S. and Munns, D. N. (1984). Effect of soil acidity factors on nodulation and growth of *Vigna unguiculata* in solution culture. *Agronomy Journal* **76**, 477–481.

Holding, A. J. and King, J. (1963). The effectiveness of indigenous populations of *Rhizobium trifolii* in relation to soil factors. *Plant and Soil* **18**, 191–198.

Holding A. J. and Lowe, J. F. (1971). Some effects of acidity and heavy metals on the rhizobium-leguminous plant association. *Plant and Soil*, Special Volume, 153–166.

Howieson, J. G. (1985). Use of an organic buffer for the selection of acid tolerant *Rhizobium meliloti* strains. *Plant and Soil* **88**, 367–376.

Howieson, J. G. and Ewing, M. (1986). Acid tolerance in the *Rhizobium meliloti-Medicago* symbiosis. *Australian Journal of Agricultural Research* **37**, 55–64.

Jensen, H. L. (1943). Nitrogen fixation in leguminous plants. IV. The influence of reaction on the formation of root nodules in *Medicago* and *Trifolium*. *Proceedings of the Linnean Society of New South Wales* **68**, 207–220.

Jensen, H. L. (1969). The distribution of lucerne and clover rhizobia in agricultural soils in Denmark. *Tidsskrift for Planteavl* **73**, 61–72.

Jones, D. G. (1966). The contribution of white clover to a mixed upland sward. II. Factors affecting density and effectiveness of *Rhizobium trifolii*. *Plant and Soil* **24**, 250–260.

Jones, D. G. and Morley, S. J. (1981). The effect of pH on host plant preference for strains of *Rhizobium trifolii* using fluorescent ELISA for strain identification. *Annals of Applied Biology* **97**, 183–190.

Jones, D. G., Hughes, R., Munro, J. M. M. and Davies, W. E. (1964). The contribution of white clover to a mixed upland sward. I. The effect of inoculation on the early development of white clover. *Plant and Soil* **21**, 63–69.

Jones, H. R. and Curnow, B. C. (1985). Nodulation of subterranean clover growing in permanent pastures on acid soils in North Central Victoria. *Australian Journal of Experimental Agriculture* **26**, 31–36.

Kamprath, E. J. (1978). Lime in relation to Al toxicity in tropical soils. *In* 'Mineral Nutrition of Legumes in Tropical and Subtropical Soils'. (C. S. Andrew and E. J. Kamprath, eds.), pp. 233–246. CSIRO, Melbourne.

Keyser, H. H. and Munns, D. N. (1979a). Tolerance of Rhizobia to acidity, aluminium and phosphate. *Soil Science Society of America Journal* **43**, 519–523.

Keyser, H. H. and Munns, D. N. (1979b). Effects of calcium, manganese and aluminium on growth of Rhizobia in acid media. *Soil Science Society of America Journal* **43**, 500–503.

Keyser, H. H. and Munns, D. N. (1981). Response of *Rhizobium* to acid and aluminium stress. *Soil Biology and Biochemistry* **13**, 115–118.

Keyser, H. H., Munns, D. N. and Hohenberg, J. S. (1979). Acid tolerance of rhizobia in culture and in symbiosis with cowpea. *Soil Science Society of America Journal* **43**, 719–722.

Kijne, J. W., Smit, G., Diaz, C. L. and Lugtenberg, B. J. J. (1985) Attachment of *Rhizobium leguminosarum* 248 to pea root hairs. *In* 'Nitrogen Fixation Research Progress'. (H. J. Evans, P. J. Bottomley and W. E. Newton, eds.), p. 260. NifhoFF.

Kim, M. K. (1985). Effects of pH and aluminium on nodulation and growth of *Trifolium subterraneum* cultivars. Ph.D. Thesis, University of Queensland.

Kim, M. K., Edwards, D. G. and Asher, C. J. (1985). Tolerance of *Trifolium subterraneum* cultivars to low pH. *Australian Journal of Agricultural Research* **36**, 569–578.

Lie, T. A. (1969). Effect of low pH on different phases of nodule formation in pea plants. *Plant and Soil* **31**, 391–406.

Loneragan, J. F. and Dowling, E. J. (1958). The interaction of calcium and hydrogen ions in the nodulation of subterranean clover. *Australian Journal of Agricultural Research* **9**, 464–472.

Loneragan, J. F., Meyer, D., Fawcett, R. G. and Anderson, A. J. (1955). Lime pelleted clover seeds for nodulation on acid soils. *Journal of the Australian Institute of Agricultural Science* **21**, 264–265.

Loos, M. A. and Louw, H. A. (1965). Influence of calcium carbonate amendments on the nodulation of white clover in acid soils of George. *South African Journal of Agricultural Science* **8**, 729–736.

Lowendorf, H. S., Baya, A. M. and Alexander, M. (1981). Survival of *Rhizobium* in acid soils. *Applied and Environmental Microbiology* **42**, 951–957.
Lowther, W. L. and Loneragan, J. F. (1968). Calcium and nodulation of subterranean clover (*Trifolium subterraneum* L.). *Plant Physiology* **43**, 1362–1366.
Lowther, W. L. and Loneragan, J. F. (1970). Calcium in the nodulation of legumes. Proceedings 11th International Grasslands Conference pp. 446–450.
Mahler, R. L. (1983). Influence of pH on yield and N and P nutrition of alfalfa grown on an Andic Mission silt loam. *Agronomy Journal* **75**, 731–735.
Mahler, R. L. and Wollum, A. G. (1981). Seasonal variation of *Rhizobium trifolii* in clover pastures and cultivated fields in North Carolina. *Soil Science* **132**, 240–246.
Mahler, R. L. and Wollum, A. G. (1982). Seasonal fluctuation of *Rhizobium japonicum* under a variety of field conditions in Northern Carolina. *Soil Science* **134**, 317–324.
Marques-Pinto, C., Yao, P. Y. and Vincent, J. M. (1974). Nodulating competitiveness amongst strains of *Rhizobium meliloti* and *R. trifolii. Australian Journal of Agricultural Research* **25**, 317–329.
Marshall, K. C. (1964). Survival of root-nodule bacteria in dry soils exposed to high temperatures. *Australian Journal of Agricultural Research* **15**, 273–281.
Munns, D. N. (1965a). Soil acidity and growth of a legume. I. Interactions of lime with nitrogen and phosphate on growth of *Medicago sativa* L. and *Trifolium subterraneum* L. *Australian Journal of Agricultural Research* **16**, 733–741.
Munns, D. N. (1965b). Soil acidity and growth of a legume. II. Reactions of aluminium and phosphate in solution and effects of aluminium, phosphate, calcium and pH on *Medicago sativa* L. and *Trifolium subterraneum* L. in solution culture. *Australian Journal of Agricultural Research* **16**, 743–756.
Munns, D. N. (1968). Nodulation of *Medicago sativa* in solution culture. I. Acid-sensitive steps. *Plant and Soil* **28**, 129–146.
Munns, D. N. (1970). Nodulation of *Medicago sativa* in solution culture. V. Calcium and pH requirements during infection. *Plant and Soil* **32**, 90–102.
Munns, D. N. (1977a). Soil acidity and related problems. *In* 'Exploiting the Legume-*Rhizobium* Symbiosis in Tropical Agriculture'. (J. M. Vincent, A. S. Whitney, and J. Bose, eds.), pp. 211–236. Miscellaneous Publication 145, University of Hawaii.
Munns, D. N. (1977b). Mineral nutrition and the legume symbiosis. *In* 'A Treatise on Dinitrogen Fixation, Section IV, Agronomy and Ecology'. (R. W. F. Hardy and A. H. Gibson, eds.), pp. 353–391. Wiley, New York.
Munns, D. N. (1978). Soil acidity and nodulation. *In* 'Mineral Nutrition of Legumes in Tropical and Subtropical Soils'. (C. S. Andrew and E. J. Kamprath, eds.), pp. 247–264. CSIRO, Melbourne.
Munns, D. N. (1984). Coping with soil acidity in legume symbioses. *Australian Institute of Agricultural Science Occasional Publication* **12**, 137–141.
Munns, D. N., Fox, R. L. and Koch, B. L. (1977). Influence of lime on nitrogen fixation by tropical and temperate legumes. *Plant and Soil* **46**, 591–601.
Munns, D. N., Keyser, H. H., Fogle, V. W., Hohenberg, J. S., Righetti, T. L., Lauter, D. L., Zaroug, M. G., Clarkin, K. L. and Whitacre, K. W. (1979). Tolerance of soil acidity in symbioses of Mung Bean with Rhizobia. *Agronomy Journal* **71**, 256–260.
Munns, D. N., Hohenberg, T. L., Righetti, T. L. and Lauter, D. J. (1981). Soil acidity tolerance of symbiotic and nitrogen-fertilized soybeans. *Agronomy Journal* **73**, 407–410.
Murphy, H. E., Edwards, D. G. and Asher, C. J. (1984). Effects of aluminium on nodulation and early growth of four tropical pasture legumes. *Australian Journal of Agricultural Research* **35**, 663–673.
Newman, E. I. and Bowen, H. J. (1974). Patterns of distribution of bacteria on root surfaces. *Soil Biology and Biochemistry*, **6**, 205–209.

Norris, D. O. (1959). The role of calcium and magnesium in the nutrition of *Rhizobium. Australian Journal of Agricultural Research* **10**, 651–698.

Norris, D. O. (1965). Acid production by *Rhizobium*: A unifying concept. *Plant and Soil* **22**, 143–166.

Norris, D. O. (1971). Seed pelleting to improve nodulation of tropical and sub-tropical legumes. 2. The variable response to lime and rock phosphate pelleting of eight legumes in the field. *Australian Journal of Experimental Agriculture and Animal Husbandry* **11**, 282–289.

Nutman, P. S. (1952). Host factors influencing infection and nodule development in leguminous plants. *Proceedings of the Royal Society of London* (Series B) **139**, 176–185.

Parker, C. A., Trinick, N. J. and Chatel, D. L. (1977). Rhizobia as soil and rhizosphere inhabitants. *In* 'A Treatise on Dinitrogen Fixation, Section IV, Agronony and Ecology'. (R. W. F. Hardy and A. H. Gibson, eds.) pp. 311–352. Wiley, New York.

Paton, D. F. (1960). Problems in legume establishment with special reference to nodulation and nutrition. *Journal of the Australian Institute of Agricultural Science* **26**, 32–37.

Pena-Cabriales, J. J. and Alexander, M. (1983). Growth of *Rhizobium* in unamended soil. *Soil Science Society of America Journal* **47**, 81–84.

Petersen, H. B. and Gooding, T. H. (1941). Nebraska Agricultural Experiment Station Bulletin 121.

Purchase, H. F. and Nutman, P. S (1957). Studies on the physiology of nodule formation. VI. The influence of bacterial numbers in the rhizosphere on nodule initiation. *Annals of Botany* **21**, 439–454.

Quispel, A. (1974). 'The Biology of Nitrogen Fixation'. North Holland Publishing Company, Amsterdam.

Rai, R. and Prasad, V. (1983). Effect of soil acidity factors on nodulation, active iron content of nodules and relative efficiency of symbiotic N_2-fixation by mutant strains of *Lens esculenta Rhizobium. Journal of Agricultural Science (Cambridge)* **100**, 607–611.

Rakotoarisoa, R. R., Tailliez, R. and Guillaume, J. B. (1981). Obtention d'un mutant de *Rhizobium meliloti* adapte aux cultures de luzerne en terres acids. *Plant and Soil* **60**, 99–110.

Ranga Rao, V. (1977). The effect of root temperature on the infection processes and nodulation in *Lotus* and *Stylosanthes. Journal of Experimental Botany* **28**, 241–259.

Rerkasem, B. (1977). Differential sensitivity to soil acidity of legume-*Rhizobium* symbioses. Ph.D. Thesis, University of Western Australia.

Rice, W. A., Penney, D. C. and Nyborg, M. (1977). Effects of soil acidity on rhizobia numbers, nodulation and nitrogen fixation by alfalfa and red clover. *Canadian Journal of Soil Science* **57**, 197–203.

Robson, A. D. (1978). Mineral nutrients limiting nitrogen fixation in legumes. *In* 'Mineral Nutrition of Legumes in Tropical and Subtropical Soils'. (C. S. Andrew and E. J. Kamprath, eds.), pp. 277–294. CSIRO, Melbourne.

Robson, A. D. and Loneragan, J. F. (1970a). Nodulation and growth of *Medicago truncatula* on acid soils. II. Colonization of acid soils by *Rhizobium meliloti.* *Australian Journal of Agricultural Research* **21**. 435–445.

Robson, A. D. and Loneragan, J. F. (1970b). Nodulation and growth of *Medicago truncatula* on acid soils. I. Effect of calcium carbonate and inoculation level on the nodulation of *Medicago truncatula* on a moderately acid soil. *Australian Journal of Agricultural Research* **21**, 427–434.

Roughley, R. J., Date, R. A. and Walker, M. H. (1966). Inoculating and lime pelleting legume seed. *Agricultural Gazette of New South Wales* **77**, 142–146.

Rovira, A. D. (1961). Rhizobium numbers in the rhizospheres of red clover and paspalum in relation to soil treatment and the numbers of bacteria and fungi. *Australian Journal of Agricultural Research* **12**, 77–83.

Rovira, A. D. and Stern, W. R. (1961). Rhizosphere bacteria in grass-clover associations. *Australian Journal of Agricultural Research* **12**, 1108–1118.

Rowe, B. A. (1982). Effects of limestone on pasture yields and the pH of two krasnozems in north-western Tasmania. *Australian Journal of Experimental Agriculture and Animal Husbandry* **22**, 100–105.

Russell, J. S. (1960). Soil fertility changes in the long term experimental plots at Kybybolite, S. A. I. Changes in pH, total nitrogen, organic carbon, and bulk density. *Australian Journal of Agricultural Research* **11**, 902–926.

Russell, P. E. and Jones, D. G. A. (1975). Immunofluorescent studies of selection of strains of *R. trifolii* by S184 white clover (*T. repens* L.). *Plant and Soil* **42**, 119–129.

Sartain, J. B. and Kamprath, E. J. (1975). Effect of liming a highly Al-saturated soil on the top and root growth and soybean nodulation. *Agronomy Journal* **67**, 507–510.

Schmidt, E. L. (1979). Initiation of plant root-microbe interactions. *Annual Review of Microbiology* **33**, 355–376.

Sewell, M. C. and Ganey, P. L. (1930). Interrelation of nutrient and soil reaction on growth and inoculation of alfalfa. *Soil Science* **30**, 291–299.

Shannon, P. W., Pringle, R. M., Mansell, G. P. and Edmeades, D. C. (1984). Effects of lime on pasture production on soils in the North Island of New Zealand. 2. The nature of seasonal pasture growth responses to lime application. *New Zealand Journal of Agricultural Research* **27**, 357–361.

Shipton, W. A. and Parker, C. A. (1967). Nodulation of lime-pelleted lupins and serradella when inoculated with peat and agar cultures. *Australian Journal of Experimental Agriculture and Animal Husbandry* **7**, 259–262.

Siman, A., Craddock, F. W. and Hudson, A. W. (1974). The development of manganese toxicity in pasture legumes under extreme climatic conditions. *Plant and Soil* **41**, 129–140.

Simmons, K. V. (1962). Establishing sub. clover on acid soils. *Agricultural Gazette of New South Wales* **73**, 183–188.

Souto, S. M. and Dobereiner, J. (1969). Manganese toxicity in tropical forage legumes. (In Portuguese) *Pesquisa Agropecuaria Brasileira* **4**, 129–138.

Spain, J. M., Francis, C. A., Howeler, R. H. and Calvo, F. (1974). Differential species and varietal tolerance to soil acidity in tropical crops and pastures. *In* 'Soil Management in Tropical America'. (E. Bornemisza and A. Alvarado, eds.), North Carolina State University, Raleigh.

Spencer, D. (1950). The effect of calcium and soil pH on nodulation of *T. subterraneum* L. clover on a yellow podsol. *Australian Journal of Agricultural Research* **1**, 374–381.

Ssali, H. (1981). The effect of level of $CaCO_3$, inoculation and lime pelleting on the nodulation and growth of beans in five acid soils. *Plant and Soil* **61**, 53–63.

Tan, I. K. P. and Broughton, W. J. (1981). Rhizobia in tropical legumes. XII. Biochemical basis of acid and alkali reactions. *Soil Biology and Biochemistry* **13**, 389–393.

Thomas, G. W. (1975). The relationship between organic matter content and exchangeable aluminium in acid soil. *Soil Science Society of America Journal* **39**, 591.

Thornton, F. C. and Davey, C. B. (1983a) Response of the clover-Rhizobium symbiosis to soil acidity and Rhizobium strain. *Agronomy Journal* **75**, 557–560.

Thornton, F. C. and Davey, C. B. (1983b). Acid tolerance of *Rhizobium trifolii* in culture media. *Soil Science Society of America Journal* **47**, 496–501.

Trinick, M. J. and Parker, C. A. (1982). Self-inhibition of Rhizobial strains and the influence of cultural conditions on microbial interactions. *Soil Biology and Biochemistry* **14**, 79–86.

Vincent, J. M. (1958). Survival of the root nodule bacteria. *In* 'Nutrition of the Legumes'. (E. G. Hallsworth, ed.), pp. 108–123. Butterworths, London.

Vincent, J. M. (1962). Influence of Ca and Mg on growth of *Rhizobium. Journal of General Microbiology* **28**, 653–663.

Vincent, J. M. (1980). Factors controlling the legume-*Rhizobium* symbiosis. *In* 'Nitrogen Fixation, Vol. II'. (W. E. Newton and W. H. Orme-Johnson, eds.), pp. 103–129. University Park Press, Baltimore.

Vincent, J. M. and Waters, L. M. (1953). The influence of the host on competition amongst clover root-nodule bacteria. *Journal of General Microbiology* **9** 357–370.

Vincent, J. M., Whitney, A. S. and Bose, J. (1977). 'Exploiting the Legume-*Rhizobium* Symbiosis in Tropical Agriculture'. College of Tropical Agriculture. Miscellaneous Publication 145, University of Hawaii.

Vose, P. B. and Jones, D. G. (1963). The interaction of manganese and calcium on nodulation and growth in three varieties of *Trifolium repens. Plant and Soil* **18**, 372–385.

Wilson, D. O. and Reisenauer, H. M. (1970). Effect of manganese and zinc ions on growth of *Rhizobium. Journal of Bacteriology* **102**, 729–732.

Wood, M., Cooper, J. E. and Holding, A. J. (1983). Method to assess the effects of soil acidity factors on legume-*Rhizobium* symbioses. *Soil Biology and Biochemistry* **15**, 123–124.

Wood, M, Cooper, J. E. and Holding, A. J. (1984a). Aluminium toxicity and nodulation of *Trifolium repens. Plant and Soil* **78**, 381–391.

Wood, M., Cooper, J. E. and Holding, A. J. (1984b). Soil acidity factors and nodulation of *Trifolium repens. Plant and Soil* **78**, 367–379.

Yao, P. Y. and Vincent, J. M. (1976). Factors responsible for the curling and branching of clover root hairs by *Rhizobium. Plant and Soil* **45**, 1–16.

Zaroug, M. G. and Munns, D. N. (1980). Screening strains of *Rhizobium* for the tropical legumes *Clitoria ternatea* and *Vigna trilobata* in soils of different pH. *Tropical Grasslands* **14**, 28–33.

4

The Effect of Soil Acidity on Microbial Activity in Soils

A. D. ROBSON and L. K. ABBOTT

I. INTRODUCTION

There has been considerable study of the effects of soil acidity on the growth and activity of soil microorganisms involved in chemical transformations within soil (Alexander, 1980). The effects of soil acidity on the interactions between plants and microorganisms has also been studied extensively (see for example review by Smiley, 1975). In these studies there are some common difficulties in experimental methodology which we will consider before discussing the effects of soil acidity on microbial activity in soil.

The first difficulty is that in acid soils, toxicities of aluminium and manganese and deficiencies of calcium and molybdenum may be as important as effects of hydrogen ions *per se* (Foy, 1984). Liming acid soils decreases the concentrations of aluminium and manganese and increases the concentrations of calcium, molybdenum and magnesium in soil solutions. In some instances, liming acid soils induces nutrient deficiencies, for example phosphorus (Robson and Loneragan, 1970). Despite the

SOIL ACIDITY AND PLANT GROWTH
ISBN 0 12 590655 2

complexity of soil acidity, relatively few studies have attempted to identify the particular component of the acid-soil complex that affects microbial activity.

The second difficulty is that many authors do not state the method used to measure soil pH. There are many procedures for measuring soil pH and the values obtained depend markedly on the method used (see review by White, 1969). To obtain the best estimate, there appears to be a strong case for the measurement of soil pH using salt solutions with ionic strengths comparable to the soil solution (Dolling and Ritchie, 1985).

Irrespective of the method used, it is difficult to equate a particular value of soil pH with values obtained either in solution or in agar media used for microbial growth. There has been relatively little study of the effect of soil acidity on the growth and activity of microorganisms in soil; most studies have been conducted in laboratory media where artefacts such as those associated with inadequate buffering may occur.

A third difficulty is that most studies have been conducted using a very restricted range of microorganisms isolated from a particular soil. The microorganisms are often originally isolated onto media of neutral pH. There may be a wide range of responses to soil acidity by microorganisms from particular genera or species. For example, isolates of *Gaeumannomyces graminis* var. *tritici* from acid soils appeared to have a lower optimum pH on agar for growth than did isolates of the same fungus from alkaline soils (Sivasithamparam and Parker, 1981).

A final consideration is that soil is extremely heterogeneous with large variations in soil pH over extremely short distances. In particular, the pH of the rhizosphere may differ from the pH of the bulk soil (Marschner *et al.*, 1982; Nye, 1981). The form of nitrogen supplied to the plant markedly influences rhizosphere pH (Smiley, 1974; Jarvis and Robson, 1983a,b). When plants are supplied with NH_4^+ the pH of the rhizosphere falls, whereas when plants are supplied with NO_3^- the pH of the rhizosphere may rise. This effect of form of nitrogen on the difference between rhizosphere pH and bulk soil pH may have important consequences for microbial activity in soil. For example, Smiley (1975) demonstrated a close association between effect of form of nitrogen and effect of soil pH on the incidence and severity of disease, particularly for those pathogens which spread along roots. The application of NH_4^+ tends to increase the severity of diseases which occur mainly in acid soils and to decrease the severity of diseases which occur mainly in neutral and alkaline soils. In contrast, the application of NO_3^- tends to decrease the severity of diseases which occur mainly on acid soils and to increase the severity of those which occur mainly on neutral and alkaline soils.

II. THE EFFECT OF SOIL ACIDITY ON MICROORGANISMS INVOLVED IN CHEMICAL TRANSFORMATIONS IN SOIL

Liming acid soils generally increases the mineralisation of organic nitrogen (Fig. 1; review by Harmsen and van Schreven, 1955; Barrow, 1965) and organic sulphur (Fig. 1; White, 1959; Barrow, 1965). The activity and growth of the heterotrophic microorganisms involved in this mineralisation appear to be limited in acid soils (see review by Schmidt, 1982). Soil acidity may also affect the availability of nutrients to plants by affecting the rates of other microbially-mediated processes in soil (e.g. denitrification). In this section we will consider effects of soil acidity on microbial involvement in the cycling of nitrogen, phosphorus and sulphur.

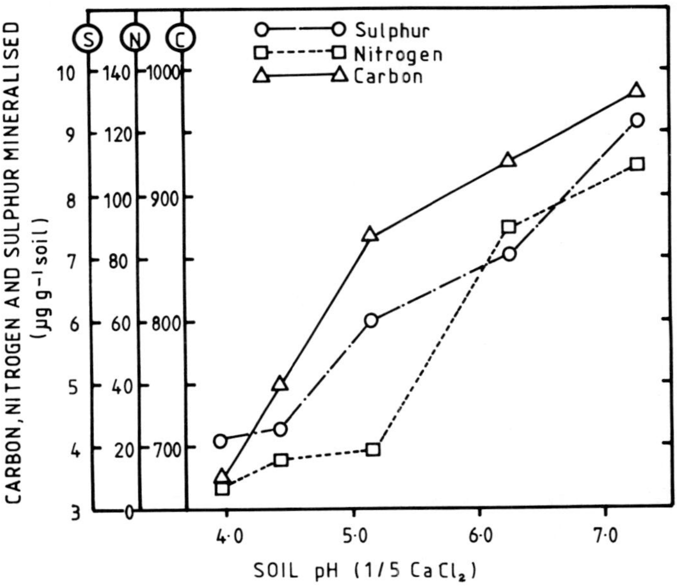

Fig. 1. Effects of soil pH on the mineralisation of carbon, nitrogen and sulphur after incubation of Coolup sand for 32 days at 27°C (after Barrow, 1965).

A. Cycling of nitrogen

Soil acidity may affect the availability of nitrogen to plants by affecting the activity of microorganisms involved in ammonification, nitrification, denitrification, immobilisation and non-symbiotic nitrogen fixation. We will focus on each of these reactions, presenting, where possible, quantitative relationships between soil pH and the rates of these reactions.

Mineralisation of organic nitrogen in soil consists of two stages: ammonification and nitrification. Ammonification appears to be much less sensitive to soil acidity than is nitrification. Acidification of a neutral clay soil (pH 6.5, (1:2 soil:water) to pH 4.0 decreased ammonification only slightly but almost completely suppressed nitrification (Cornfield, 1953). It is thus not surprising that the effects of soil acidity on the mineralisation of organic nitrogen may be less and more temporary than the effects on nitrification (Nyborg and Hoyt, 1978).

Nitrification, which involves the oxidation of ammonium to nitrite and the oxidation of nitrite to nitrate, is very sensitive to soil acidity (Fig. 2 and review by Schmidt, 1982). Liming acid soils (pH 5.0–5.2), (1:2.5 soil:water) increased the number of both ammonium oxidisers and nitrite oxidisers (Edmeades *et al.*, 1981).

Several studies (Dancer *et al.* 1973; Bhat *et al.*, 1980; Gilmour, 1984) have demonstrated that nitrification does not occur when soil pH (1:2.5 soil:water) is less than 4.1. The rate of nitrification increased linearly with increasing soil pH. There is some dispute as to the pH required for maximum nitrification (Fig. 3). Earlier workers (Weber and Gainey, 1962; Morrill and Dawson, 1967; Saranthchandra, 1978) considered that the rate reached a maximum at pH 6 and was independent of soil pH in the range 6 to 8. However, later studies have indicated that higher values of pH (6.5, Bhat *et al.*, 1980; 7.2, Gilmour, 1984) are required for the maximum rate of nitrification.

Nitrification proceeds in acid soils with pH values far below the limits observed for the growth of nitrifying bacteria in pure culture. In pure culture, the growth of *Nitrobacter agilis* was greatest at pH 8 with marked decreases in growth on more acidic media (Aleem and Alexander, 1960). In soil perfusion studies, the optimum pH values for the growth of *Nitrobacter* and *Nitrosomonas* were greater than 6.6 and 7.6 respectively (Morrill and Dawson, 1962). There are at least two explanations for the discrepancy between pH effects on growth of nitrifying bacteria and nitrification in acid soils. Within acid soils there may be microsites which are less acid in which nitrification may proceed. Alternatively, the growth of strains of nitrifying bacteria present in acid soils may be more tolerant of acidity than that of the bacteria studied by Aleem and Alexander (1960) or Morrill and Dawson (1962).

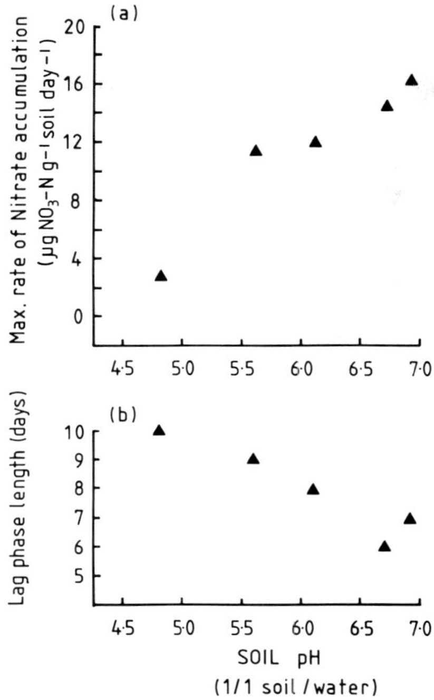

Fig. 2. The relationship between soil pH before incubation and (A) the maximum rate of nitrate accumulation and (B) the length of the lag phase for production of nitrate for soils treated with 100 μg NH_4^+-N (as $(NH_4)_2SO_4$) per g soil and incubated at 23°C for 30 days. Data are from Dancer *et al.*, (1973) for treatments receiving annual applications of 84 kg N/ha^{-1}.

It has been suggested (Brar and Giddens, 1968) that nitrification is limited by high concentrations of aluminium in acid soils. Additionally, the rate of nitrification was sensitive to high concentrations of manganese (Premi and Cornfield, 1969). However, in a range of acid soils where liming markedly increased nitrification, no relationship existed between the amounts of aluminium or manganese soluble in $CaCl_2$ and nitrification, indicating that nitrification was apparently limited by high concentrations of H^+ (Nyborg and Hoyt, 1978).

Denitrification is also limited by soil acidity (Fig. 3; Müller *et al.*, 1980; review by Firestone, 1982). Liming an acid peat which raised soil pH from 3.5 to 6.5 (method not stated) increased the rate of N_2O production at least tenfold (Klemedtsson *et al.*, 1977). In an earlier study, increasing the initial pH from 4.9 to 6.0 (1 : 1 soil:water) increased the rate of denitrification (Wijler and Delwiche, 1954). Further increasing soil pH to 7.3 did not affect the total denitrification rate but changed the

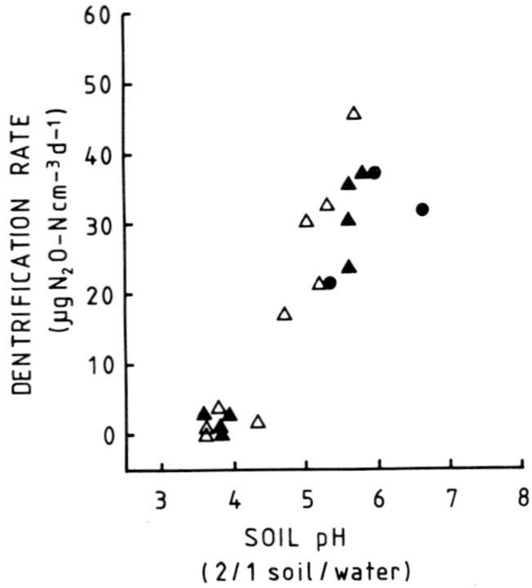

Fig. 3. The relationship between rate of denitrification and soil pH for surface horizons of mire (▲), forest ((△) and agricultural (●) soils (data from Müller *et al.*, 1980).

products. Below pH 7.0, N_2O was the major product of denitrification, whereas above this value N_2 was predominant. It is not clear which component of the soil acidity complex limits denitrification (see for example Bremner and Shaw, 1958). It is also not clear whether the effects of soil acidity on denitrification are related to effects on microbial growth or to effects on enzyme activity. There may be some selection for denitrifiers well adapted to the prevailing soil pH. The optimum pH for denitrification activity in an acid soil was approximately 3.9 (1:1 soil:water), whereas in an adjacent limed soil the optimum pH was 6.3 (Fig. 4; Parkin *et al.*, 1985). These values were close to the initial pH values for the two soils (4.1, 6.0).

Non-symbiotic nitrogen fixation may also represent a significant input of nitrogen into some ecosystems (see for example Parker, 1957; Mullaly *et al.*, 1967). There has been little study on the effect of soil pH on the rates of non-symbiotic nitrogen fixation. The available evidence would suggest that, like most microbially-mediated processes, non-symbiotic nitrogen fixation would be limited in acid soils (see La Rue, 1977; Granhall, 1981).

B. Mineralisation of organic phosphorus and organic sulphur

Liming acid soils (pH 5.0–5.5, 1:1 soil:water) stimulated the mineralisation of organic phosphorus (Halstead *et al.*, 1963). However, for a range of soils, differences in the rate of mineralisation of organic phosphorus did not appear to be related to differences in soil pH (Fig. 5; Thompson

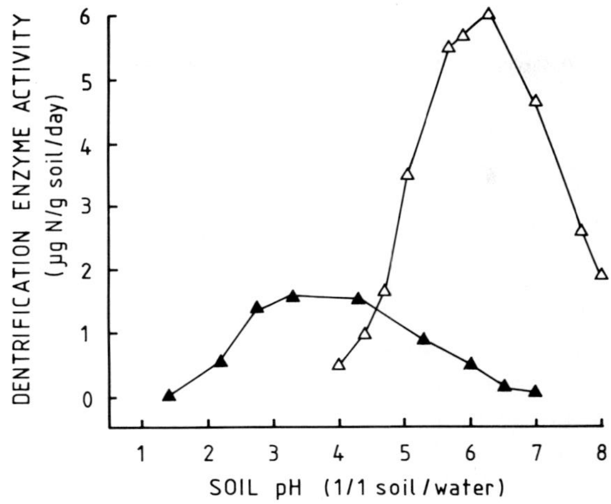

Fig. 4. The effect of soil pH on denitrification enzyme activity in adjacent soils with (△) and without (▲) previous lime application which increased soil pH from 4.1 to 6.0 (1 : 1 soil : water) (data from Parkin *et al.*, 1985).

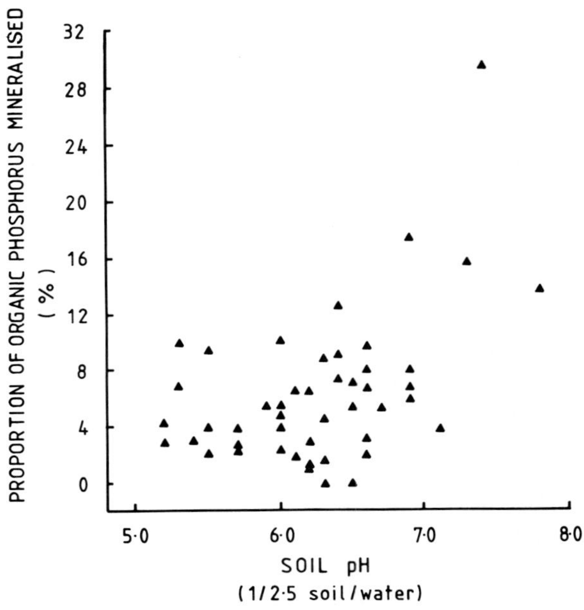

Fig. 5. The mineralisation of organic phosphorus in relation to soil pH in 25 pairs of virgin and cultivated soils (data from Thompson *et al.*, 1954).

et al., 1954). The incorrect interpretation of the data by the authors has often been cited to indicate that soil acidity limited the mineralisation of organic phosphorus (see for example Cosgrove, 1967; Jackson, 1967; Foy, 1984).

The mineralisation of some forms of organic phosphorus in soil may be more sensitive to acidity than that of other forms. For example, liming which increased soil pH from 5.8 to 7.2 (method not stated) increased the mineralisation of phytin more than it increased the mineralisation of nucleic acids (Pearson *et al.*, 1941).

Soil acidity could limit the mineralisation of organic phosphorus in several ways. First, calcium phytates are more soluble than iron and aluminium phytates which predominate in acid soils (Cosgrove, 1967). It is therefore likely that calcium phytates would be more assessible for mineralisation than iron and aluminium phytates. Second, liming may stimulate the growth of microorganisms involved in the mineralisation of organic phosphorus (Halstead *et al.*, 1963). Finally, the activity of phosphatases may depend on soil pH (Halstead, 1964).

There is little data to suggest that effects of soil acidity on the mineralisation of organic phosphorus differ from effects on the mineralisation of nitrogen or sulphur. However, the ratios of total nitrogen to organic phosphorus were greater in alkaline soils than in acid soils (Williams *et al.*, 1960; Williams and Steinbergs, 1958); as were the ratios of non-sulphate sulphur to organic phosphorus. These findings suggest that the mineralisation of organic phosphorus in acid soils is more limited than is the mineralisation of nitrogen and sulphur.

Liming acid soils may increase markedly the concentration of sulphate in soil (White, 1959; Williams, 1967; Nelson, 1964). However, this can be due to effects other than those of lime in stimulating the activity of microorganisms (see review by Freney, 1967). For instance, sulphate can be added as a contaminant in the calcium carbonate, it can be desorbed from anion exchange sites and it can be released from organic matter by chemical hydrolysis. Increasing soil pH from 4 to 5 (1:1 soil:water) increased acetate-extractable sulphate threefold, but only after the soil was incubated for 3 months (Nelson, 1964). Effects of acidity on the mineralisation of sulphur may be restricted to quite acid soils because further increasing soil pH beyond 5 only slightly affected acetate-extractable sulphate (Nelson, 1964).

Effects of acidity on the mineralisation of sulphur appear to be similar to those on the mineralisation of nitrogen (White, 1959). However, in a field survey the ratio of non-sulphate sulphur to nitrogen changed with soil pH (Williams and Steinbergs, 1958). Those soils with a pH > 7 (1:5 soil:water) contained a higher ratio of non-sulphate sulphur to nitrogen than did soils with a pH < 7. Within the acid soils (pH 5.2–7), there appeared to be no effect of soil pH on the proportion of non-sulphate sulphur to nitrogen.

III. THE EFFECT OF SOIL ACIDITY ON THE INTERACTIONS BETWEEN PLANTS AND MICROORGANISMS

The associations between plants and microorganisms may be beneficial to both organisms (e.g. mycorrhizas) or may be to the sole advantage of the microorganism (e.g. plants colonised by pathogens). In both cases, four processes should be considered when examining the effect of soil acidity on interactions between plants and microorganisms. These are: the growth of the microorganism in the soil and on the root; the infection of the plant by the microorganism; the physiology of the interaction; and the growth of the host plant.

In most studies of the effect of soil acidity on mycorrhizas and plant/pathogen interactions, there has been little attempt to separate effects of soil acidity on the growth of the mycorrhizal fungus or pathogen from effects on initial infection, on the spread of infection, and on the physiological consequences of this infection for plant growth. Relatively few studies have compared the effect of soil acidity on the growth of the plant in the absence and presence of the mycorrhizal fungus or pathogen (see, however, Trolldenier, 1981). These studies would provide a basis for identifying the component of the interaction between plants and microorganisms which is most sensitive to soil acidity. For example, if the growth of the host plant is more sensitive to soil acidity than any of the additional components associated with disease, enhanced disease in acid soils may merely reflect indirect effects of soil acidity on host plant growth.

In this section of the review, we will consider the effects of soil acidity on the formation and functioning of mycorrhizas and on disease.

A. Vesicular-arbuscular (VA) mycorrhizas

1. Variation in species response to soil acidity

Among the large number of species of VA mycorrhizal fungi, there is considerable variation in response to soil acidity. This variation is demonstrated by different relationships between soil pH and the distribution and abundance of different fungal species. It is also demonstrated by contrasting responses in both the formation and functioning of mycorrhizas for different species of fungus. Finally, the effect of soil pH on the germination of spores, on hyphal growth from spores and hyphal growth from mycorrhizal roots can be quite different for different species. Each of these areas will be considered.

Some species of VA mycorrhizal fungi are restricted to acid soils. Other species are restricted to alkaline soils, whereas others occur in both acid and alkaline soils (Table I). For example, spores of *Acaulospora laevis* were not observed in Western Australian soils of pH greater than 6.4 (1:5 soil:0.01 M $CaCl_2$) and the greatest abundance of spores was found in more acid soils (pH 4.5–4.9) (Porter *et al.*, 1987a). Similarly, spores of *A. laevis* were the only spores of VA mycorrhizal fungi recovered from soils of pH 4.0–4.5 (method not stated) in central Florida (Nicolson and Schenck, 1979). In contrast, spores of two undescribed species of *Glomus* (isolates WUM 2 and WUM 3) were restricted in Western Australian soils to soils with a pH > 6.5 (1:5 soil:0.01 M $CaCl_2$) (Porter *et al.*, 1987a). For other species (e.g. *Glomus monosporum*) soil pH had no marked effects on their distribution and abundance (Porter *et al.*, 1987a). In fields cropped to cereals in Oregon, there was also a strong association between soil pH and the distribution of species of VA mycorrhizal fungi (Young *et al.*, 1985). In soils with pH < 4.9 (1:2 soil:water), virtually all spores were those formed by *A. laevis*, whereas in soils with pH > 6.0 almost all spores were formed by *G. mosseae.*

It is therefore not surprising that there is little variation in the extent of mycorrhizal infection in natural vegetation growing on soils over a wide range of pH (Read *et al.*, 1976; Sparling and Tinker, 1978); however, the species of fungi that form mycorrhizas in each soil may be quite different. In a long-term liming experiment, increasing soil pH

Table I. The effect of soil pH on the distribution of three species of mycorrhizal fungi in the Merredin area of Western Australia (after Porter *et al.*, 1987a). Percentage of samples containing spores of each fungus.

pH range (1/5 0.01 M $CaCl_2$)	*Glomus* sp. WUM 3	*Acaulospora laevis*	*G. monosporum*
4.5–4.9	0	100	80
5.0–5.4	0	0	67
5.5–5.9	0	43	86
6.0–6.4	0	67	100
6.5–6.9	100	0	0
7.0–7.4	50	0	50
7.5–7.9	50	25[a]	25
8.0–8.4	100	0	0

[a] One spore was found in 1 soil.

from 4.5 to 7.5 (1:2.5 soil:water) did not affect the length of root converted to VA mycorrhizas (Wang *et al.*, 1984). But the species of fungi responsible for the mycorrhizas were not the same over this range. In the most acid soil, roots were only infected by fungi with fine hyphae, whereas in the least acid soil there was no infection by fungi with fine hyphae.

The effect of liming on the formation of VA mycorrhizas depends on the fungal species (Table II). Liming an acid soil (pH 5.3, 1:5 soil:0.01 M $CaCl_2$) did not affect the percentage of root length of subterranean clover infected by *G. fasciculatum* but enabled *Glomus* sp. (WUM 16) to form

Table II. The effect of soil pH on the infection of roots by VA mycorrhizal fungi.

Fungus	Response	Reference
Acaulospora laevis	Increasing pH from 4 to 5.6 either did not affect or decreased infection.	Mosse (1972)
Gigaspora margarita	At moderate phosphorus supply no effect of pH (from 5.5 to 6.4, 1/5 water) on infection.	Siqueira *et al.* (1984)
Glomus sp. (isolate WUM 16)	Liming soil from pH 5.3 to 7.0 (1/5 0.01 M $CaCl_2$) increased infection from 2 to 42%.	Abbott and Robson (1985)
Glomus fasciculatum	Liming soil from pH 4.4 to 6.0 (method not stated) did not affect infection.	Mosse *et al.* (1973)
" "	No effect of increasing pH 5.3 to 7.5 (1/5 0.01 M $CaCl_2$) on infection.	Abbott and Robson (1985)
" "	Increasing pH from 5.1 to 6.0 (method not stated) increased colonisation; further increasing pH decreased colonisation.	Davis *et al.* (1983)
Glomus mosseae	Maximum infection at pH 6.1 (1/5 water). Increasing or decreasing pH decreased infection.	Siqueira *et al.* (1984)
" "	Maximum infection at pH 7.4–7.6.	Davis *et al.* (1983)
" "	No infection at pH 4 (method not stated); considerable infection at pH 5.6.	Mosse (1972)

mycorrhizas (Abbott and Robson, 1985). In another study, the addition of lime to a soil of pH 5.3 (saturated paste) decreased the extent of mycorrhizal infection in roots of lucerne by indigenous fungi (Kucey and Diab, 1984); but the extent of infection was increased when the soil was inoculated with a mixed inoculum of VA mycorrhizal fungi isolated from alkaline soils (pH 7.2–7.7).

In general, infection by *A. laevis* appears to be greater in acid soils and is decreased by increasing pH. Infection by *G. fasciculatum* and *Gigaspora margarita* appears to be unaffected by soil pH over a wide range, whereas infection by *G. mosseae* and some other *Glomus* spp. (e.g. WUM 16) appears to be greater in neutral or alkaline soils and to be increased by liming.

2. *Influence of soil acidity on formation and function*

Soil pH could affect the ability of species of VA mycorrhizal fungi to colonise and persist in soils by affecting survival and germination of spores, growth of hyphae in soil, penetration and infection of the root and propagule formation. Most attention has been paid to effects of pH on the germination of spores (Fig. 6, Tables III and IV) and on the growth of hyphae from those spores (Table IV). Usually effects of pH on the growth of hyphae from spores paralleled effects on spore germination (Siqueira *et al.*, 1984; Porter *et al.*, 1987b). Effects of soil pH on infection (Table II) are closely related to effects on spore germination (Table III) and hyphal growth including the growth of hyphae in soil from infected roots (Abbott

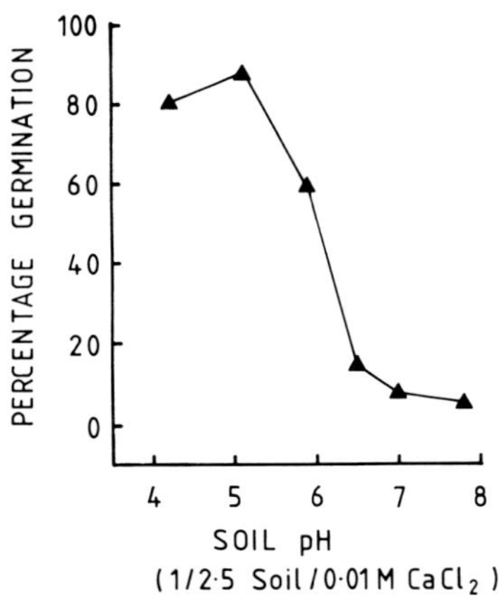

Fig. 6. Germination of spores of *Acaulospora laevis* in limed Sawyers soil in relation to soil pH (after Hepper, 1984).

Table III. The effect of pH on the germination of spores of VA mycorrhizal fungi.

Fungus	Medium	Response	Reference
Acaulospora laevis	Soil	Increasing pH beyond 5 (1/2.5 0.01 M $CaCl_2$) decreased germination.	Hepper (1984)
" "	Soil	Liming soil from pH 4.6 to 7.4 (1/5 0.01 M $CaCl_2$) decreased germination from 30% to 1%. Acidifying soil from pH 7.6 to 4.9 increased germination from 4 to 23%.	Porter *et al.* (1987b)
Gigaspora margarita	Agar	No effect of pH in range 5.8 to 7.8.	Pons and Gianinazzi-Pearson (1984)
" "	Agar	Little effect of pH between 4.5 and 6.0.	Siqueira *et al.* (1982)
" "	Agar	Liming soil from pH 5.5 to 6.1 (1/5 water) increased germination from 50 to 63%.	Siqueira *et al.* (1984)
Gigaspora heterogama	Agar	Optimum pH 5 for germination	Green *et al.* (1976)
Gigaspora corallordea	Agar	Optimum pH 6 for germination	Green *et al.* (1976)
Glomus mosseae	Agar	Increasing pH from 5 to 7 increased germination from 2 to 65%, further increasing pH decreased germination.	Green *et al.* (1976)
" "	Agar	Increasing pH from 5.8 to 6.8 increased germination, further increasing pH did not affect germination.	Pons and Gianinazzi-Pearson (1984)
" "	Soil	Liming acid soil from 5.5 to 5.9 increased germination from 43 to 60%, no further increase with additional lime.	Siqueira *et al.* (1984)
Glomus epigaeus	Agar	Germination at pH 5.8 greater than that at 6.2 or 6.8 but similar to that at 7.5.	Pons and Gianinazzi-Pearson (1984)
	Agar	Optimum pH 6–8, germination decreased slightly at pH values outside range.	Daniels and Trappe (1980)
Glomus sp. (WUM 2)	Soil	Liming soil from pH 4.6 to 7.4 (1/5 0.01 M $CaCl_2$) increased germination from 4 to 67%. Acidifying soil from pH 7.6 to 4.9 decreased germination from 64 to 1%.	Porter *et al.* (1987b)
Glomus caledonium	Agar	No effect of pH in range 5.3 to 7.4 on germination. Hyphal growth increased from pH 5.3 to 6.6, further increasing pH decreased hyphal growth.	Carr and Hinkley (1985)

Table IV. The effect of soil pH on the germination of spores and hyphal growth of *Acaulospora laevis* and *Glomus* sp. WUM 3 (after Porter *et al.*, 1987b).

Species	Soil[a]	Treatment	pH[b]	Spore germination (%)	Hyphal growth (mm hyphae per germinated spore)
A. laevis	Ulva sandy loam	Nil	4.6	30	28
		$CaCO_3$	7.4	1	1
		$CaSO_4$	4.6	15	6
		Nil	7.6	4	2
		H_2SO_4	4.9	23	8
		$CaSO_4$	7.5	2	1
Glomus sp. WUM 3	Merredin sandy clay loam	Nil	4.6	4	2
		$CaCO_3$	7.4	67	75
		$CaSO_4$	4.6	5	6
		Nil	7.6	64	111
		H_2SO_4	4.9	1	8
		$CaSO_4$	7.5	55	64

[a] Soil descriptions from Bettenay and Hingston (1961)
[b] (1/5, 0.01 M $CaCl_2$)

and Robson, 1985). Thus it is likely that acidity limits mycorrhizal infection mainly by decreasing the growth of the fungi in soil.

It is not clear whether the effects of soil acidity in depressing spore germination, hyphal growth and infection result from toxic effects of H^+ ions *per se* or from aluminium and manganese. Acidifying soil solutions extracted from neutral soils decreased spore germination and hyphal outgrowth for *Glomus* sp. (WUM 2), indicating that H^+ ions were the most likely toxic factor (Porter *et al.*, 1987b). Siqueira *et al.* (1984) concluded that liming an acid soil increased germination of *G. mosseae* spores by decreasing aluminium concentrations. However, this conclusion appears to be based on a correlation between spore germination and acid-soluble (0.5 N H_2SO_4 and 0.025 N HCl) aluminium. Unfortunately, acid-soluble aluminium will not be closely correlated with aluminium concentration in soil solutions because aluminium solubility increases markedly with increasing acidity. Nevertheless, increasing the concentration of both aluminium and manganese in solutions applied to acid-washed sand at pH 4.5 depressed the formation of mycorrhizas in oats by *G. caledonium* (Wang *et al.*, 1984).

There are also many reports that the effects of inoculation with mycorrhizal fungi on plant growth depend upon soil pH (see for example

Mosse, 1972; Skipper and Smith, 1979; Yawney *et al.*, 1982). Soil acidity could influence the functioning of VA mycorrhizas in several ways apart from effects on mycorrhizal formation. In at least one study, effects of lime application on the efficiency with which VA mycorrhizas increased plant growth were related to effects on the formation of mycorrhizas (Abbott and Robson, 1985). However, in many other studies it is not possible to determine whether these effects operated on the formation or function of mycorrhizas because no data were presented on the extent of mycorrhizal infection (see for example Skipper and Smith, 1979; Graw, 1979).

Decreasing soil acidity could influence the response to inoculation with a mycorrhizal fungus by: (1) affecting the availability of phosphorus to plants; (2) altering the form of phosphorus in soil (see later discussion on ectomycorrhizas); (3) alleviating a deficiency or toxicity of nutrients other than phosphorus thus enabling a greater response to phosphorus; or (4) mitigating the effects of aluminium or manganese toxicity on plant growth. To evaluate some of these possibilities, the response to phosphorus of mycorrhizal and non-mycorrhizal plants must be compared on limed and unlimed soils using horizontal rather than vertical comparisons (see Abbott and Robson, 1984). The effects of soil pH on the development of infection can only be determined satisfactorily by harvesting plants at several times.

The effects of lime application on the availability of phosphorus to plants are complex (see review by Haynes, 1982). Adding lime to an acid soil has resulted in increased (Stewart and Pearson, 1952; MacLean and Cook, 1955) or decreased (Pierre and Browning, 1935; Rossiter, 1965; Robson and Loneragan, 1970) phosphate uptake by plants. These different results have usually been explained in relation to the effects of lime on reactions of phosphate in soil. However, liming acid soils may decrease the formation of mycorrhizas by fungal species poorly adapted to neutral or alkaline soils and thus restrict the uptake of phosphorus.

It has been suggested that mycorrhizas may be particularly important when root growth is restricted by unfavourable soil conditions (Bowen, 1983). However, inoculation with *G. fasciculatum* did not alleviate aluminium toxicity in subterranean clover grown in an acid soil (White, 1984). In this experiment adding aluminium sulphate to an acid soil (pH 4.7, 1/5 0.01 M $CaCl_2$) depressed plant growth but did not depress the formation of mycorrhizas.

Involvement of mycorrhizal fungi in uptake and/or assimilation of nitrate and ammonium would mean that infection of roots could modify the magnitude and spatial distribution of H^+ or OH^- release by roots. The way in which internal pH is controlled varies in different organisms as well as with different sources of nitrogen (see Raven and Smith, 1976).

Hence fungal involvement in pH regulation could alter the ratio between cations and inorganic anions, and the accumulation of organic anions in plant tissues as well as the release of H^+ or OH^- to the rhizosphere Raven *et al.*, 1978; Smith, 1980; Smith and Smith, 1983). Buwalda *et al.* (1983) showed that mycorrhizal plants took up more inorganic anions than non-mycorrhizal plants without an accompanying increase in cation uptake. This could be explained in terms of a reduction in organic acid synthesis as a means of pH regulation (Raven and Smith, 1976; Smith, 1980). If that is the case, then increased OH^- efflux from roots would presumably have occurred although soil pH was not measured.

Decrease of rhizosphere pH when ammonium is the nitrogen source would be expected, whether roots were mycorrhizal or not, as all organisms deal with the excess H^+ generated in ammonium assimilation by efflux into the soil (see ectomycorrhizas below). Suggestions that mycorrhizal fungi might exude organic acids into the rhizosphere require experimental investigation.

B. Ectomycorrhizas and ericoid mycorrhizas

1. Formation and fungal growth

Despite generalisations such as ‘ectotrophic and ericalean mycorrhizae seem in general to be well adapted to acid soils’ (Mosse *et al.*, 1981) and ‘decrease in mycorrhizal production in alkaline soils’ (Bowen and Theodorou, 1973), there appears to have been no systematic study of the relation of soil pH to the distribution and abundance of fungi that form either ectomycorrhizas or ericoid mycorrhizas.

In many situations the formation of ectomycorrhizas appears to be little affected by soil pH over a wide range (see for example Theodorou and Bowen, 1969; Marx and Zak, 1965), although mycorrhiza formation has been shown to be depressed in some alkaline soils (Richards, 1965; Theodorou and Bowen, 1969). Species of fungi appear to differ in their ability to form ectomycorrhizas in soils of different pH (Theodorou and Bowen, 1969; Marx and Zak, 1965).

It has been suggested that increasing soil pH may indirectly decrease the formation of ectomycorrhizas by increasing the supply of nitrate to the plant (Richards, 1965). Indeed, as discussed earlier (Section II,A), increasing soil pH from 4 to 7 (1 : 5 soil : water) generally increases rates of nitrification. Additionally, nitrate can greatly inhibit the formation of ectomycorrhizas (Bigg, 1981; cited by Alexander, 1983). It is possible that effects of nitrate in depressing the formation of ectomycorrhizas may be associated with increases in rhizosphere pH, which in turn is due to associated with hydroxyl extrusion as a means of regulating internal pH. Theodorou and Bowen (1969) compared the colonisation of roots of *Pinus radiata* and glass fibres by *Rhizopogon luteolus* at pH 5 and 8, and with two levels of nitrate (Table V). Increasing pH from 5 to 8 markedly decreased colonisation of both fibres and roots; increasing nitrate supply at pH 5 decreased colonisation of roots but increased colonisation of glass fibres. In contrast to roots, there will be no nitrate uptake and hence no increase in rhizosphere pH with glass fibres.

The germination of spores of some ectomycorrhizal fungi and the growth of ectomycorrhizal fungi on agar can be affected by pH of the medium (see extensive review by Hung and Trappe, 1983); the effect of pH depended on the species and even the isolate of fungus. There has been little study of the effects of pH on the growth of ericoid mycorrhizal fungi. However, the growth of an ericoid mycorrhizal fungus in a liquid culture was markedly depressed by acidity; maximum growth was attained with an initial pH of 6.6 (Pearson and Read, 1975). Despite the possible difficulties in relating data obtained on agar to effects occurring in soil (see discussion by Giltrap and Lewis, 1981; Harley and Smith, 1983; Hung and Trappe, 1983), there does not appear to be any investigation of the effect of soil pH on the germination of spores and hyphal growth for ectomycorrhizal fungi or ericoid mycorrhizal fungi in soil.

The difficulties of extrapolating from agar to soil for ectomycorrhizal fungi are largely associated with effects of the fungal growth on the pH of the medium and with effects of chemicals used for pH buffering on fungal growth. The need for buffers is related to effects of fungal growth on pH of the media. The extent of pH change is probably related to the amount of fungal growth and the nitrogen source of the medium. Therefore these problems are likely to be more important for ectomycorrhizal fungi and ericoid mycorrhizal fungi than for VA mycorrhizal fungi because of

Table V. The effects of soil pH and nitrate concentration on the colonisation of roots of seedlings of *Pinus radiata* and of glass fibres by *Rhizopogon luteolus* (after Theodorou and Bowen, 1969).

pH	Nitrate conc. (μg.g^{-1} soil)	Percentage colonised		Length colonised (mm)	
		Fibres	Seedlings	Fibres	Seedlings
5.0	12	55	95	4.8	26.5
5.0	115	70	88	8.5	16.0
8.0	12	15	0	0.8	0.0
8.0	115	25	0	1.4	0.0

differences in the amount of growth on agar. It may be possible to define accurately the effects of pH on growth of ectomycorrhizal and ericoid mycorrhizal fungi by using small amounts of inoculum in relation to the volume of medium and assessing growth before marked changes in the pH of the medium occur.

2. Ectomycorrhizal infection, plant growth and soil acidity

There have been relatively few studies of the effect of soil pH on the growth response of plants to infection by ectomycorrhizal fungi. However, in aseptic culture, effects of inoculation with three fungi in increasing dry weight of seedlings of slash pine were more marked at pH 4.0 and 4.6 than at pH 5.6 and 6.6 (Marx and Zak, 1965). In this study, increasing the pH of the medium increased the growth of uninoculated seedlings more than the growth of inoculated seedlings. Acidifying a soil from 6.5 to 4.5 (1:2.5 soil:water), which decreased the formation of ectomycorrhizas in *Pinus caribaea*, decreased the response in growth to inoculation with soil containing ectomycorrhizal fungi (Hart *et al.*, 1980). Acidification of the soil decreased growth of the inoculated seedlings without affecting the growth of uninoculated seedlings.

As for endomycorrhizas, the effects of pH on growth response to inoculation with ectomycorrhizal fungi could arise from (1) effects of soil acidity on the availability of nutrients for uptake by plants, as well as from (2) effects of soil acidity on the formation of ectomycorrhizas (see Section III,A). In addition, pH may affect the uptake of nutrients from solution by mycorrhizal and non-mycorrhizal roots in different ways. The optimum pH for ammonium uptake from solution by ectomycorrhizal roots of spruce and Douglas firs was 6.3; whereas for non-mycorrhizal roots, ammonium uptake was increased by increasing pH of the solution from 6.3 to 7.1 (Rygiewicz *et al.*, 1984a). Mycorrhizal roots released fewer hydrogen ions per ammonium ion taken up (Rygiewicz *et al.*, 1984a) and fewer hydroxyl ions per nitrate ion taken up (Rygiewicz *et al.*, 1984b) than did non-mycorrhizal roots. Hence ectomycorrhizal roots may affect rhizosphere pH less than non-mycorrhizal roots.

C. Disease

The effect of soil acidity on plant disease has received considerable study (see for example a general account by Palti (1981) and a review on diseases associated with *Phytophthora* species by Schmitthener and Canaday (1983)). However, it is difficult to establish general principles about the effects of soil pH on disease largely because few studies have identified which component of the interaction between the plant and the pathogen is the most sensitive to soil acidity (see introduction to this chapter and to Section III). Nevertheless, it is likely that, as for mycorrhizal associations, effects of soil pH on disease can operate through effects on spore survival, on spore germination and on infection of roots.

Almost all of the reports of effects of soil pH on disease are for root disease. This is somewhat surprising as one may have expected that effects of soil pH on the availability of essential nutrients may have affected the incidence and severity of leaf diseases as well.

The incidence and severity of several diseases are affected by soil pH (Table VI), but the generalisations made are not without exceptions. For example, most reports suggest *Phytophthora* root rots are decreased by acidifying soil and increased by lime application (see review by Schmitthener and Canaday (1983)). On the other hand, there are also situations where liming decreased the incidence (e.g. Boughton *et al.*, 1978) or severity (e.g. Fernandes-Falcon *et al.*, 1984) of *Phytophthora* root rots.

For some pathogenic fungi, the effect of soil pH on disease severity depends on its influence on the growth of the host plants. The main effect

Table VI. Effect of soil acidity on disease.

Disease	Pathogen	Reference
Diseases decreased by acidification of alkaline soil or increased by liming acid soils.		
Scab of potato	*Streptomyces scabies*	Martin (1920)
Take-all in wheat	*Gaeumannomyces graminis* var. *tritici*	Trolldeiner (1981)
Phytophthora root rots	Various species	Schmitthener and Canaday (1983)
Pythium root rots of pines	*Pythium irregulare*	Roth and Riker (1943)
Verticillium wilt of sunflowers	*Verticillium albo-atrum*	Orellana *et al.* (1975)
Diseases increased by acidification of alkaline soils or decreased by liming acid soils.		
Fusarium wilt of tomato	*Fusarium oxysporum* f.sp *lycopersici*	Jones and Overman (1971)
Club root of brassicas	*Plasmodiophora brassicae*	Garrett (1970)
Damping off in medics	*Pythium ultimum* *Pythium irregulare*	Kleinig (1965); MacKenzie *et al.* (1972)
Bacterial canker of peach	*Pseudomonas syringae*	Weaver and Wehunt (1975)

of unfavourable pH in these cases may be to prolong the stage of growth during which the plant is most susceptible to infection. For example, for *Pythium* root rots, the disease is most severe in those interactions in which the host plant is under the greatest stress. Hence for plants whose growth is sensitive to soil acidity (for example *Medicago* species), liming may decrease the severity of *Pythium* root rots (Kleinig, 1965; MacKenzie *et al.*, 1972). With other plant species that are well adapted to acid soils (for example conifers), liming may increase the incidence or severity of *Pythium* roots rots (Roth and Riker, 1943).

The effect of soil pH on the incidence and severity of disease associated with other more specialised parasites appears to be the same for different host plants (for example the vascular wilt fungi). Hence infection by *Fusarium* species of a wide range of plants varying in their tolerance to soil acidity is greatest in acid soils and is decreased by liming (see reviews by Garrett, 1956, 1970). Garrett (1970) considered that these effects of soil acidity are not related to direct effects on the fungus before infection, but to effects on the internal progress of infection.

For tomato wilt, due to *Fusarium oxysporum* f. *lycopersici*, plants grown in acid conditions before inoculation were more severely affected than those grown in alkaline conditions before inoculation and then transferred to acid conditions after inoculation (Foster and Walker, 1947). There have been relatively few studies of the effect of soil acidity on the growth of these fungi in soil. However, growth and sporulation of

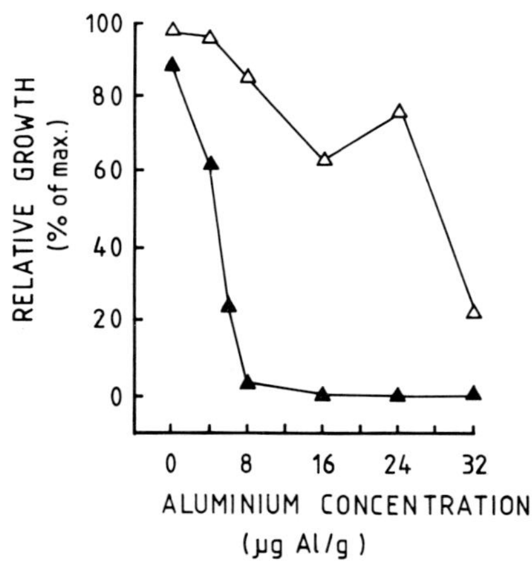

Fig. 7. The effect of soluble aluminium on the relative growth of *Verticillium albo-atrum* (▲) and *Whetzelinia sclerotiorum* (△) in a defined nutrient medium at pH 4.7. Growth is expressed as a percentage of the weight of each fungus in Al-free medium maintained at pH 6.3 (data after Orellana *et al.*, 1975).

Fusarium oxysporum f. *cubense* was greatest in acid soils and was decreased by liming (Stover, 1956).

It is not clear which component of soil acidity predisposes plants to infection by *Fusarium* species. For fungal species associated with vascular wilts (*Verticillium albo-atrum*), marked effects of aluminium in decreasing growth in laboratory media suggest that increased severity of disease following liming may be due to a reduction in concentrations of aluminium in the soil solution (Fig. 7, Table VII; Orellana *et al.*, 1975). The growth of *Whetzelinia sclerotiorum* was less affected by aluminium than the growth of *V. albo-atrum*; liming an acid soil from pH 4.4 to 5.4 (method not stated) decreased the severity of disease associated with infection of sunflowers by *W. sclerotiorum* (Orellana *et al.*, 1975).

Increased incidence of root diseases may be associated with the decline in clover content of pastures which accompanies the gradual acidification of soils (Stovold, 1974). However, there is little evidence to support this suggestion. In old pastures on acid soils, root rots of subterranean clover associated with infection by both *Pythium* and *Fusarium* species have been observed (Burgess *et al.*, 1973; McGee and Kellock, 1974; Stovold, 1974). Acidity would be expected to increase the severity of disease associated with these fungi (see above). Indeed, liming acid soils increased the establishment of *Medicago* species by decreasing the incidence of 'damping off' associated with infection by *Pythium irregulare* (MacKenzie *et al.*, 1972). Moreover, coating seeds with lime increased emergence of *M. truncatula* on an acid soil (pH 6.1, method not stated) in the absence of a fungicide but not in the presence of fungicide

Table VII. The effect of the addition of $CaCO_3$ on the fresh weight of sunflowers and disease severity when inoculated or not with *Verticillium albo-atrum* and *Whetzelinia sclerotiorum* (data from Orellana *et al.*, 1975).

$CaCO_3$ added to soil ($\mu g.g^{-1}$)	Soil pH[c]	Inoc.	*V. albo-atrum* FW[b] shoots	*V. albo-atrum* DSI[a]	*W. sclerotiorum* FW whole plant	*W. sclerotiorum* DSI
750	4.4	−	1.97	0	3.05	0
3000	5.4	−	2.63	0	5.40	0
750	4.4	+	1.10	2.9	2.53	4.3
3000	5.4	+	0.70	4.2	4.43	0.8

[a] DSI Disease severity index
[b] FW Fresh weight
[c] Method not stated

(Kleinig, 1965). The benefits of increasing pH were most marked under adverse conditions of temperature and rainfall.

Other studies have indicated that *Pythium* and *Fusarium* species may be less important pathogens of legumes than *Phytophthora clandestina* (Greenhalgh and Taylor, 1985). No association has been reported between soil pH and the incidence or severity of disease associated with *P. clandestina*. There is thus a clear need to examine the effects of lime application and fungicides on the incidence of root rots in relation to the growth and persistence of pasture legumes on acid soils.

REFERENCES

Abbott, L. K. and Robson, A. D. (1984). The effect of VA mycorrhizae on plant growth. *In* 'VA Mycorrhiza'. (C. Ll. Powell and D. J. Bagyaraj, eds.), pp. 113–130. CRC Press Inc., Boca Raton.

Abbott, L. K. and Robson, A. D. (1985). The effect of soil pH on the formation of VA mycorrhizas by two species of *Glomus*. *Australian Journal of Soil Research* **23**, 253–261.

Aleem, M. and Alexander, M. (1960). Nutrition and physiology of *Nitrobacter agilis*. *Applied Microbiology* **8**, 80–84.

Alexander, I. (1983). The significance of ectomycorrhizas in the nitrogen cycle. *In* 'Nitrogen as an Ecological Factor'. (J. A. Lee, S. McNeill and I. N. Rorison, eds.), pp. 69–93. Blackwell Scientific Publications, Oxford.

Alexander, M. (1980). Effect of acidity on microorganisms and microbial processes in soil. *In* 'Effects of Acid Precipitation on Terrestrial Ecosystems'. (T. Hutchinson and M. Harvas, eds.), pp. 363–364. Plenum Publishing Corp., New York.

Barrow, N. J. (1965). Further investigations of the use of lime on established pastures. *Australian Journal of Experimental Agriculture and Animal Husbandry* **5**, 442–449.

Bettenay, E. and Hingston, F. J. (1961). The soils and land use of the Merredin area, Western Australia. CSIRO (Aust.) Soil and Land Use Series No. 41.

Bhat, K. K. S., Flowers, T. H. and O'Callagan, J. R. (1980). A model for the simulation of the fate of nitrogen in farm wastes on land application. *Journal of Agricultural Science* **94**, 183–193.

Bigg, W. L. (1981). Ph.D. thesis, University of Aberdeen.

Boughton, T. J., Malajczuk, N. and Robson, A. D. (1978). Suppression of the infection of jarrah roots by *Phytophthora cinnamomi* with lime application. *Australian Journal of Botany* **26**, 611–615.

Bowen, G. D. (1983). Microbial determinants of plant nutrient uptake. *In* 'Soils: An Australian Viewpoint'. pp. 693–710 (CSIRO Division of Soils, Melbourne/Academic Press, London).

Bowen, G. D. and Theodorou, C. (1973). Growth of ectomycorrhizal fungi around seeds and roots. *In* 'Ectomycorrhizae, their Ecology and Physiology'. (G. C. Marks and T. T. Kozlowski, eds.), pp. 107–150. Academic Press, New York.

Brar, S. S. and Giddens, J. (1968). Inhibition of nitrification in Bladen grassland soil. *Soil Science Society of America Proceedings* **32**, 821–823.

Bremner, J. M. and Shaw, K. (1958). Denitrification in soil. II. Factors affecting denitrification. *Journal of Agricultural Science* **51**, 39–52.

Burgess, L. W., Ogle, H. J., Edgerton, J. P., Stubbs, L. W. and Nelson, P. E. (1973). The biology of fungi associated with root rot of subterranean clover in Victoria. *Proceedings of the Royal Society of Victoria* **86**, 19–28.

Buwalda, J. G., Stribley, D. P. and Tinker, P. B. (1983). Increased uptake of anions by plants with vesicular-arbuscular mycorrhizas. *Plant and Soil* **71**, 463–467.

Carr, G. R. and Hinkley, M. A. (1985). Germination and hyphal growth of *Glomus caledonium* on water agar containing benomyl. *Soil Biology and Biochemistry* **17**, 313–316.

Cornfield, A. H. (1953). The mineralization of nitrogen in a soil acidified with sulphur, aluminium sulphate or ferrous sulphate. *Journal of the Science of Food and Agriculture* **4**, 298–301.

Cosgrove, D. J. (1967). Metabolism of organic phosphates in soil. *In* 'Soil Biochemistry'. (A. D. MacLaren and G. H. Peterson, eds.), pp. 216–228. Edward Arnold (Publishers) Ltd., London.

Dancer, W. S., Peterson, L. A. and Chesters, G. (1973). Ammonification and nitrification of N as influenced by soil pH and previous N treatments. *Soil Science Society of America Proceedings* **37**, 67–69.

Daniels, B. A. and Trappe, J. M. (1980). Factors affecting spore germination of the vesicular-arbuscular mycorrhizal fungus *Glomus epigaeus. Mycologia* **72**, 457–471.

Davis, E. A., Young, J. L. and Linderman, R. G. (1983). Soil lime level (pH) and VA mycorrhiza effects on growth response of sweet gum seedlings. *Soil Science Society of America Journal* **47**, 251–256.

Dolling, P. J. and Ritchie, G. S. P. (1985). Estimates of soil solution ionic strength and the determination of pH in West Australian soils. *Australian Journal of Soil Research* **23**, 309–314.

Edmeades, D. C., Judd, M. and Sarathchandra, S. U. (1981). The effect of lime on nitrogen mineralisation as measured by grass growth. *Plant and Soil* **60**, 177–186.

Fernandes-Falcon, M., Fox, R. L. and Trujillo, E. E. (1984). Interactions of soil pH, nutrients and moisture on phytophthora root rot of avocado. *Plant and Soil* **81**, 165–176.

Firestone, M. K. (1982). Biological denitrification. *In* 'Nitrogen in Agricultural Soils'. (F. J. Stevenson, ed.), pp. 289–326. American Society of Agronomy, Madison, Wisconsin.

Foster, R. E. and Walker, J. C. (1947). Pre-disposition of tomato to Fusarium wilt. *Journal of Agricultural Research* **74**, 165–185.

Foy, C. D. (1984). Physiological activities of hydrogen, aluminium and manganese toxicities in acid soil. *In* 'Soil Acidity and Liming'. (F. Adams, ed.), pp. 57–97. American Society of Agronomy, Madison, Wisconsin.

Freney, J. R. (1967). Sulfur-containing organics. *In* 'Soil Biochemistry'. (A. D. MacLaren and G. H. Peterson, eds.), pp. 229–259. Edward Arnold (Publishers) Ltd., London.

Garrett, S. D. (1956). 'Biology of Root-Infecting Fungi'. Cambridge University Press, Cambridge.

Garrett, S. D. (1970). 'Pathogenic Root-Infecting Fungi'. Cambridge University Press, Cambridge.

Gilmour, J. T. (1984). The effects of soil properties on nitrification and nitrification inhibition. *Soil Science Society of America Journal* **48**, 1262–1266.

Giltrap, N. J. and Lewis, D. H. (1981). Inhibition of growth of ectomycorrhizal fungi in culture by phosphate. *New Phytologist* **89**, 669–675.

Granhall, U. (1981). Biological nitrogen fixation in relation to environmental factors and functioning of natural ecosystems. *In* 'Terrestrial Nitrogen Cycles'. (F. E. Clark and T. Rosswall, eds.), Ecol. Bull. **33**, 1131–144, (Stockholm).

Graw, D. (1979). The influence of soil pH on the efficiency of vesicular-arbuscular mycorrhiza. *New Phytologist* **82**, 687–695.

Green, N. E., Graham, S. O. and Schenck, N. C. (1976). The influence of pH on the germination of vesicular-arbuscular mycorrhizal spores. *Mycologia* **68**, 919–934.

Greenhalgh, F. C. and Taylor, P. A. (1985). *Phytophthora clandestina*, cause of severe tap-root rot of subterranean clover in Victoria, Australia. *Plant Disease* **69**, 1002–1004.

Halstead, R. L. (1964). Phosphatase activity of soils as affected by lime and other treatments. Canadian Journal of Soil Science **44**, 137–144.

Halstead, R. L., Lapensee, J. M. and Ivarson, K. C. (1963). Mineralisation of soil organic phosphorus with particular reference to the effect of lime. *Canadian Journal of Soil Science* **43**, 97–106.

Harley, J. L. and Smith, S. E. (1983). 'Mycorrhizal Symbiosis'. Academic Press, London.

Harmsen, G. W. and van Schreven, D. A. (1955). Mineralisation of organic nitrogen in soil. *Advances in Agronomy* **7**, 299–398.

Hart, P. B. S., Widdowson, J. P., Watts, H. M. and Chu-Chou, M. (1980). Response of *Pinus caribaea* var. *hondurensis* seedlings to mycorrhizal inoculum, phosphorus and pH. *Australian Forest Research* **10**, 389–396.

Haynes, R. J. (1982). Effects of liming on phosphate availability in acid soils. *Plant and Soil* **68**, 289–308.

Hepper, C. M. (1984). Regulation of spore germination of the vesicular-arbuscular mycorrhizal fungus, *Acaulospora laevis*, by soil pH. *Transactions of the British Mycological Society* **83**, 54–56.

Hung, L. L. and Trappe, J. M. (1983). Growth variation between and within species of ecto-mycorrhizal fungi in response to pH *in vitro*. *Mycologia* **75**, 234–241.

Jackson, W. A. (1967). Physiological effects of soil acidity. *In* 'Soil Acidity and Liming'. (R. W. Pearson and F. Adams, eds.), pp. 43–124. American Society of Agronomy, Madison, Wisconsin.

Jarvis, S. C. and Robson, A. D. (1983a). The effects of nitrogen nutrition of plants on the development of acidity in Western Australian soils. I. Effects with subterranean clover grown under leaching conditions. *Australian Journal of Agricultural Research* **34**, 341–353.

Jarvis, S. C. and Robson, A. D. (1983b). The effect of nitrogen nutrition of plants on the development of acidity in Western Australian soils. II. Effects of differences in cation/anion balance between plant species grown under nonleaching conditions. *Australian Journal of Agricultural Research* **34**, 355–365.

Jones, J. P. and Overman, A. J. (1971). Control of Fusarium wilt of tomato with lime and soil fumigants. *Phytopathology* **61**, 1415–1417,

Kleinig, C. R. (1965) Emergence of *Medicago tribuloides* on moderately acid soils. *Australian Journal of Agricultural Research* **16**, 311–319.

Klemedtsson, L., Svensson, B. H., Lindbert, T. and Rosswall, T. (1977). The use of acetylene inhibition of nitrous oxide reductase in quantifying denitrification in soils. *Swedish Journal of Agricultural Research* **7**, 179–185.

Kucey, R. M. N. and Diab, G. E. S. (1984). Effects of lime, phosphorus, and addition of vesicular-arbuscular (VA) mycorrhizal fungi on indigenous VA fungi and on growth of alfalfa in a moderately acidic soil. *New Phytologist* **98**, 481–486.

La Rue, T. A. (1977). The bacteria. *In* 'A Treatise on Dinitrogen Fixation, Section III, Biology'. (R. W. F. Hardy and W. S. Silver, eds.), pp. 19–62. John Wiley and Sons Inc., New York.

McGee, D. C. and Kellock, A. W. (1974). *Fusarium avenaceum*, a seed borne pathogen of subterranean clover roots. *Australian Journal of Agricultural Research* **25**, 549–557.

MacKenzie, J., Glasby, T. and Diatloff, A. (1972). The role of nodulation and damping-off in lucerne establishment on acidic sandy soils on the Darling Downs. *Australian Journal of Experimental Agriculture and Animal Husbandry* **12**, 4428–4432.

MacLean, A. J. and Cook, R. L. (1955). The effect of soil reaction on availability of phosphorus for alfalfa in some Eastern Ontario soils. *Soil Science Society of America Proceedings* **19**, 311–314.

Marschner, H., Romheld, V. and Ossenberg-Neuhans, H. (1982). Rapid method for measuring changes in pH and reducing processes along roots of intact plants. *Zeitschrift für Pflanzenphysiologie* **105**, 407–416.

Martin, W. H. (1920). The relation of sulphur to soil acidity and to the control of potato scab. *Soil Science* **9**, 393–409.

Marx, D. H. and Zak, B. (1965). Effect of pH on mycorrhizal formation of slash pine in aseptic culture. *Forest Science* **11**, 66–75.

Morrill, L. G. and Dawson, J. E. (1962). Growth rates of nitrifying chemoautotrophs in soil. *Journal of Bacteriology* **83**, 205–206.

Morrill, L. G. and Dawson, J. E. (1967). Patterns observed for the oxidation of ammonium to nitrate by soil organisms. *Soil Science Society of America Proceedings* **31**, 757–760.

Mosse, B. (1972). The influence of soil type and *Endogone* strain on the growth of mycorrhizal plants in phosphate-deficient soils. *Revue d'Ecologie et de Biologie du Sol* **3**, 529–537.

Mosse, B., Hayman, D. S. and Arnold, D. J. (1973). Plant growth responses to vesicular-arbuscular mycorrhizas. V. Phosphate uptake by three plant species from P-deficient soils labelled with ^{32}P. *New Phytologist* **72**, 809–815.

Mosse, B., Stribley, D. P. and Le Tacon, F. (1981). Ecology of mycorrhizae and mycorrhizal fungi. *Advances in Microbial Ecology* **5**, 137–210.

Mullaly, J. V., McPherson, J. B., Mann, A. P. and Rooney, D. R. (1967). The effect of length of legume and non-legume leys on gravimetric soil nitrogen at some locations in the Victorian wheat areas. *Australian Journal of Experimental Agriculture and Animal Husbandry* **7**, 568–571.

Müller, M. M., Sundman, V. and Skujins, J. (1980). Denitrification in low pH podsols and peats determined with the acetylene inhibition method. *Applied and Environmental Microbiology* **40**, 235–239.

Nelson, L. E. (1964). The effect of pH on the acetate-soluble sulfur content of a mayhew soil in Mississippi before and after incubation. *Soil Science Society of America Proceedings* **28**, 290–291.

Nicolson, T. N. and Schenck, N. C. (1979). Endogonaceous mycorrhizal endophytes in Florida. *Mycologia* **71**, 178–198.

Nyborg, M. and Hoyt, P. B. (1978). Effects of soil acidity and liming on mineralization of soil nitrogen. *Canadian Journal of soil Science* **58**, 331–338.

Nye, P. H. (1981). pH changes across the rhizosphere induced by roots. *Plant and Soil* **61**, 7–26.

Orellana, R. G., Foy, C. D. and Fleming, A. L. (1975). Effect of soluble aluminium on growth and pathogenicity of *Verticillium albo-atrum* and *Whetzelinia sclerotiorum* from sunflower. *Phytopathology* **65**, 202–205.

Palti, J. (1981). 'Cultural Practices and Infectious Crop Diseases'. Springer Verlag, Berlin, p. 29.

Parker, C. A. (1957). Non-symbiotic nitrogen-fixing bacteria in soil. III. Total nitrogen changes in a field soil. *Journal of Soil Science* **8**, 48–59.

Parkin, T. B., Sexstone, A. J. and Tiedje, J. M. (1985). Adaption of denitrifying populations to low soil pH. *Applied and Environmental Microbiology* **49**, 1053–1056.

Pearson, R. W., Norman, A. G. and Ho, C. (1941). The mineralization of the organic phosphorus of various compounds in soil. *Soil Science Society of America Proceedings* **6**, 168–175.

Pearson, V. and Read, D. J. (1975). The physiology of the mycorrhizal endophyte of *Calluna vulgaris. Transactions of the British Mycological Society* **64**, 1–7.

Pierre, W. H. and Browning, G. M. (1935). The temporary injurious effect of excessive liming of acid soils and its relation to phosphate nutrition of plants. *Journal of American Society of Agronomy* **27**, 742–751.

Pons, F. and Gianinazzi-Pearson, V. (1984). Influence du phosphore, du potassium, de l'azote et du pH sur le comportement *in vitro* de champignons endomycorrhizogenes a vesicles et arbuscles. *Cryptogamie Mycologie* **5**, 87–100.

Porter, W. M., Robson, A. D. and Abbott, L. K. (1987a). Field survey of the distribution of VA mycorrhizal fungi in relation to soil pH. *Journal of Applied Ecology* **24**, 659–662.

Porter, W. M., Robson, A. D. and Abbott, L. K. (1987b). Factors controlling the distribution of VA mycorrhizal fungi in relation to soil pH. *Journal of Applied Ecology* **24**, 663–672.

Premi, P. R. and Cornfield, A. H. (1969). Effects of addition of copper, manganese, zinc and chromium compounds on ammonification and nitrification during incubation of soil. *Plant and Soil* **31**, 345–352.

Raven, J. A. and Smith, F. A. (1976). Nitrogen assimilation and transport in vascular land plants in relation to intracellular pH regulation. *New Phytologist* **76**, 415–431.

Raven, J. A., Smith, S. E. and Smith, F. A. (1978). Ammonium assimilation and the role of mycorrhizas in climax communities in Scotland. *Transactions of the Proceedings of the Botanical Society*, Edinburgh **43**, 27–35.

Read, D. J., Koucheki, H. K. amd Hodgson, J. (1976). Vesicular-arbuscular mycorrhiza in natural vegetation systems. I. The occurrence of infection. *New Phytologist* **77**, 641–653.

Richards, B. N. (1965). Mycorrhiza development of loblolly pine seedlings in relation to soil reaction and the supply of nitrate. *Plant and Soil* **22**, 187–199.

Robson, A. D. and Loneragan, J. F. (1970). Growth of annual *Medicago* species on an acid soil. 11th International Grassland Conference, University of Queensland Press, Brisbane, pp. 424–428.

Rossiter, R. C. (1965). Phosphorus toxicity in subterranean clover and oats grown on Muchea sand and the modifying effects of lime and nitrate-nitrogen. *Australian Journal of Agricultural Research* **3**, 227–242.

Roth, L. F. and Riker, L. F. (1943). Influence of temperature, moisture and soil reaction on the damping-off of red pine seedlings by *Pythium* and *Rhizoctonia. Journal of Agricultural Research* **7**, 273–293.

Rygiewicz, P. T., Bledsoe, C. S. and Zasoski, R. J. (1984a). Effects of ectomycorrhizae and solution pH on [^{15}N] ammonium uptake by conifer seedlings. *Canadian Journal of Forest Research* **14**, 885–892.

Rygiewicz, P. T., Bledsoe, C. S. and Zasoski, R. J. (1984b). Effects of ectomycorrhizae and solution pH on [^{15}N] nitrate uptake by coniferous seedlings. *Canadian Journal of Forest Research* **14**, 893–899.

Saranthchandra, S. V. (1978). Nitrification activities and the changes in the populations of nitrifying bacteria in soil perfused at two different H-ion concentrations. Plant and Soil 50, 99–111.

Schmidt, E. L. (1982). Nitrification in soil. *In* 'Nitrogen in Agricultural Soils'. (F. J. Stevenson, ed.), pp. 253–288. American Society of Agronomy, Madison, Wisconsin.

Schmitthener, A. F. and Canaday, C. H. (1983). Role of chemical factors in development of *Phytophthora* diseases. *In* '*Phytophthora* — its Biology, Taxonomy, Ecology and Pathology'. (D. C. Erwin, S. Bartnicki-Garcia and P. H. Tsao, eds.) pp. 189–196. American Phytopathological Society, St. Paul, Minnesota.

Siqueira, J. O., Hubbell, D. H. and Schenck, N. C. (1982). Spore germination and germ tube growth of a vesicular-arbucular mycorrhizal fungus *in vitro. Mycologia* **74**, 952–959.

Siqueira, J. O., Hubbell, D. H. and Mahmud, A. W. (1984). Effect of liming on spore germination, germ tube growth and root colonization by vesicular-arbuscular mycorrhizal fungi. *Plant and Soil* **76**, 115–124.

Sivasithamparam, K. and Parker, C. A. (1981). The physiology and nutrition of *Gaeumannomyces graminis* in culture. *In* 'Biology and Control of Take-All'. (P. J. Shipton and M. J. C. Asher, eds.), pp. 125–150. Academic Press, London.

Skipper, H. D. and Smith, G. W. (1979). Influence of soil pH on the soybean-endomycorrhiza symbiosis. *Plant and Soil* **53**, 559–563.

Smiley, R. W. (1974). Rhizosphere pH as influenced by plants, soils and nitrogen fertilizers. *Soil Science Society of America Proceedings* **38**, 795–799.

Smiley, R. W. (1975). Forms of nitrogen and the pH in the root zone and their importance to root infection. *In* 'Biology and Control of Soil Borne Pathogens'. (G. W. Bruehl, ed.), pp. 55–62. American Phytopathological Society, St. Paul, Minnesota.

Smith, S. E. (1980). Mycorrhizas of autotrophic higher plants. *Biological Review* **55**, 475–510.

Smith, S. E. and Smith, F. A. (1983). Could mycorrhizas be involved in changes in soil pH? Proceedings of the 7th Australian Nodulation Conference, AIAS, Occ. Publication.

Sparling, G. P. and Tinker, P. B. (1978). Mycorrhizal infection in Pennine grassland. I. Levels of infection in the field. *Journal of Applied Ecology* **15**, 943–950.

Stewart, E. H. and Pearson, R. W. (1952). Utilization of phosphorus by fertilizer placement and liming. *Agronomy Journal* **44**, 501–502.

Stover, R. H. (1956). Studies on Fusarium wilt of bananas. I. The behaviour of *F. oxysporum* f. *cubense* in different soils. *Canadian Journal of Botany* **34**, 927–942.

Stovold, G. E. (1974). Root rot caused by *Pythium irregulare* Buissman, an important factor in the decline of established subterranean clover pastures. *Australian Journal of Agricultural Research* **25**, 537–548.

Theodorou, C. and Bowen, G. D. (1969). The influence of pH and nitrate on mycorrhizal associations of *Pinus radiata* D. Don. *Australian Journal of Botany* **17**, 59–67.

Thompson, L. M., Black, C. A. and Zoellner, J. A. (1954). Occurrence and mineralization of organic phosphorus in soils, with particular reference to associations with nitrogen, carbon and pH. *Soil Science* **77**, 185–196.

Trolldenier, G. (1981). Influence of soil moisture, soil acidity and nitrogen source on take-all in wheat. *Phytopathology* **102**, 163–177.

Wang, G., Stribley, D. P., Tinker, P. B. and Walker, C. (1984). Soil pH and vesicular-arbuscular mycorrhiza. Proceedings of the 6th North American Conference on Mycorrhizae, p. 289.

Weaver, D. J. and Wehunt, E. J. (1975). Effect of soil pH on susceptibility of peach to *Pseudomonas syringae. Phytopathology* **65**, 984–989.

Weber, D. F. and Gainey, P. L. (1962). Relative sensitivity of nitrifying organisms to hydrogen ions in soils and in solutions. *Soil Science* **94**, 138–145.

White, J. G. (1959). Mineralization of nitrogen and sulphur in sulphur-deficient soils. *New Zealand Journal of Agricultural Research* **2**, 255–258.

White, P. F. (1984). The effect of vesicular arbuscular mycorrhizas on the growth of subterranean clover grown under conditions of aluminium toxicity. Honours Thesis, School of Agriculture, The University of Western Australia.

White, R. E. (1969). Measurement of soil pH. *Journal of the Australian Institute of Agricultural Science* **35**, 3–15.

Wijler, J. and Delwiche, C. C. (1954). Investigations on the denitrifying process in soil. *Plant and Soil* **5**, 155–169.

Williams, C. H. (1967). Some factors affecting the mineralization of organic sulphur in soils. *Plant and Soil* **26**, 206–223.

Williams, C. H., Williams, E. G. and Scott, N. M. (1960). Carbon, nitrogen, sulphur and phosphorus in some Scottish soils. *Journal of Soil Science* **11**, 334–346.

Williams, C. H. and Steinbergs, A. (1958). Sulphur and phosphorus in some Eastern Australian soils. *Australian Journal of Agricultural Research* **9**, 483–491.

Yawney, W. J., Schultz, R. C. and Kormanik, P. P. (1982). Soil phosphorus and pH influence the growth of mycorrhizal sweetgum. *Soil Science Society of America Journal* **46**, 1316–1320.

Young, J. L., Davis, E. A. and Rose, S. L. (1985). Endomycorrhizal fungi in breeder wheats and triticale cultivars field-grown on fertile soil. *Agronomy Journal* **77**, 219–224.

5

Selection of Genotypes Tolerant of Aluminium and Manganese

B. J. SCOTT and J. A. FISHER

I. INTRODUCTION

Species vary in their reaction to soil acidity. However, the last 25 years has seen the recognition that differences also exist among genotypes within a species (Ouellette and Dessureaux, 1958; Neenan, 1960). Differential tolerances to aluminium within species have been reported for barley, wheat, triticale, rice, alfalfa, tomato, soybean, perennial ryegrass, snapbean, cotton, maize, rye, sunflower, pea and sweet potato (Foy, 1974; 1983a). Reports of differential tolerances to manganese within plant species include wheat, apple, triticale, soybean, cotton and flax (Foy, 1983a).

The existence of such differences has raised the possibility of breeding, within agriculturally important plants, for improved tolerance

SOIL ACIDITY AND PLANT GROWTH
ISBN 0 12 590655 2

to soil acidity. These differences have been extensively and frequently reviewed (Wright and Ferrari, 1976; Foy and Fleming, 1978; Foy, 1984), with emphasis on the physiological and biochemical basis for toxicity (Haug, 1984) and tolerance, particularly with reference to aluminium.

Views diverge widely on the role of plant tolerance in the restoration of productivity on acid soils. Some authors enthusiastically endorse the usefulness of tolerance (e.g. Vose, 1981; 1983; Frey, 1985), while others express doubts. Fox (1980) describes the use of tolerance as like 'the hazard of flying a low-powered aircraft up a narrow canyon. The course starts easily and the scenery is beautiful, but options run out very quickly and to continue is to invite disaster'. Such a view seems to envisage tolerance as an alternative to liming. In this chapter we have explored the relationship between liming and plant tolerance to acidity.

We believe that using the plant's tolerance to acidity is an important, complementary approach to soil amendment (e.g. liming) and this is particularly true in the extensive agriculture practised in most of Australia. The adoption of liming is limited by its high cost in an agricultural system which has relatively low cost inputs and low profitability per hectare. Using crops tolerant of acid soils may prove to be an essential aspect of management of soil acidity as in South America (Sanchez and Salinas, 1982). Indeed, it has been suggested that sensitivity to mineral stress in the wheat and rice cultivars of the 'green revolution' may have been a major shortcoming in the adaptation of these cultivars (Foy, 1983a).

In this review we emphasise several areas of interest which have not been reviewed recently. We aim to meet the information requirements of a plant breeder attempting to produce cultivars more tolerant of soil acidity. We use examples relevant to temperate Australian agriculture with emphasis on cereals, partly because much is known about these species and partly because this is the area in which the authors are actively involved.

II. TOXICITY SYMPTOMS AND TOLERANCE MECHANISMS

A. Aluminium

1. Toxicity

The effects of aluminium are primarily on the root system, causing reduced root elongation and, at high aluminium levels, the death of the growing point. The lateral roots appear to be more sensitive to damage than primary roots but all root growth is damaged as aluminium levels are increased. This gives the typical 'stubby root' appearance with few, fine-branching roots. As a result the root system is inefficient in absorbing nutrients and water.

However, symptoms of aluminium toxicity in tops are reasonably diverse and not easily identified. This is not surprising as aluminium interferes with the uptake, transport and use of essential elements such as calcium, magnesium, phosphorus, potassium and iron (Foy, 1983b) and an aluminium-induced deficiency of any of these elements may affect the symptoms shown.

The effects of aluminium toxicity on root elongation and death of the growing point are rapid, occurring in a matter of hours. This observation on roots is the basis of many of the rapid tests for aluminium tolerance in plants.

2. *Tolerance*

Tolerance to aluminium can operate by preventing or reducing aluminium uptake by plants or by detoxifying aluminium once it is in the plant tissue. Exclusion of aluminium from the plant may operate in several ways.

First, the pH in the root zone may be increased and this decreases the solubility and toxicity of aluminium. This process involves the plant accumulating more anions than cations and excreting hydroxyl ions or Bronsted-Lowry bases (Helyar, 1978). This process has been demonstrated in wheat (Foy and Fleming, 1978; Fleming, 1983). Varieties of wheat tolerant of aluminium tended to increase the pH of the nutrient solution in which they were grown while aluminium-sensitive varieties lowered the pH. The mechanism appeared to be that tolerant plants were able to take up some nitrate in the presence of ammonium while sensitive plants took up relatively more ammonium.

Second, low cation exchange capacity of the roots has been suggested as being associated with aluminium tolerance (Vose and Randall, 1962). This would be expected to reduce the diffusion of trivalent aluminium and also to reduce the severity of the effects of aluminium on cell wall expansion as there would be fewer, free carboxyl radicals in cell walls to which the aluminium could bond (Helyar, 1978).

Third, some resistance by the plasmalemma to aluminium damage has been suggested. This would then restrict aluminium diffusion into the cytoplasm (Henning, 1975; Foy, 1983b).

Fourth, morphological change under aluminium stress has been suggested as a mechanism for tolerance (Hecht-Buchholz and Foy, 1981). These authors suggested that a mucigel was produced to protect the root from aluminium damage. Hecht-Buchholz (1983) has reviewed morphological damage and adaptation to a range of mineral deficiencies and toxicities.

Once aluminium has penetrated the plant tissue, other tolerance mechanisms may operate. First, in certain woody species and in 'accumulator' species, cytoplasmic aluminium is complexed by organic molecules and so does not disrupt the cell metabolism (Jones, 1961; Jackson, 1967). This aluminium, once inactivated, may be held in the cytoplasm or deposited in old xylem vessels or cell walls (Helyar, 1978). The detoxification of aluminium by tea (*Camellia sinensis* L. Ktze), an aluminium accumuluator, appears to be accomplished by this mechanism (Sivasubramaniam and Talibudeen, 1972; Memon *et al.*, 1981).

The importance of this internal tolerance in distinguishing between genotypes of a species is unclear. It may be unimportant in the broad range of agricultural plants, although some examples of a relationship between tolerance and aluminium accumulation in tissue can be found in the literature. Differential internal tolerance has been reported with rye (Aniol *et al.*, 1980), snapbean (Naidoo, 1976) and in the callus cell cultures of tomato (Meredith, 1978a,b).

Second, differential distribution of tissue aluminium between tops and roots has been reported as a possible tolerance mechanism (Foy, 1983b).

Third, as aluminium disrupts the uptake, transport and use of a range of essential nutrient elements, 'efficiencies' in uptake or utilisation of nutrients have been suggested as negating the disruptive effect of aluminium toxicity. In potato and maize, magnesium uptake was greater in varieties tolerant of aluminium toxicity than in sensitive varieties (Foy, 1983b). The ability of plants to tolerate low internal concentrations of calcium, phosphorus and iron have been suggested as the basis for aluminium tolerance (Baumgartner *et al.*, 1976; Foy and Fleming, 1978).

B. Manganese

1. *Toxicity*

Manganese toxicity primarily affects plant tops rather than roots and produces specific symptoms in the tops of some species. With soybean grafts, Heenan and Carter (1976) reported that the genotype of the scion determined the response to manganese toxicity. However, with alfalfa genotypes, Ouellette and Dessureaux (1958) concluded that the genotype of the root determined the response.

The most common symptom of manganese toxicity is the chlorosis of leaf tip and margins followed by necrosis, particularly on the older leaves. This can be accompanied by chlorotic or necrotic flecking or mottling on the leaf margins or whole leaf. Under greater stress the symptoms appear on progressively younger plant tissue. 'Puckering', 'crinkling' or 'cupping' has been described in soybean (Heenan and Carter, 1977), and for a range of crops including cotton, rape, kale, lettuce and snapbean (Foy, 1983

Generally the damaged plant tissue (i.e. leaf margins, interveinal spo or flecks on older tissue) has a higher concentration of manganese than the surrounding tissue. This accumulation appears to be due to low phloem mobility of manganese which leads to manganese accumulation at the extremities of the xylem system (leaf margins and interveinal leaf tissue; Vlamis and Williams, 1973). Interveinal chlorosis has been shown to be due to manganese-induced iron deficiency in some but not all instances (Hewitt, 1963).

Browning of roots has been reported in nutrient cultures high in manganese (Morris and Pierre, 1949; Foy, 1983b). E. A. Polle (pers. comm.) drew attention to a symptom of 'banding' on the roots of seedling wheat grown in dilute solutions of manganese. This symptom was reported in 1927 (Barton and Trelease, 1927). In Polle's experience and ours, the appearance of this symptom was not related to manganese tolerance although varieties did vary in their reaction, from no apparent effect to a clear 'banding' (see Fig. 1).

Helyar (1978) has noted the role of manganese as a cofactor in the activity of many enzymes and has suggested that plant damage with high concentrations of manganese in tissues is due to loss of control of at least some of these enzymes activated by manganese. Manganese has a role in phosphate transfer reactions, the Krebs cycle, a number of oxidases, nitrate reductase and in photosynthesis.

Effects of manganese toxicity on some of these processes have been noted (Sirkar and Amin, 1974; Aoba *et al.*, 1977; Scott, Burke and Bostrom, 1987). The circumstantial evidence, then, is that effects of manganese toxicity on these plant functions are likely to be the primary damage to the plant. However, with such a diverse set of sites for metabolic disruption under manganese stress, it is likely that secondary effects occur and that the concentrations of many metabolites are likely to differ from the normal level.

2. *Tolerance*

Plant tolerance to high levels of soil manganese may operate by exclusion, by restriction of manganese transport to plant tops, and by tolerance of shoots to high internal concentrations of manganese.

Exclusion of manganese as a tolerance mechanism may improve plant performance in a manganese-toxic soil but would also make the plant more susceptible to deficiency in other situations. There is

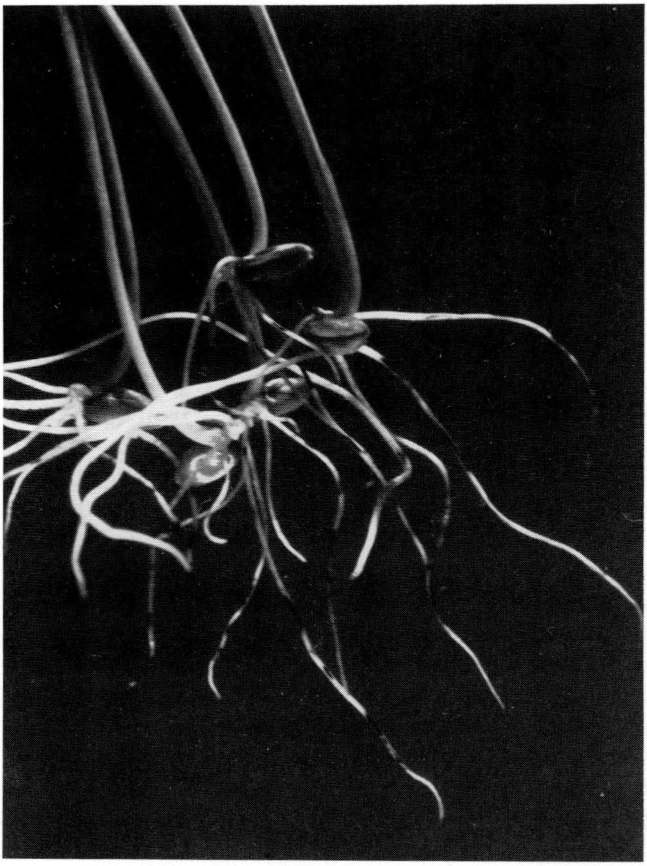

Fig. 1. Clear 'banding' in the roots of the Brazilian cultivar. Carazinho, four days after germination and growth in a dilute solution (0.2 mM) of manganese chloride. Similar results were obtained with manganese sulphate solutions.

evidence of this mechanism operating among species and some evidence of differential uptake at a cultivar level (Helyar, 1978).

Restricted movement of manganese from roots to tops has been suggested as a basis of tolerance in lucerne (Ouellette and Dessureaux, 1958), and a range of other species (Andrew and Hegarty, 1969; Robson and Loneragan, 1970; Edwards and Asher, 1982). Such a mechanism proposes that manganese can be isolated or 'inactivated' in the root and that this does not damage the root. For example, some key metabolic processes (e.g. photosynthesis), disrupted by high levels of tissue manganese, occur in the tops rather than in the roots.

Tolerance to high levels of manganese in shoots may be the major mechanism operating in a wide range of species. Heenan and Carter (1976), in a grafting experiment with soybean, demonstrated that tolerance was related to the genotype of the shoots rather than the genotype of the roots. Edwards and Asher (1982) thought this was the most important mechanism for manganese tolerance shown by a wide range of species tested by them. However, the first two mechanisms were also operating. Similar conclusions were reached by Temple-Smith and Koen (1982) with a range of crop types and by Smith (1979) with a group of tropical grass species.

At the level of cultivars within a species, tolerance to manganese in shoots has been reported as important in soybean (Brown and Devine, 1980; Heenan and Campbell, 1981), wheat (Foy *et al.*, 1973; authors' unpublished data), cotton (Foy *et al.*, 1981), siratro (*Macroptilium atropurpureum*, Hutton *et al.*, 1978) and cowpea (Horst, 1980; 1983).

The basis of tissue tolerance to manganese has been attributed to: (1) the formation of metabolically inactive, organic manganese complexes (Foy, 1984); (2) binding to cell walls and/or deposition in vacuoles (Helyar, 1978); and (3) tolerance by some vital enzyme systems of high concentrations of ionic manganese (Scott, Burke and Bostrom, 1987).

Localisation of manganese in plant tissue has been described by Memon *et al.* (1980), who noted that the highest concentrations of manganese occurred in the epidermis, collenchyma and bundle sheath cells in the manganese-accumulator species *Acanthopanax sciadophylloides*. Foy (1984) has suggested that this localisation may be a tolerance mechanism that keeps manganese from key metabolic sites. By contrast, Horst and Marschner (1978) and Horst (1983) ascribed the greater manganese tolerance of beans, when silicon was added to solutions, to the more homogeneous distribution of manganese in the leaf. High concentrations of manganese in local areas of leaf were associated with symptoms of brown spots.

III. TOLERANCE AND LIMING

Concern is expressed that the introduction of a range of tolerant cultivars would negate the need for liming and, in acidifying agricultural systems, acidity problems would intensify unchecked. We believe this view is based on reports in the literature of the non-response of tolerant cultivars to lime application (Martini *et al.*, 1977). In this study there was no response to liming for 19 Brazilian wheat cultivars tolerant of aluminium toxicity (mean yield 2110 kg. ha^{-1} when unlimed, 1820 kg. ha^{-1} when limed) compared to the response of 49 cultivars from a range of other countries (mean yield 570 kg. ha^{-1} when unlimed, 1210 kg. ha^{-1} when limed).

There are, however, other data in the literature, which indicate that the relationship between tolerance to aluminium and response to lime is frequently different from this. In fact, in another study reported by Martini *et al* (1977), increased grain yield in a tolerant cultivar of Brazilian wheat (IAS-52) is reported following liming. We believe that the broader evidence indicates that both liming and tolerance have a role to play in increasing production from acid soils. A range of such views has been discussed by Foy (1983a).

In order to consider the relationship between tolerance and lime use, it is necessary to describe the responses of a tolerant and a sensitive cultivar to the application of lime.

A. Relationship of tolerance and response to lime — a proposed model

The proposed model is illustrated in Figure 2. A serious aluminium problem is corrected by liming. The points marked A to D on this figure

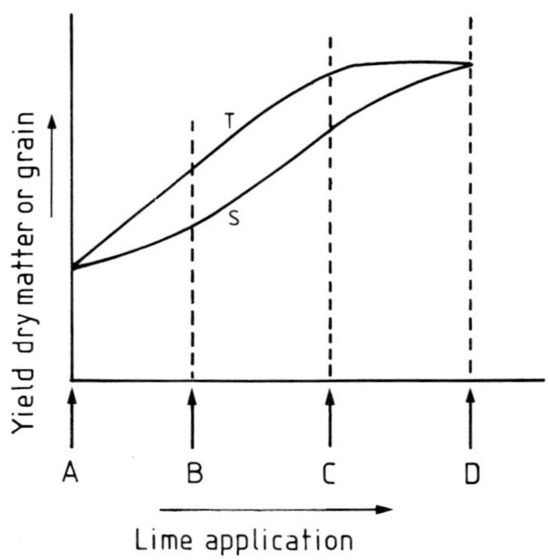

Fig. 2. The idealised response to lime application by an aluminium tolerant (T) and a sensitive (S) cereal cultivar grown in a soil with toxic levels of aluminium. At point A aluminium toxicity limits both tolerant and sensitive cultivar; at point B the tolerant cultivar outyields the sensitive cultivar; between B and C both cultivar react similarly and between C and D the tolerant cultivar does not respond but the sensitive cultivar does.

are described here. At point A there is a serious aluminium toxicity which restricts the growth of both the tolerant and the sensitive cultivar. When lime is applied, the tolerant cultivar gives a greater response up to point B, but the sensitive cultivar shows little or no response. Between B and C both cultivars react similarly i.e. the yield increase (dry matter or grain) is the same in absolute terms. Tolerance here results in higher yield but not in a differential response by the cultivars. Between C and D the tolerant cultivar does not respond to further lime application but the sensitive cultivar continues to respond. D is the point at which aluminium toxicity is removed from the plant root environment of both cultivars and in this example the cultivars have been assumed to give the same yield under these circumstances.

The above description is based on absolute yield and if ratios of yield (nil or low lime/high lime) are used to correct for yield differences between cultivars, the essential features of the model are obscured. We suggest that ratios are inappropriate as they appear to show the sensitive as more responsive only because of their lower yield under stress. Liming beyond point D may give yield reductions under some circumstances and again ratios need to be interpreted with care.

This model would apply where aluminium toxicity was the dominant factor limiting yield and cultivars were reasonably adapted to the area. If these conditions are not met, low yields may be due to factors other than aluminium sensitivity. We were unable to find data on which to base a model for liming, manganese toxicity and manganese tolerance.

B. Evidence in support of the model

This model is supported by some examples from experiments using wheat and barley (Fig. 3). The data presented describe the responses to liming of tolerant or sensitive cultivars of wheat and barley in experiments in pots (‘Bladen’ and ‘Tatum’ soils) and in the field. The aluminium tolerances referred to have been determined independently of these results.

In a pot experiment using the aluminium toxic ‘Bladen’ soil (Foy *et al.*, 1973), the responses of the wheat cultivars Atlas 66 (aluminium-tolerant) and Monon (aluminium-sensitive) to liming were determined (Fig. 3A). The range of approximately parallel response (B to C, Fig. 2) is apparent in the range from nil lime to 1500 $\mu g.g^{-1}$. The range of non-response of the tolerant and continued response of the sensitive (C to D) is present in the range between 1500 to 3000 $\mu g.g^{-1}$ of lime. At the highest rate of lime the plants would probably not be exposed to an aluminium stress.

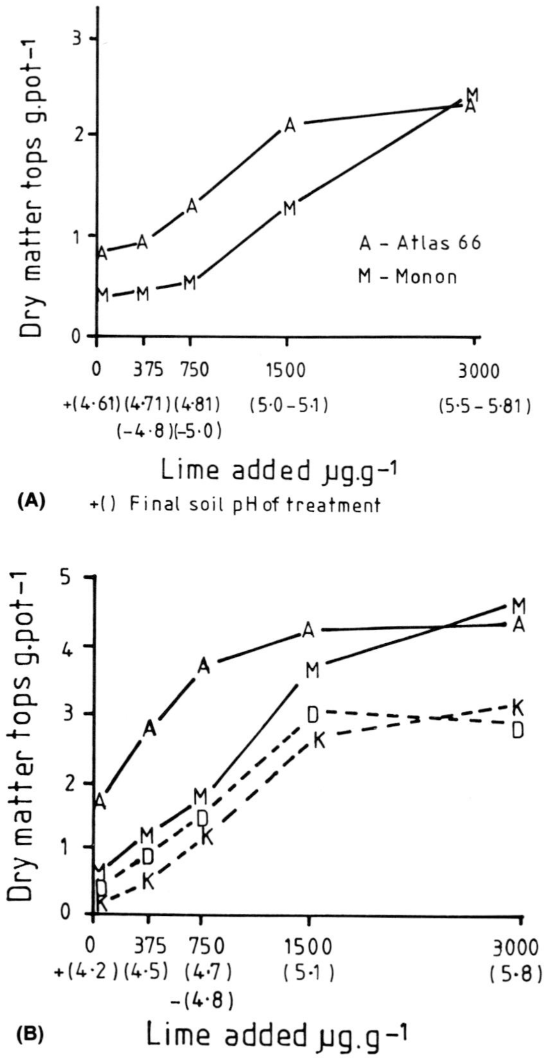

Fig. 3.(A) Response in dry matter yield by the wheat cultivars Atlas 66 (A; aluminium tolerant) and Monon (M; aluminium sensitive) to the application of lime. Plants were grown in a pot experiment using aluminium toxic 'Bladen soil; (after Foy *et al.*, 1973).

(B) The response in dry matter yield by wheat and barley to lime application.Plants were grown in pots using aluminum toxic 'Tatum' soil. The wheat cultivars (solid lines) were Atlas 66 (A) and Monon (M) and the barley cultivars (broken lines) were Dayton (D; aluminium tolerant) and Kearney (K; aluminium sensitive); (after Foy *et al.*, 1965).

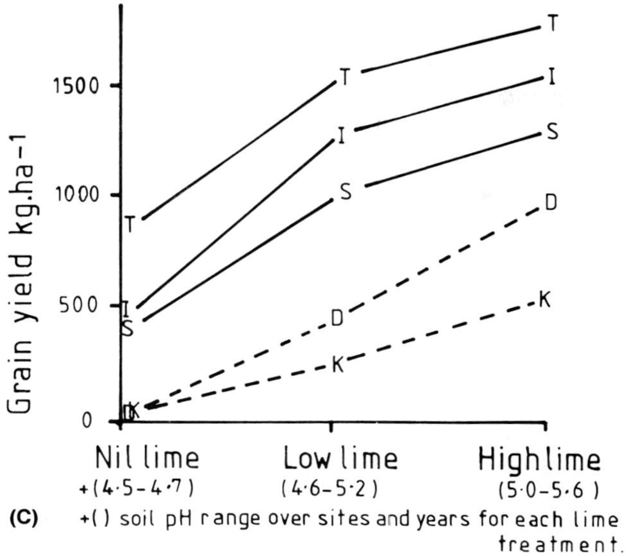

(C)

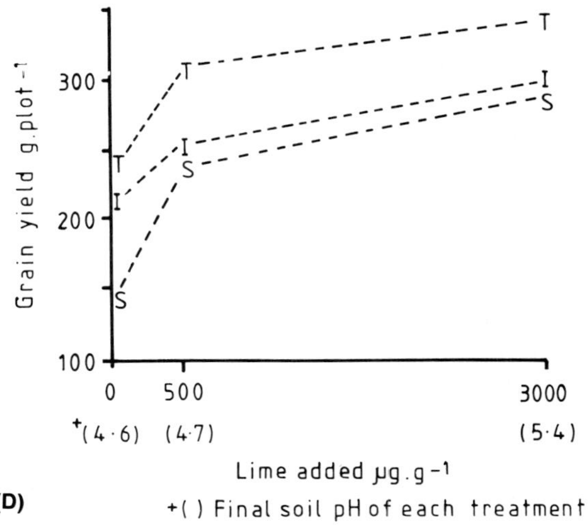

(D)

(C) Effects of lime on grain yield in the field, averaged over three seasons, for wheat (solid lines) and barley (broken lines). The wheat cultivars were grouped, (as described in the text) into a tolerant group (T), a sensitive group (S) and a group with intermediate reaction (I) to high aluminium. The barley cultivars are Dayton (D) and Kearney (K); (after Lafever *et al.*, 1977).

(D) Effects of lime application on the grain yield of barley cultivars in the field in one season. The barley cultivars have been grouped on the basis of a greenhouse pot experiment into the 10 most tolerant of aluminium (T), the 10 most sensitive (S) and the remaining 10 intermediate lines (I); (after Reid *et al.*, 1969).

In another experiment conducted in pots using the aluminium-toxic 'Tatum' soil (Foy *et al.*, 1965), Atlas 66 and Monon wheat were again tested as well as the barley cultivars Dayton (aluminium-tolerant) and Kearney (aluminium-sensitive). Results for both the wheat and barley cultivars (Fig. 3B) were similar to those presented in Figure 3A. The higher rates of lime applied in this experiment depressed yields.

In the same paper (Foy *et al.*, 1965), the two barley cultivars were grown on the 'Bladen' soil. Liming increased dry matter production per pot by Dayton from 2.8 to 3.2 g and increased that by Kearney from 0.4 to 0.5 g. We believe that this illustrates the range A to B in Figure 2, as the lime rate used was low and the '+ lime' treatment only attained pH 4.7 (method not stated). Compared to Kearney, the yield of Dayton was higher and its yield response to liming was greater in absolute terms.

Lafever *et al.* (1977) reported responses of cereal grain yields to liming in the field (Fig. 3C). The 23 cultivars of wheat (the authors' group 1) have been split into three classes based on the results of the solution culture tests reported in the same paper. The seven cultivars which were most aluminium-tolerant in solution culture were grouped as the tolerant group (T). The seven most sensitive were grouped (S) and the remaining nine formed the intermediate tolerance group (I). The grain yield data were averaged over three seasons and, as the experiment was not on the same plots each year and lime was applied at different rates, the pH ranges of the surface soil are given and the 'low' and 'high' lime nomenclature is used.

We believe that the section B to C of Figure 2 is illustrated in the wheat yields (Fig. 3C). All groups responded similarly to liming and tolerance was an advantage irrespective of liming treatment at this site. Aluminium toxicity was not eliminated and the authors emphasise this point. The sub-surface soil was acid and so surface liming did not ameliorate aluminium toxicity in the rooting profile of the crop. In other words point D of Figure 2 (no aluminium toxicity) was not attained.

Correlation coefficients between an independent measure of aluminium tolerance for each wheat cultivar (the ratio of root length at nil aluminium to root length at 8 $\mu g.mL^{-1}$ of aluminium in solution) and the grain yield in the field for nil, low and high lime rates were calculated from the data given by Lafever *et al.* (1977). The coefficients were 0.68, 0.69 and 0.62 respectively and all were significiant ($p < 0.01$). Clearly, tolerance to aluminium was associated with higher grain yield in the field even when high rates of lime had been applied.

At 'high' rates of lime, the sensitive group produced 345 per cent of the yield achieved when no lime was applied. Similarly, the tolerant group produced 217 per cent of the yield achieved when no lime was applied. The sensitive cultivars then appear to be more responsive, but in fact the absolute increase in yield was 956 $kg.ha^{-1}$ for the sensitive group and 959 $kg.ha^{-1}$ for the tolerant group. The tolerant group then was equally responsive, in terms of absolute yield increases, and the tolerant group had the higher yield at all lime rates. Using ratios to compare yields at high lime with those at nil lime is misleading.

In the same experiment (Fig. 3C), the yield of the barley cultivar Dayton (aluminium-tolerant) increased more than that of Kearney (aluminium-sensitive) and matched the range A to B in Figure 2. Compared to the wheat cultivars which covered the range B to C, the barley was under a greater stress. We suggest that this is due to the greater sensitivity of barley to aluminium. The response of these barley cultivars is an example of greater response by an aluminium-tolerant cultivar to lime application compared to the response from a sensitive cultivar.

Reid *et al.* (1969) reported the response of 30 cultivars of barley to liming in the field. Their grain yields are given in Figure 3D with the barley cultivars grouped into the 10 most tolerant of aluminium (T), the 10 intermediate (I) and the 10 most sensitive (S) lines. The basis of this grouping is the tolerance ranking (yield at low lime to yield at high lime), reported in the same paper, from a greenhouse experiment using an aluminium-toxic soil. This tolerance index correlated well ($r = 0.92$; $p < 0.001$) with that obtained for 13 of the lines in a subsequent solution culture experiment (Reid *et al.*, 1971), in which the index used was root length at nil aluminium/root length at 8 $\mu g.mL^{-1}$ aluminium.

All tolerance groups responded to lime application (Fig. 3D). The sensitive cultivars gave a greater increase in yield than the tolerant cultivars (98 $g.plot^{-1}$ cf. 51 $g.plot^{-1}$). The data are in the range B to greater than C in Figure 2. Aluminium toxicity was probably not eliminated at this site as the tolerant group out-yielded the sensitive group even at the high lime rate. A significant correlation ($r = 0.39$; $p < 0.05$) was present between the tolerance index derived in the greenhouse (described above) and absolute grain yield in the field at the high lime rate, suggesting that the plants were still experiencing aluminium toxicity.

C. Implications and discussion of the model

It should not be expected that sensitive cultivars will always be more responsive to liming than tolerant ones. The responsiveness will vary with the initial soil condition and the extent to which that has been changed by liming, as well as the relative tolerance range of the cultivars being tested. In some situations (Fig. 2, A to B) liming may only be economic if a tolerant cultivar is used. In all other situations a tolerant variety will exceed or equal the yield and hence the economic return obtained from growing a sensitive cultivar.

It is our experience in New South Wales that tolerance to acid soil conditions can play an important role as a component of acid soil management. In two situations on acid soils (Doyle and Bradley, 1982; Scott, 1982) higher yields were obtained with tolerant cereals than with sensitive cereals. Additionally both tolerant and sensitive cereals responded to liming. In practice, the complete elimination of acidity from a soil profile, for the duration of crop growth, is a situation which is difficult to achieve for a number of reasons.

First, we have already drawn attention to acidity in the profile which is below the depth of lime incorporation. Second, variability of the acidity problem across a site results in insufficient lime being applied in some very acid areas following treatment at a single lime rate. Third, poor spreading or incorporation of lime may leave areas or portions of soil unlimed. Fourth, liming just prior to sowing may leave young plants exposed to an acidity stress as the reaction of lime with the soil may be incomplete. Finally, the farmer's practice of 'hedging' or reducing fertilizer rates to less than those recommended may leave acidity problems only partially amended. The variability problems are described in greater detail in Chapter 6 this monograph.

Using ratios of yield with and without lime to compare cultivars of varying tolerance obscures these economic implications of tolerance. Ratios can be used as an index of tolerance only if liming produces point D in Figure 2 and the cultivars show a range of responses under the stress. If point D is not achieved then the ranking of cultivars will depend on the range of stress covered.

IV. BREEDING FOR TOLERANCE

A. Requirements for breeding

A frequently quoted example of successful breeding for tolerance is that of wheat breeding in Brazil. Since 1919 in that country, breeders have selected and later bred wheat for tolerance to 'crestamento', a leaf 'burn' condition in wheat (Da Silva, 1976). In retrospect it appears that selection in this way was for tolerance to soil acidity, in particular to aluminium toxicity. While such an experience shows what is possible in breeding for tolerance, the breeding was done on an *ad hoc* basis. Devine (1976) has listed the techniques and information required for the formulation of a deliberate breeding program. These requirements are that: (1) techniques are available for identifying tolerance; (2) there is a reasonable range of variability within the species; (3) the character is heritable and some estimate of heritability is available; (4) there are no strong, undesirable genetic correlations with tolerance; (5) an estimate can be made of the level of improvement attainable in the field.

It is generally acknowledged that these requirements need to be met by plant nutritionists, biochemists, physiologists and others before plant breeders will be able or motivated to undertake planned breeding programs aimed at the regular production of improved, acid-tolerant cultivars. While much has been achieved in this direction, crops and pastures with greater tolerance to soil acidity remain, in general, a promise yet to be fulfilled.

While claims have been made of tolerance to acid soil by species and cultivars, these claims are based on field observations rather than being the result of a conscious breeding effort. At Wagga Wagga in southern New South Wales, we are committed to a project on the breeding of a wheat cultivar tolerant of acid soils (Fisher and Scott, 1983). We are not alone as Mexican workers, cooperating with other workers In Brazil, have already produced aluminium-tolerant lines for further testing (Anon., 1984).

B. Screening for tolerance

There are three stages of selection in a breeding program, and the requirements of a test for tolerance differ in these stages.

The first stage is identifying parental lines. Tests at this stage must be precise and frequently multiple tests would be used, often in series, to finally identify the best parental lines. In our program at Wagga Wagga, three tests are used in sequence, a rapid laboratory test, growing plants in nutrient solution and finally testing the lines in field trials. In the final choice of parental lines, it is not essential to be able to test large numbers of lines. Field trials are used, not only to confirm the tolerance of the chosen lines, but also to identify faults in the cultivars which may need correction in the breeding program. If different tolerance mechanisms exist, a knowledge of the basis of tolerance in a particular cultivar would be useful and may allow different types of tolerance to be combined. Hopefully, different tolerance mechanisms combined in this way would be additive.

The second stage of selection is testing in segregating populations. The requirements of the selection test will depend on the aim and methods of the breeding program. At one extreme, if the aim is to move one dominant gene from a tolerant cultivar to an otherwise well adapted cultivar, then the population size required would be small (20 plants per generation) and relatively slow tests would be adequate. At the other extreme, where inheritance is complex and it is necessary to seek improvement in other characters simultaneously, it will be essential to screen many thousands of plants. This latter case is the more common in plant breeding, so it is desirable that a selection test be very rapid. Sometimes a less precise test may be acceptable if it allows a larger population to be handled. In this case the selected population would need to be retested with a more precise test. In backcrossing or complex crosses, where the selected line is to be used immediately in the crossing program, it is essential that the results of testing are available before the end of flowering.

The final stage of testing is confirmation that the selected lines perform as required. The requirements at this stage are similar to selecting parental lines. Rapidity of tests is not as important as precision. It is essential to do field tests in the environment in which the cultivar will be grown.

Plants may be screened for their tolerance to acid soil conditions by testing in soil either in the field or glasshouse. Alternatively sand or solution culture systems can be used. All of these approaches have their advantages and disadvantages.

1. Field screening

Field screening has been the basis of the success of the Brazilian wheat breeding effort. The testing relied on the presence or absence of the condition known as ‘crestamento’. This condition is described as causing premature death in severe situations (Da Silva, 1976) but is more usually described as a leaf ‘burn’ with associated poor root development and ‘underdeveloped, yellow-green plants with short, thin stems and narrow leaves’ (Martini *et al.*, 1977). Campbell and Lafever (1976) also drew attention to the stunting of wheat on acid soils and suggested that differential stunting could be used to separate lines which differed greatly in their tolerance.

Howeler and Cadavid (1976) have successfully screened rice in the field for tolerance but found better differentiation of their lines when they used a low rate of lime, suggesting that the unamended soil was too acid relative to the range of tolerance available in their material. Their selection criterion was absolute grain yield in the field.

Kang and Fox (1980) used lime and sulphur and/or sulphuric acid to establish a pH gradient on a manganiferous oxisol. Cowpea (*Vigna unguiculata*) cultivars were sown along the pH gradient with the aim of testing cultivars for their manganese tolerance. They were able to establish tolerance differences by this 'continuous function experimental design'.

The testing of tolerance in this way evaluates plants in the soil and climate in which they are to be grown commercially. The disadvantages of field testing are: the time taken (usually one growing season); problems of soil variability; the confusing effects of differential resistances of material under test to diseases and pests; and the vulnerability of the material to environmental hazards such as drought, flood and lodging (Howeler and Cadavid, 1976). All of these factors may make tolerance to a soil factor difficult to identify with certainty and precision.

2. Glasshouse screening with soil

Studies here can be divided into two types: (1) those which use a problem field soil; and (2) those in which the soil is used only as a screening medium and is not directly associated with the acidity problems under study.

The soil in category 2 is chosen carefully to produce either aluminium or manganese toxicity in the species to be evaluated. The approach is that of a screening test and not that of studying a field problem in the glasshouse.

Surprisingly few studies fall into the first category. Some examples would be those of Mugwira *et al.* (1978), Camargo and Oliveira (1981), Long *et al.* (1973) and Culvenor *et al.* (1986a). The advantage is that plants are screened for their suitability on a relevant soil, i.e. while it may be suggested that aluminium or manganese toxicity is involved, a precise diagnosis of the problem, while helpful, is not essential.

The use of soil as a screening medium for aluminium and manganese tolerance has been discussed by Foy (1976). Andrew (1976) has outlined the use of soil as a medium for screening for manganese tolerance in legumes. The problems are those of producing a toxicity of the correct type and of a magnitude appropriate to the range of tolerances under test. The soil chosen should produce a single stress of either aluminium toxicity or manganese toxicity, which does not interact with other nutritional deficiencies (e.g. calcium, phosphorus, molybdenum, magnesium).

The problem with a soil as a screening medium is variability caused by the site of collection itself, time of collection and storage conditions for the soil (particularly with manganese). The stress chosen can be obtained by acidifying a soil, liming or adding manganese. Additions of basal nutrients are aimed at preventing nutrients other than the desired stress from affecting growth. In practice, experience is required both with the soil and with the plant species being tested, and care is needed in the interpretation of results. Using cultivars of known reaction to the planned toxicity in each experiment is an advantage.

Despite these precautions, Foy (1976) points to surprising results in some experiments and gives examples of molybdenum deficiencies and combined aluminium and manganese toxicities operating in some experiments where aluminium stress alone was planned.

3. Screening with nutrient solutions

The advantage of solution culture experiments in research on plant tolerance to aluminium and manganese is that greater control can be exerted over the level of exposure to the toxic element, the level of other nutrients in the solution and pH.

The major concern is that the rankings for tolerance in solution culture cannot be reproduced in a soil of the same toxicity. For this reason there are numerous papers which document cultivar tolerance in both a solution culture system and in a pot experiment using soil. Examples for aluminium tolerance are in barley (MacLeod and Jackson, 1967; Reid *et al.*, 1971), wheat (Foy *et al.*, 1965), rice (Nelson, 1983), soybean (Sartain and Kamprath, 1978) and sunflower (Foy *et al.*, 1974). Similar research has been conducted on manganese tolerance e.g. in triticale (Mugwira *et al.*, 1981), wheat (Foy *et al.*, 1973) and cotton (Foy *et al.*, 1969). The general thrust of these papers is that solution culture results gave similar cultivar rankings to those obtained in soil. However some variability has been reported (Foy, 1976) and may be due to the greater availability of nutrients such as phosphorus in nutrient solutions compared to availability in a soil, particularly to a plant with a root system damaged by aluminium toxicity.

However, there are several factors (particularly those related to the control of the toxic element stress imposed) which need to be considered to ensure a successful solution culture experiment.

a Aluminium: The factors controlling the solubility and speciation of aluminium in solution are discussed in detail in Chapter 1 of this monograph. The pH and concentration of phosphorus in solution should be low in order to impose and control the aluminium stress. It is necessary to control these factors either by frequent monitoring and adjustment and/or by frequent solution changes. Flowing culture systems which require only low phosphorus levels in solution for plant growth and have continuously monitored pH are ideal but complex. The control of nitrate and ammonium concentration offers some scope for stabilising the solution pH.

The level of calcium and magnesium in solution can also influence the severity of aluminium toxicity. This has been exploited by Rhue and Grogan (1977), who varied the calcium and magnesium in solution to control the plant damage caused by a constant aluminium level. A comprehensive approach to the design of a short-term (14 to 17 days) screening system for testing aluminium tolerance in sorghum has been described by Furlani and Clark (1981). These authors varied concentrations of aluminium, calcium, potassium, magnesium and phosphorus, source of nitrogen (nitrate, or ammonium, or a combination of both) and temperature regime in establishing levels of each for a screening test. High phosphorus, calcium or magnesium protected the plants against aluminium damage. Potassium level, in the range tested, had no effect on severity of aluminium toxicity, and plant damage by aluminium increased with increasing temperature in the range (day/night) of 23/18 to 33/28°C. Increasing the temperature during both day and night from 5 to 35°C also increased plant damage (Aniol, 1983).

b Manganese: The expression of manganese toxicity can be altered by a number of factors, other than the level of manganese in solution. These factors need to be controlled in order to separate reliably cultivars or species on the basis of their manganese tolerance.

Increased concentrations of calcium, magnesium, and ammonium in solution as well as lowered pH decreased the uptake of manganese by plants (Jackson, 1967; Robson and Loneragan, 1970; Heenan and Carter, 1975; Shuman and Anderson, 1976; Osawa and Ikeda, 1980). The beneficial effects of ammonium may be due partly to the lowering of external pH when plants absorb ammonium (Foy, 1983b). An increase in phosphate in solution has been reported as increasing manganese uptake and toxicity (Foy, 1983b).

Increasing temperature appears to increase uptake of manganese by plants but also enhances the plant's tolerance of high levels of tissue manganese. These effects have been reported for soybean (Heenan and Carter, 1977), tobacco (Rufty *et al.*, 1979) and rice (Nelson, 1982).

Addition of silicon to nutrient solutions has been found to alleviate manganese toxicity in barley (Williams and Vlamis, 1957), grasses (Vlamis and Williams, 1967) and beans (Horst and Marschner, 1978; Horst, 1983). This effect appears to be due to a more homogeneous distribution of manganese in the leaf tissue rather than to lower manganese concentrations. Silicon prevented the accumulation of manganese in local areas of the leaf, which otherwise resulted in necrotic spots (Horst and Marschner, 1978). However there is also some evidence of silicon reducing manganese uptake (Foy and Fleming, 1978; Foy, 1983b).

4. Rapid screening

Soil or solution culture studies, taking 4 to 8 weeks to complete, are generally used in experiments on aluminium tolerance in plants. However, the development of more rapid laboratory or glasshouse screening techniques for the detection of tolerance offers many benefits. The search for improved tolerance is easier as more material can be screened, and heritability studies can be more comprehensive. From the plant breeder's viewpoint, such rapid testing enables large numbers of breeding lines or individual plants to be tested routinely.

In all rapid laboratory tests for aluminium tolerance, the basis for predicting plant performance in soil is the effect of aluminium on roots over a short time span. The correspondence between the two is acceptable in general terms only (Reid *et al*, 1971; Campbell and Lafever, 1976; Howeler and Cadavid, 1976). Sartain and Kamprath (1978) drew attention to some lack of agreement between ranking of soybeans for aluminium tolerance in a soil experiment in pots and a short-term assessment of root growth. They noted that the short-term studies of root elongation accounted for the effects of aluminium on cell elongation and cell division while the pot experiment integrated many other effects over a longer time. These included the continued effect of aluminium on top and root growth and nutrient uptake. Hanson and Kamprath (1979) reported similar experiments and noted that some expressions of seedling tolerance do not carry over as tolerance in adult plants.

In most rapid tests, root length is measured (or estimated) and compared with a control at nil aluminium. Such methods have been described for wheat by Campbell and Lafever (1976) and Kerridge *et al.* (1971), for barley by Reid (1976), Kaminska (1981) and McNeilly (1982), for phalaris by Culvenor *et al.* (1986a) and for a range of crops by Konzak *et al.* (1976) and Mugwira *et al.* (1978).

These methods are most successful when the treatment and the nil aluminium control can be run concurrently. This may only be possible when testing the tolerance of established cultivars in self-fertilizing species. In cross-pollinating species the variability within a cultivar increases the experimental variability. When segregating breeding material is used, it is not possible to run a nil aluminium control, although the parents of the cross may be tested.

If it is not possible to run a concurrent control, the experiment should be run for a relatively long time so that slow root growth due to other factors is not confused with aluminium sensitivity (e.g. Reid, 1976 for 18 days; McNeilly, 1982 for 13 days). The disadvantage of these systems is that they must be maintained with frequent pH corrections and a decision must be made either to include or exclude phosphorus. In some systems, however, the pH is only adjusted at the beginning of the test (e.g. Mugwira *et al.*, 1978) and in others only aluminium is added to the solution, and no nutrients are present (Kaminska, 1981).

With wheat, a number of rapid tests have been developed. The most convenient is that using a haematoxylin stain (Polle *et al.*, 1978). This test can demonstrate aluminium tolerance in four-day-old seedlings and therefore with such small plants minimal space is required. Because of the short duration of the test, no phosphorus is required. The haematoxylin stain test, with some modifications, is successful with other species. This strategy — the testing of young seedlings — is needed in other crop and pasture species and for other mineral stresses (e.g. manganese tolerance). No test is available for the rapid evaluation of manganese tolerance in cereals.

At Wagga Wagga we are developing a test for aluminium tolerance which has some features in common with that of Polle *et al.* (1978). It is an enhancement of the approach described by Moore *et al.* (1976) and refined by Ali (1973) and Rhue (1976). This enhancement is the addition of a stain which permits a rapid visual scoring of test results, in a way similar to the haematoxylin test, and avoids measuring root length.

The method is identical to the haematoxylin test for the first three days. The aluminium exposure is for 48 hours rather than 16 hours as in the haematoxylin test and the aim is to cause irreversible damage (i.e. death) of the growing points of the roots in sensitive cultivars as outlined by Moore *et al.* (1976). At the completion of the aluminium exposure, the roots are stained, simply to mark which ones were exposed to the aluminium. The plants remain on the rafts during staining as described by Polle *et al.* (1978) for the haematoxylin staining. The rafts are then floated on a nil aluminium, recovery solution for three days. Fig. 4 shows the results of a test run with a sensitive and a tolerant cultivar of wheat. Death of the growing point of the root has occurred with the sensitive but not with the tolerant cultivar and this is made obvious by the unstained root growth. We used a peroxidase stain but we have not fully evaluated the effects of the stain on root growth. However, neutral red may be suitable (Schumacher *et al.*, 1983).

Apart from rapid visual scoring, the test shares a number of advantages with the haematoxylin test. These are that seed and not seedlings are handled during the test, the total duration of the test is short (seven days cf. four days in the haematoxylin test) and the duration of the aluminium exposure is short. This means that first, phosphorus need not

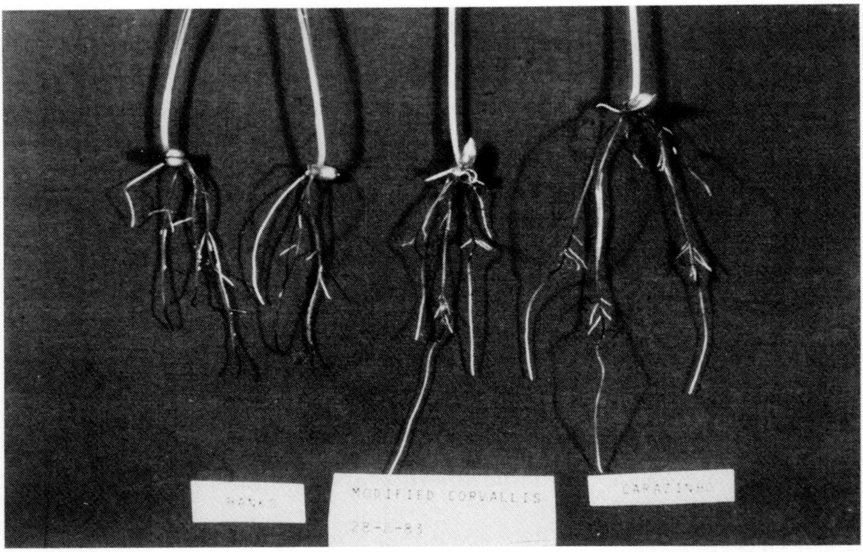

Fig. 4. Seven-day-old seedlings of the wheat cultivars 'Banks' (Australian, aluminium-sensitive) and 'Carazinho' (Brazilian, aluminium-tolerant) after three days growth with no aluminium in solution, two days exposure to aluminium, followed immediately by staining and a three day recovery with no aluminium in solution.
The growth in the recovery solution is unstained and shows that the growing point of the root of the sensitive cultivar was killed but that of the tolerant cultivar continued to grow.

be added, although it could be added to the recovery solution if necessary, and second, the pH and nutrient solution control is only vital for the time of the aluminium exposure. Indeed, the recovery solution need only be controlled broadly and it has been our practice to permit recovery to take place over a weekend without supervision.

The major advantage of our tests over other tests is that it appears to be suitable for a wider range of crops as the death of the growing point of the root under aluminium stress is probably a universal effect on plants. Also death of the growing point is probably not essential and differential root elongation could be the basis of cultivar differentiation. In establishing the test in other species, there would need to be changes in the aluminium levels tested, the nutrient solution at aluminium exposure, the duration of exposure and recovery, and the stain used. We have had success with wheat and barley.

Aniol *et al.* (1980) also used a similar methodology on rye, triticale and wheat. The roots were held in place between chromatography paper and a lucite plate as described by Konzak *et al.* (1976), and the root tip position was marked on the lucite plate after an aluminium exposure.

Staining may also be useful in replacing the first measurement in tests such as that of Hanson and Kamprath (1979) in which root growth was measured twice in 48 hours of aluminium exposure. Staining at the beginning of the aluminium exposure and measurement from the area of stain to the root tip 48 hours later would give the root growth during the aluminium exposure.

Tests available for assessing manganese tolerance are not as well researched as those for aluminium. As a result they are largely confined to relatively long duration (four to eight weeks) experiments in a manganese-toxic soil or in sand or solution culture. The only alternative method of identifying tolerance seems to be that of Horst (1982; 1983) who described the direct application of manganese to the petioles of intact plants of cowpea and found the symptoms induced to be related to tolerance information obtained from a sand culture experiment.

All the rapid tests described here are non-destructive and use scores of toxicity symptoms, stain patterns or root length measurements or estimates. This permits the plants tested to be grown to maturity. For a breeder these tests mean that plants can be assessed for tolerance and used in a crossing program or grown for seed.

C. Sources of tolerance

1. Within species

The most convenient sources of tolerance to use are those that occur within the same species, provided that an adequate level of tolerance can be achieved. Differences within species for tolerance to the various facets of soil acidity have been reviewed extensively (Wright and Ferrari, 1976; Foy and Fleming, 1978; Foy, 1983a). The identification of the most tolerant lines for use in a breeding program usually involves screening large numbers of collections. Collections expressing tolerance are more likely to be identified from areas with acid soil, like the lateritic soils of Brazil. Takagi (1984) reported that in 686 common wheat cultivars from 15 countries, only four highly tolerant lines were identified and these were from Brazil. Slootmaker and Reid (unpublished data, cited by Reid *et al.*, 1980) tested 1795 barley lines in the USDA world collection and identified 180 lines to use in a composite cross. Barley composite cross XXXIV provides a diverse gene pool made up of segregates from all known lines tolerant of aluminium and/or acid soils (Reid *et al.*, 1980). Howeler (1973) tested approximately 1000 cultivars of rice from various parts of the world for aluminium tolerance when searching for suitable parental material.

2. Related species

If adequate tolerance is not available within a crop species, it may be possible to introgress the genes for tolerance from related species. Breeding highly acid-tolerant cultivars of the legume *Centrosema pubescens* appeared to be impossible, however selections from crosses between *C. pubescens* and *C. macrocarpum* provided a basis for breeding highly acid-tolerant lines with desirable agronomic characters (Hutton, 1981).

However, the expression of genes from related species is often modified when they are transferred. Differences within the Triticinae have been reported by Slootmaker (1974) and Takagi (1984). The best tolerance to aluminium is present in rye and this tolerance is expressed in triticales. Lopez-Benitez (1977) has investigated the possibility of transferring this tolerance to wheat and found that the rye tolerance was not additive to the tolerance already found in wheat. Modifier genes from the wheat genotype influenced the expression of the full level of tolerance to aluminium contributed by the rye parent in the amphidiploids. Choudry (1980) has studied the addition of a single pair of rye chromosomes to wheat and their effect on wheat tolerance. No transfer of a single chromosome pair markedly improved tolerance in the wheat to which it had been added, suggesting that the rye tolerance is probably multigenic. Aniol and Gustafson (1984) tested rye additions and substitutions in different wheat cultivars and found a reduced expression of the rye tolerance in the wheat background.

The CIMMYT wheat cultivar Alondra (Anon., 1984) or Alondra 'S' (Anon., 1978) has been observed to grow well on the acid soils in Brazil, although it appears to be sensitive to aluminium in laboratory testing, presumably using the haematoxylin stain test. This caused interest as the line carried a rye translocation from the wheat cultivar Weique-Red Mace (Anon., 1978). These reports have attributed the success of Alondra in the field to its resistance to *Septoria* and/or its improved efficiency in phosphorus uptake.

At Wagga Wagga we have tested the tolerance of Alondra (Australian Wheat Collection number 21831) to aluminium and manganese in solution culture and found it to be highly tolerant of both (Figs. 5 and 6) and thus predict that it would perform well on acid soils. The source of the seed was originally Brazil but it arrived in the Australian collection from Corvallis, Oregon (Syme, pers. comm.) and it is believed to be Alondra 4546. There are several possible explanations for the difference

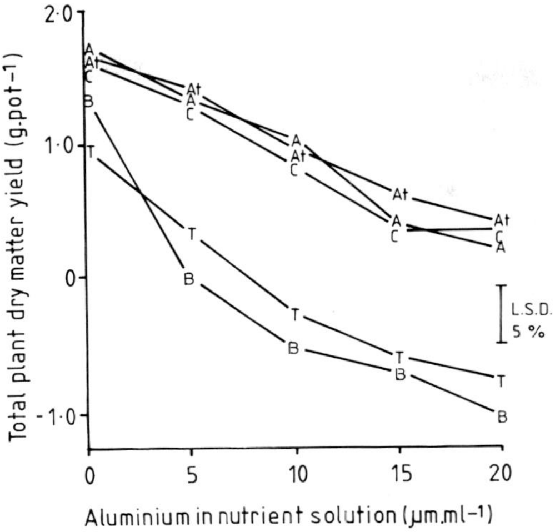

Fig. 5. The effect of aluminium concentrations in solution on the dry matter yield ($\log_e$) of five cultivars of wheat rested at Wagga Wagga, New South Wales. The plants were grown for four weeks in a sub-irrigated gravel bed system similar to that described by Hutton *et al.* (1978). The wheat cultivars are Carazinho (C), Atlas 66 (At), Alondra (A; Aus 21831), Teal (T) and Brevor (B). The first three cultivars are clearly tolerant of aluminium, Teal is intermediate and Brevor is aluminium sensitive.

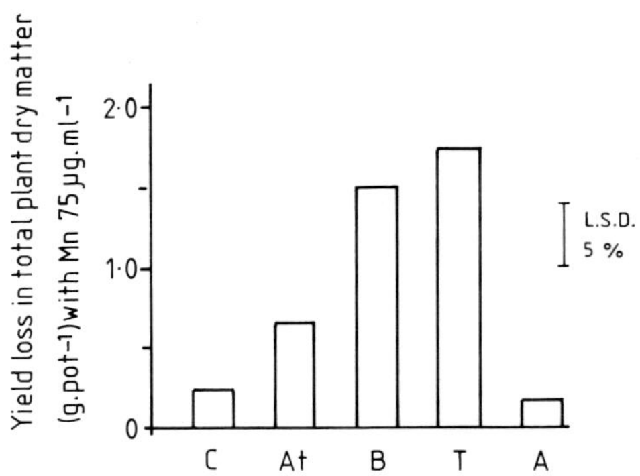

Fig. 6. In the same experiment as described for Figure 5, a treatment with manganese at 75 μg.mL^{-1} was tested. The total dry matter yield losses ($\log_e$ yield at Mn_o — $\log_e$ yield at Mn_{75}) are presented. The wheat cultivars are Carazinho (C), Atlas 66 (At), Alondra (A; Aus 21831), Teal (T) and Brevor (B). Teal was the most manganese-sensitive cultivar (largest yield loss), Brevor was also sensitive, Atlas 66 was intermediate and Carazinho and Alondra were tolerant of the manganese level tested.

between this observation and those of CIMMT. These are that Alondra (Aus 21831) is different to the material tested by CIMMYT, despite the similar name, or that the cultivar has an anomalous reaction with the haematoxylin test.

3. Mutation breeding

Mutation breeding offers the possibility of developing novel properties in existing crop plants. In crops where the desired trait is already available from other sources, mutation breeding offers the prospect of producing a simply modified cultivar (produced maybe by one or two generations of inbreeding) which might then be released very rapidly (Mayo, 1980). However, because of the low frequency of desirable mutants, a very efficient screening system is essential for mutation breeding to be successful.

4. Somaclonal variation

In recent years a new source of variability, somaclonal variation, has become available to breeders. The value of somaclonal variation to plant improvement depends on the ease with which a given plant species can be cultured and plants regenerated from cell lines. Somaclonal variation may find its greatest application for plant improvement in concert with selection for desirable mutations at the cellular level (Nabors, 1976; Scowcroft and Larkin, 1982).

Genotypic differences in aluminium tolerance have been demonstrated in cultures of tomato cells (Meredith, 1978a) and it was possible to select aluminium resistant variants. However, it was not possible to regenerate plants from the cell cultures (Meredith, 1978b). Meredith's experiments show that aluminium tolerance is not exclusively a whole-plant phenomenon even though most of the proposed mechanisms for tolerance operate at the whole-plant level (Section II, A).

5. Recombinant DNA

Recombinant DNA technologies also have the potential for introducing tolerance to acid soils into a cultivar. These techniques require detailed understanding of the genetics and biochemistry of tolerance, as well as a suitable vector for introducing the new DNA sequence into the target species. As yet, nobody has reported the use of this new technology in breeding plants for acid soils.

D. Heritability of tolerance

Breeders will aim, wherever possible, to use major genes in preference to polygenic variation. Selection of traits determined by a few genes of pronounced effect is much more effective in achieving genetic gains than selection for traits determined by many genes with small effects (Mayo, 1980). Rarely is a trait controlled by one class of gene only. More commonly a trait is controlled by a few genes of major effect, a larger number of genes with moderate effects and many genes whose effects are too small to be identified individually.

If selection for a few genes with major effects can give an adequate level of response, then a breeding strategy such as backcrossing can be adopted. This breeding program can be relatively small and the probability of producing the desired cultivar is high. If it is necessary to select for minor genes, large populations are required and the rate of progress and probability of success are lower. When both classes of genes are selected for, it is often advantageous to ensure that the population is homozygous for the major genes before selecting for the minor genes.

1. Aluminium

a Major genes The existence of major genes for tolerance to aluminium has been reported in a number of species. The wheat cultivar, Druchamp, has a major dominant gene compared to the susceptible cultivar, Brevor (Kerridge and Kronstad, 1968). Sensitivity to aluminium in Redcoat and Arthur compared to Seneca and Thorne was conditioned by a single recessive gene (Campbell and Lafever, 1978). Camargo (1981) reported a single dominant gene for tolerance in BH-1146, which was effective at 10 $\mu g.m^{-1}$ of aluminium in solution culture, and two dominant genes in Atlas 66 which were less effective at high levels of aluminium.

There are a number of reports on the location of genes controlling aluminium tolerance in wheat. Chromosome 4D carries a gene controlling aluminium response (Polle *et al.*, 1978). Chinese Spring has genes for aluminium tolerance located on chromosome arms 4Ds and 2Ds (Takagi, 1984). However, Aniol and Gustafson (1984) using Chinese Spring nullisomic-tetrasomic and ditelosomic lines concluded that aluminium tolerance was located on chromosome arms 6A1, 7As, 2D1, 3D1, 4D1, 4B1 and chromosome 7.

Reid (1970) showed that the difference in aluminium tolerance between Dayton and Smooth Awn 86, a winter barley, was controlled by a single dominant gene. Stolen and Anderson (1978) reported that in barley a single dominant gene located on chromosome 4 controlled tolerance to low soil pH. They suggested that the gene may be synonymous with that reported by Reid (1970) for aluminium tolerance.

In maize a single major gene for aluminium tolerance was reported by Rhue *et al.* (1978) and, to account for the observed range in tolerance among inbred lines, they suggested a multiple allelic series at a single locus. Mortinez-Racines (1977) indicated that tolerance in Monolaya and Bluebonnet 50 rice cultivars, was conditioned by two recessive genes. In grain sorghum, Boye-Goni (1982) concluded that the aluminium tolerant trait was simply inherited. In phalaris, reaction to aluminium can be largely explained by a two-gene model, with tolerance requiring at least one dominant allele at each locus (Culvenor *et al.*, 1986b).

b Quantitative inheritance While the existence of these major genes has been noted, most of the authors also indicated that the major genes account for only part of the observed variation. Campbell and Lafever (1978) found that, in wheat, inheritance of aluminium tolerance was more complex than a single gene with incomplete dominance. To achieve a level of tolerance equivalent to the older Ohio cultivars would require selection for the minor genes affecting aluminium tolerance. Heritability estimates reported for Redcoat/Thorne and Redcoat/Seneca were 0.41 and 0.79. However, these estimates would be affected by the segregation of the major gene for susceptibility. Camargo *et al.* (1980) also examined the quantitative inheritance of aluminium tolerance in wheat and reported that the parent-progeny regression estimate for aluminium tolerance was high (b = 0.9). In southern New South Wales, it appears that backcrossing a major gene for aluminium tolerance into a susceptible cultivar may not achieve the desired improvement in tolerance, and selection for minor genes will also be necessary (Fisher and Scott, unpublished data).

Culvenor *et al.* (1986b) obtained heritability estimates between 0.48 and 0.75 for relative root extension after eliminating the sensitive class which lacked one or both of the genes needed for tolerance. However, heritability estimates were low for shoot growth in a field site high in aluminium, possibly due to site variability.

2. *Manganese*

Tolerance to high levels of exchangeable manganese in lettuce was controlled by from one to four genes in different cultivars (Eenink and Garretsen, 1977). In lucerne, Dessureaux (1959) reported that genes controlling manganese tolerance were additive and the heritability of tolerance was high. Brown and Devine (1980) concluded that control of tolerance to excess manganese in soybeans was multigenic rather than controlled by a single locus. Reciprocal differences in progeny suggested that cytoplasmic inheritance also influenced manganese tolerance. Heenan *et al.* (1981) also concluded that minor genes were important in the inheritance of manganese tolerance in soybeans, although it was suggested that a single gene may influence tolerance to high levels of manganese in the cross Bragg x Amredo. In wheat, a few genes appear to account for a substantial proportion of the variation in manganese tolerance, and these can be selected successfully in a backcrossing program (Scott and Fisher, unpublished data).

F. Genetic correlations with acid soil tolerance

There is a fear that plants tolerant of acid soils may be essentially low yielding and, in addition, may be poorly adapted to non-acid soils. The notion that cultivars tolerant of acid soils are low yielding occurred because the traditional cultivars grown in acid soil areas were both acid soil tolerant and low yielding when compared with the modern, high yield potential 'green revolution' cultivars. These newer cultivars had not been selected for acid soil tolerance.

In wheat, the tall Brazilian cultivars were the most tolerant of acid soils (Takagi, 1984). In upland rice, the tall traditional cultivars were, in general, rather tolerant while the new dwarf cultivars were more susceptible (Howeler, 1973). As a result it was assumed that these tall, low-yielding characteristics were likely to be linked with aluminium tolerance. This has not proved to be the case in wheat (Anon., 1980a; Camargo *et al.*, 1980). The genetic correlations between aluminium tolerance and plant height or grain yield were not significant.

In Australia some cultivars of wheat, reasonably tolerant of aluminium, have been bred and selected in areas of non-acid soils e.g. Timgalen and Gamenya from northern New South Wales. Should aluminium tolerance be a disadvantage on a non-acid soil, it would be expected that selection pressure would be towards aluminium sensitive cultivars in non-acid areas. The example given suggests that this is not occurring. Similarly with manganese tolerance in wheat, 'Sunstar', a cultivar with considerable manganese tolerance, has been selected in northern New South Wales in the absence of manganese toxicity (authors' unpublished data).

With lucerne, selections made on an acid soil performed better at all pH levels and not just at low pH. These selections were also more responsive to phosphorus application than other germplasm (Bouton and

Sumner, 1983). In soybeans, inefficiency in absorption or utilisation of iron is independent of tolerance to manganese toxicity (Brown and Devine, 1980)

In aluminium-tolerant plants, yield increases have been obtained with the addition of small amounts of aluminium in solution experiments. This has reinforced the claim that tolerant plants may be low yielding in non-acid situations. These increases have been summarised by Foy (1984), but there is no evidence that aluminium is an essential plant nutrient.

In summary, while there is much speculation on adverse correlations with acid soil tolerance, we conclude that there is little evidence in the literature of such problems.

F. Examples of breeding programs

A joint breeding program was commenced by CIMMYT and some Brazilian research institutes in 1973 to transfer tolerance to aluminium into high-yielding semi-dwarf wheats. The tolerance was located primarily in tall, low-yielding, poor agronomic types of wheat. The selection program used a 'shuttle breeding' technique with field tests for tolerance to acid soils in Brazil, laboratory tests for aluminium tolerance in Mexico and selection for disease resistance and other aspects of agronomic performance in Mexico. This program led to Alondra 4546 being released in Parana State of Brazil and many elite lines are now available in the Aluminium Screening Nursery (Anon. 1978, 1980a, 1980b, 1984).

In addition to our program at Wagga Wagga, breeding for tolerance to acid soils in wheat is also being undertaken in Western Australia (Barclay, 1985). These two Australian breeding programs aim at the release of commercial cultivars.

Germplasm tolerant of acid soils has been registered for a number of crops, for example barley (Composite Cross XXXIV; Reid *et al.*, 1980), crownvetch (*Coronilla varia* L.; Virginia Synthetic; Miller, 1980) and sorghum (GPIR; Duncan, 1981).

REFERENCES

Ali, S. M. E. (1973). Influence of cations on aluminium toxicity in wheat (*Triticum aestivum* Vill Host). Ph. D. thesis, 102 pp. Soils Department, Orgeon State University, Corvallis, Oregon.

Andrew, C. S. (1976). Screening tropical legumes for manganese tolerance. *In* 'Plant Adaptation to Mineral Stress in Problem Soils'. (M. J. Wright and A. S. Ferrari, eds.), pp. 329–340. Proceedings of a Workshop, Beltsville. Cornell University Press, Ithaca, New York.

Andrew, C. S. and Hegarty, M. P. (1969). Comparative responses to manganese excess of eight tropical and four temperate legume species. *Australian Journal of Agricultural Research* **20**, 687–696.

Aniol, A. (1983). Aluminium uptake by roots of two winter wheat varieties of different tolerance to aluminium. *Biochemie und Physioligie der Pflanzen* **178**, 11–20.

Aniol, A. and Gustafson, J. P. (1984). Chromosome location of genes controlling aluminium tolerance in wheat, rye and triticale. *Canadian Journal of Genetics and Cytology* **26**, 701–705.

Aniol, A., Hill, R. D. and Larter, E. N. (1980). Aluminium tolerance of spring rye inbred lines. *Crop Science* **20**, 205–208.

Anon. (1978). 'CIMMYT Report on Wheat Improvement, 1978'. Centro International De Mejoramiento de Maiz Y Trigo, El Batan, Mexico.

Anon. (1980a). 'CIMMYT Report on Wheat Improvement, 1979'. Centro International De Mejoramiento de Maiz Y Trigo, El Batan, Mexico.

Anon. (1980b). 'CIMMYT Report on Wheat Improvement, 1980'. Centro International De Mejoramiento de Maiz Y Trigo, El Batan, Mexico.

Anon. (1984). 'CIMMYT Report on Wheat Improvement, 1981'. Centro International De Mejoramiento de Maiz Y Trigo, El Batan, Mexico.

Aoba, K., Sekiya, K. and Aime, R. (1977). Studies on metallic trace elements in orchards. III. Physiological studies on excess absorption of manganese by satsumas. Bulletin of the Fruit Tree Research Station. Series A. (Hiratsuka) **4**, 1–16. (English summary).

Barclay, I. R. (1985). Breeding wheat varieties for acid soils. *Journal of Agriculture of Western Australia* **25**, 145.

Barton, L. V. and Trelease, S. F. (1927). Stimulation, toxicity and antagonism of calcium nitrate and manganese chloride as indicated by growth of wheat roots. *Bulletin of the Torrey Botanical Club* **54**, 559–577.

Baumgartner, J. G., Haag, H. P., Oliveira, G. D. and Perecin, D. (1976). Tolerance of tomato (*Lycopersicon esulentum* Mill.) cultivars to aluminium and manganese. *Anais da escola superior de agricultura 'Luis de Queiroz'* **33**, 513–541.

Bouton, J. H. and Sumner, M. E. (1983). Alfalfa, *Medicago sativa* L., in highly weathered, acid soils. V. Field performance of alfalfa selected for acid tolerance. *Plant and Soil* **74**, 431–436.

Boye-Goni, S. R. (1982). Combining ability and inheritance of aluminium tolerance in grain sorghum (*Sorghum bicolor* (L.) Moench.) Ph. D. thesis University of Arizona.

Brown, J. C. and Devine, T. E. (1980). Inheritance of tolerance or resistance to manganese toxicity in soybeans. *Agronomy Journal* **72**, 898–904.

Camargo, C. E. O. (1981). Melhoramento do trigo. I. Hereditariedade da tolerância à toxicidade do aluminio. *Bragantia* **40**, 33–45.

Camargo, C. E. O. and Oliveira, O. F. (1981). Tolerancia de cultivares de trigo a diferentes niveis de alumino em solucao nutritava e no solo. *Bragantia* **40**, 21–31.

Camargo, C. E. O., Kronstad, W. E. and Metzger, R. J. (1980). Parent-progeny regression estimates and association of height level with aluminium toxicity and grain yield in wheat. *Crop Science* **20**, 355–358.

Campbell, L. G. and Lafever, H. N. (1976). Correlation of field and nutrient culture techniques of screening wheat for aluminium tolerance. *In* 'Plant Adaptation to Mineral Stress in Problem Soils'. (M. J. Wright and A. S. Ferrari, eds.), pp. 277–286. Proceedings of a Workshop, Beltsville. Cornell University Press, Ithaca, New York.

Campbell, L. G. and Lafever, H. N. (1978). Inheritance of aluminium tolerance in wheat. *Canadian Journal of Genetics and Cytology* **20**, 355–364.

Choudry, M. A. (1980). Ph. D. thesis, Alabama Agricultural and Mechanical University, Normal, Alabama.

Culvenor, R. A., Oram, R. N. and Fazekas de St. Groth, C. (1986a). Variation of tolerance in *Phalaris aquatica* L. and related species to aluminium in nutrient solution and soil. *Australian Journal of Agricultural Research* **37**, 383–395.

Culvenor, R. A., Oram, R. N. and Wood, J. T. (1986b). The inheritance of aluminium tolerance in *Phalaris aquatica* L. *Australian Journal of Agricultural Research* **37**, 397–408.

Da Silva, A. R. (1976). Application of the plant genetic approach to wheat culture in Brazil. *In* 'Plant Adaptation to Mineral Stress in Problem Soils'. (M. J. Wright and A. S. Ferrari, eds.), pp. 223–231. Proceedings of a Workshop, Beltsville, Cornell University Press, Ithaca, New York.

Dessureaux, L. (1959). Heritability of tolerances to manganese toxicity in lucerne. *Euphytica* **8**, 260–265.

Devine, T. E. (1976). Aluminium and manganese toxicities in legumes. *In* 'Plant Adaptation to Mineral Stress in Problem Soils'. (M. J. Wright and A. S. Ferrari, eds.), pp. 65–72. Proceedings of a Workshop, Beltsville. (Cornell University Press, Ithaca, New York.

Doyle, A. D. and Bradley, J. (1982). Lime for cereals on acid soils in northern New South Wales. Proceedings of the Second Australian Agronomy Conference. Wagga Wagga, July 1982.

Duncan, R. R. (1981). Registration of GPIR acid soil tolerant sorghum germplasm population (Reg. No. GP73). *Crop Science* **21**, 637.

Edwards, D. C. and Asher, C. J. (1982). Tolerance of crop and pasture species to manganese toxicity. Proceedings of the 9th International Plant Nutrition Colloquium. pp. 145–150. Warwick, U. K.

Eenink, A. H. and Garretsen, G. (1977). Inheritance of insensitivity of lettuce to a surplus of exchangeable manganese in steam sterilized soils. *Euphytica* **26**, 47–53.

Fisher, J. A. and Scott, B. J. (1983). Breeding wheats for tolerance to acid soils. Proceedings of the Australian Plant Breeding Conference, Adelaide, February 1983, p. 333.

Fleming, A. L. (1983). Ammonium uptake by wheat varieties differing in Al tolerance. *Agronomy Journal* **75**, 726–730.

Foy, C. D. (1974). Effects of aluminium on plant growth. *In* 'The Plant Root and Its Environment'. (E. W. Carson, ed.), pp. 601–642. University of Virginia Press, Charlottesville'.

Foy, C. D. (1976). General principles involved in screening plants for aluminium and manganese tolerance. *In* 'Plant Adaptation to Mineral Stress in Problem Soils'. (M.J. Wright and A.S. Ferrari, eds.), pp. 255–267. Proceedings of a Workshop, Beltsville. Cornell University Press, Ithaca, New York.

Foy, C.D. (1983a). Plant adaptation to mineral stress in problem soils. *Iowa State Journal of Research* **57**, 339–354.

Foy, C.D. (1983b). The physiology of plant adaptation to mineral stress. *Iowa State Journal of Research* **57**, 355–391.

Foy, C.D. (1984). Physiological effects of hydrogen, aluminium and manganese toxicities in acid soils. *In* 'Soil Acidity and Liming'. 2nd edition. (F. Adams, ed.), pp. 57–98. American Society of Agronomy Inc. Monograph, No. 12. Madison, Wisconsin.

Foy, C.D. and Fleming, A. L. (1978). The physiology of plant tolerance to excess available aluminium and manganese in acid soils. *In* 'Crop Tolerance to Suboptimal Land Conditions'. (G. A. Jung, ed.), pp. 301–328. American Society of Agronomy Special Publication, No. 32, Madison, Wisconsin.

Foy, C. D., Armiger, W. H., Briggle, L. W. and Reid, D. A. (1965). Differential aluminium tolerance of wheat and barley varieties in acid soils. *Agronomy Journal* **57**, 413–417.

Foy, C. D. Fleming, A. L. and Armiger, W. H. (1969). Differential tolerance of cotton varieties to excess manganese. *Agronomy Journal* **61**, 690–694.

Foy, C. D., Fleming A. L. and Schwartz, J. W. (1973). Opposite aluminium and manganese tolerances in two wheat varieties. *Agronomy Journal* **65**, 123–126.

Foy, C. D., Orellana, R. G., Schwartz, J. W. and Fleming, A. L. (1974). Responses of sunflower genotypes to aluminium in acid soil and nutrient solutions. *Agronomy Journal* **66**, 293–296.

Foy, C. D., Webb, H. W. and Jones, J. E. (1981). Adaptation of cotton genotypes to an acid, manganese toxic soil. *Agronomy Journal* **73**, 107–111.

Fox, R. L. (1980). Soils with variable charge: agronomic and fertility aspects. *In* 'Soils With Variable Charge (B. K. G. Theng, ed.), pp. 195–224. New Zealand Society of Soil Science, Lower Hutt.

Frey, K. J. (1985). The unifying force in agronomy — biotechnology. *Agronomy Journal* **77**, 187–189.

Furlani, P. P. and Clark, R. B. (1981). Screening sorghum for aluminium tolerance in nutrient solutions. *Agronomy Journal* **73**, 587–594.

Hansom, W. D. and Kamprath, E. J. (1979). Selection for aluminium tolerances in soybeans based on seedling root growth. *Agronomy Journal* **71**, 581–586.

Haug, A. (1984). Molecular aspects of aluminium toxicity. *CRC Critical Reviews in Plant Sciences* **1**, 345–373.

Hech-Buchholz, Ch. (1983). Light and electron microscopic investigations of the reactions of various genotypes to nutritional disorders. *Plant and Soil* **72**, 151–165.

Hecht-Buchholz, Ch. and Foy, C. D. (1981). Effect of aluminium toxicity on root morphology of barley. *Plant and Soil* **63**, 93–95.

Heenan, D. P. and Campbell, L. C. (1981). Soybean nitrate reductase activity influenced by manganese nutrition. *Plant and Cell Physiology* **21**, 731–736.

Heenan, D. P. and Carter, O. G. (1975). Response of two soya bean cultivars to manganese toxicity as affected by pH, and calcium levels. *Australian Journal of Agricultural Research* **26**, 967–974.

Heenan, D. P. and Carter, O. G. (1976). Tolerance of soybean cultivars to manganese toxicity. *Crop Science* **16**, 389–391.

Heenan, D. P. and Carter, O. G. (1977). Influence of temperature on the expression of manganese toxicity in two soybean varieties. *Plant and Soil* **47**, 219–227.

Heenan, D. P., Campbell, L. C. and Carter, O. G. (1981). Inheritance of tolerance to high Mn supply in soybeans. *Crop Science* **21**, 626–627.

Helyar, K. R. (1978). Effects of aluminium and manganese toxicity on legume growth. *In* 'Mineral Nutrition of Legumes in Tropical and Subtropical Soil'. (C. S. Andrew and E. J. Kamprath, eds.), pp. 207–231. CSIRO Melbourne.

Henning, S. J. (1975). Aluminium toxicity in the primary meristem of wheat roots. Ph.D. thesis, Oregon State University, Corvallis, Oregon.

Hewitt, F. J. (1963). The essential nutrient elements: requirements and interactions in plants. *In* 'Inorganic Nutrition of Plants' Plant Physiology Vol. III'. (F. C. Steward, ed), pp. 137–360. Academic Press, London.

Horst, W. J. (1980). Genotypic differences in Mn tolerance of cowpea (*Vigna unguiculata*). *Angewandte Botanik* **54**, 377–392 (English summary).

Horst, W. J. (1980). Quick screening of cowpea genotypes for manganese tolerance during vegetative and reproductive growth. *Zeitschrift für Pflanzenernahrung und Bodenkunde* **145**, 423–435.

Horst, W. J. (1983). Factors responsible for genotypic manganese tolerance in cowpea (*Vigna unguiculata*). *Plant and Soil* **72**, 213–218.

Horst, W. J. and Marschner, H. (1978). Effect of silicon on manganese tolerance of bean plants (*Phaseolus vulgaris L.*) *Plant and Soil* **50**, 287–303.

Howeler, R. H. (1973) Aluminium toxicity in upland rice. *Agronomy Abstracts* **193**.

Howeler, R. H. and Cadavid, L. F. (1976). Screening of rice cultivars for tolerance to Al-toxicity in nutrient solutions as compared with a field screening method. *Agronomy Journal* **68**, 551–555.

Hutton, E. M. (1981). Breeding *Centrosema pubescens* better adapted to the acid soils of South America. *In* 'Proceedings of the XlV International Grassland Congress, Lexington, Kentucky, U.S.A.' (J. A. Smith and V.W. Hays, eds.), Westview Press, Boulder, Colorado.

Hutton, E. M., Williams, W. T. and Andrew, C. S. (1978). Differential tolerance to manganese in introduced and bred lines of *Macroptilium atropurpureum. Australian Journal of Agricultural Research* **29**, 67–79.

Jackson, W. A. (1967) Physiological effects of soil acidity. *In* 'Soil Acidity and Liming'. (R. W. Pearson and F. Adams, eds.) , pp. 43–124. American Society of Agronomy, Madison, Wisconsin.

Jones, L. H. (1961). Aluminium uptake and toxicity in plants. *Plant and Soil* **13**, 293–310.

Kaminska, J. (1981). Tolerance of some barley varieties of substrate acidity. II. Methods for testing resistance of different varieties of barley to aluminium and manganese. *Biuletyn Instytutu Hodowli i Aklimatyzacjii Roslin* **143**, 43–54.

Kang, B. T. and Fox, R. L. (1980). A methodology for evaluating the manganese tolerance of cowpea (*Vigna unguiculata*) and some preliminary results of field trials. *Field Crops Research* **3**, 199–210.

Konzak, C. F., Polle, E. and Kittrick, J. A. (1976). Screening several crops for aluminium tolerance. *In* 'Plant Adaptation to Mineral Stress in Problem Soils'. (M. J. Wright and A. S. Ferrari, eds.), pp. 311-327. Proceedings of a Workshop, Beltsville. Cornell University Press, Ithaca, New York.

Kerridge, P. C. and Kronstad, W. E. (1968). Evidence of genetic resistance to aluminium toxicity in wheat (*Triticum aestivum*) Vill. Host). *Agronomy Journal* **60**, 710–712.

Kerridge, P. C., Dawson, M. D. and Moore, D. P. (1971). Separation of degrees of aluminium tolerance in wheat. *Agronomy Journal* **63**, 586–591.

Lafever, H. N., Campbell, L. G. and Foy, C. D. (1977). Differential response of wheat cultivars to Al. *Agronomy Journal* **69**, 563–568.

Long, F. L., Langdale, G. W. and Myhre, D. L. (1973). Response of an Al-tolerant and an Al-sensitive genotype to lime, P, and K on three Atlantic coast flatwoods soils. *Agronomy Journal* **65**, 30–34.

Lopez-Benitez, A. (1977). Influence of aluminium toxicity in intergeneric crosses of wheat and rye. Ph. D. Dissertation. Oregon State University, Corvallis.

MacLeod , L. B. and Jackson, L. P. (1967). Aluminium tolerances of two barley varieties in nutrient solution, peat, and soil culture. *Agronomy Journal* **59**, 359–363.

McNeilly, T. (1982). A rapid method for screening barley for aluminium tolerance. *Euphytica* **31**, 237–239.

Martini, J. A., Kochhann, R. A., Gomes, E. P. and Langer, F. (1977). Response of wheat cultivars to liming in some high Al oxisols of Rio Grande do Sul, Brazil. *Agronomy Journal* **69**, 612–616.

Mayo, 0. (1980). 'Theory of Plant Breeding'. Clarendon Press, Oxford.

Memon, A. R., Chino, M., Takeoka, Y., Hora, K. and Yatazawa, M. (1980). Distribution of manganese in leaf tissues of manganese accumulator: *Acanthopanax sciadophylloides* as revealed by electronprobe X-ray microanalyzer. *Journal of Plant Nutrition* **2**, 457–476.

Meredith, C. P. (1978a). Response of cultured tomato cells to aluminium. *Plant Science Letters* **12**, 17–24.

Meredith, C. P. (1978b). Selection and characterization of aluminium resistant variants from tomato cell cultures. *Plant Science Letters* **12**, 25–34.

Miller, J. D. (1980). Registration of Virginia Synthetic A, crownvetch germplasm (Reg. No. GP29). *Crop science* **20**, 418.

Moore, D. P., Kronstad, W. E. and Metzger, R. J. (1976). Screening wheat for aluminium tolerance. *In* 'Plant Adaptation to Mineral Stress in Problem Soils' (M. J. Wright and A. S. Ferrari, eds.) , pp. 287–295. Proceedings of a Workshop, Beltsville. Cornell University Press, Ithaca, New York.

Morris, H. D. and Pierre, W. H. (1949). Minimum concentrations of manganese necessary for injury to various legumes in culture solutions. *Agronomy Journal* **41**, 107–112.

Mortinez-Racines, C. P. (1977). Aluminium toxicity studies in rice (Oryza sativa L.). Ph. D. thesis, Oregon State University.

Mugwira, L. M., Elgawhary, S. M. and Patel, S. U. (1978). Differential tolerance in triticale, wheat and rye as measured by root growth characteristics and aluminium concentrations. *Plant and Soil* **50**, 683–693.

Mugwira, L. M., Floyd, M. and Patel, S. U. (1981). Tolerances of triticale lines to manganese in soil and nutrient solution. *Agronomy Journal* **73**, 119–322.

Nabors, N. W. (1976). The use of plant tissue cultures to produce altered varieties of agriculturally useful plants. *In* 'Plant Adaptation to Mineral Stress in Problem Soils'. (M. J. Wright and A. S. Ferrari, eds.), pp. 369–372. Proceedings of a Workshop, Beltsville. Cornell University Press, Ithaca, New York.

Naidoo, G. (1976). Aluminium toxicity in two snapbean varieties. Ph. D. thesis. University of Tennessee, Knoxville.

Neenan, M. (1960). Effects of soil acidity on the growth of cereals with particular reference to differential reaction of varieties thereto. *Plant and Soil* **12**, 324–338.

Nelson, L. E. (1982). The effect of temperature regime and substrate Mn on growth and manganese concentrations in rice. *Journal of Plant Nutrition* **5**, 1241–1257.

Nelson, L. E. (1983). Tolerances of 20 rice cultivars to excess Al and Mn. *Agronomy Journal* **75**, 134–138.

Osawa, T. and Ikeda, H. (1980). Heavy metal toxicities in vegetable crops. VIII. Effects of nitrogen form supplied and pH levels of the nutrient solution on manganese toxicities in vegetable crops. *Journal of the Japanese Society of Horticultural Science* **49**, 197–202.

Ouellette, G. J. and Dessureaux, L. (1958). Chemical composition of alfalfa as related to degree of tolerance to manganese and aluminium. *Canadian Journal of Plant Science* **38**, 206–214.

Polle, E. A., Konzak, A. F. and Kittrick, J. A. (1978). Visual detection of aluminium tolerance levels in wheat by haematoxylin staining of seedling roots. *Crop Science* **18**, 823–827.

Reid, D. A. (1970). Genetic control of reaction to aluminium in winter barley. *In* 'Barley Genetics II: Proceedings of the 2nd International Barley Genetics Symposium'. (R. A. Nilan, ed.), pp. 409–413. Washington State University Press, Pullman.

Reid, D. A. (1976). Screening barley for aluminium tolerance. *In* 'Plant Adaptation to Mineral Stress in Problem Soils'. (M. J. Wright and A. S. Ferrari, eds.) pp. 269–276. Proceedings of a Workshop, Beltsville. Cornell University Press, Ithaca, New York.

Reid, D. A., Jones, G. D., Armiger, W. H., Foy, C. D., Koch, E. J. and Starling, T. M. (1969). Differential aluminium tolerance of winter barley varieties and selections in associated greenhouse and field experiments. *Agronomy Journal* **61**, 218–222.

Reid, D. A., Fleming, A. L. and Foy, C. D. (1971). A method for determining aluminium response of barley in nutrient solution in comparison to response in Al-toxic soil. *Agronomy Journal* **63**, 600–603.

Reid, D. A., Slootmaker, L. A. and Craddock, J. C. (1980). Registration of barley Composite Cross XXXIV. *Crop Science* **20**, 416–417.

Rhue, R. D. (1976). The time-concentration interaction of Al toxicity in wheat meristems. Ph. D. thesis. Oregon State University, Corvallis.

Rhue, R. D. and Grogan, C. O. (1977). Screening corn for Al tolerance using different Ca and Mg concentrations. *Agronomy Journal* **69**, 755–760.

Rhue, R. D., Grogan, C.O., Stockmeyer, E. W. and Everett, H. L. (1978) Genetic control of aluminium tolerance in corn. *Crop Science* **18**, 1063–1067.

Robson, A. D. and Loneragan, J. F. (1970). Sensitivity of annual Medicago species to manganese toxicity as affected by calcium and pH. *Australian Journal of Agricultural Research* **21**, 223–232.

Rufty, T. W., Miner, G. S. and Raper, C. D. (1979). Temperature effects on growth and manganese tolerance in tobacco. *Agronomy Journal* **71**, 638–644.

Sanchez, P. A. and Salinas, J. G. (1982). Low input technology for managing oxisols and ultisols in tropical America. *In* 'Advances in Agronomy 34' (N. C. Brady, ed.), pp. 280–406. American Society of Agronomy, Madison, Wisconsin.

Sartain, J. B. and Kamprath, E. J. (1978). Aluminium tolerance of soybean cultivars based on root elongation in solution culture compared with growth in acid soil. *Agronomy Journal* **70**, 17–20.

Schumacher, T. E., Smucker, A. J. M., Eshel, A. and Eurry, R. B. (1983). Measurement of short-term root growth by prestaining with neutral red. *Crop Science* **25**, 1212–1214.

Scott, B. J. (1982). Responses to lime by cereals. Proceedings of the Second Australian Agronomy Conference, Wagga Wagga July, 1982 p. 260.

Scott, B. J., Burke D. G. and Bostrom, T. E. (1987). Australian research on tolerance to toxic manganese. *In* 'Genetic aspects of plant Mineral nutrition' (H. W. Gabelman and B. C. Loughman, eds.), pp. 153–163. Martinus Nijhoff, Dordrecht.

Scowcroft, W. R. and Larkin, P. J. (1982). Somaclonal variation: a new option for plant improvement. *In* 'Plant Improvement and Somatic Cell Genetics' (I. K. Vasil, W. R. Scowcroft and K. J. Frey, eds.), pp. 159–175. Academic Press, New York.

Shuman, L. M. and Anderson, O. E. (1976). Interactions of manganese with other ions in wheat and soybeans. *Communications in Soil Science and Plant Analysis* **7**, 547–557.

Sirkar, S. and Amin, J. V. (1974). The manganese toxicity of cotton. *Plant Physiology* **54**, 539–543.

Sivasubramaniam, S. and Talibudeen, O. (1972). Effects of aluminium on the growth of tea (*Camellia sinensis*) and its uptake of potassium and phosphorus. *Tea* **43**, 4–13.

Slootmaker, L. A. J. (1974). Tolerance to high soil acidity in wheat related species, rye and triticale. *Euphytiaca* **23**, 505–513.

Smith, F. W. (1979). Tolerance of seven tropical pasture grasses to excess manganese. *Communications in Soil Science and Plant Analysis* **10**, 853–867.

Stolen, O. and Anderson, S. (1978). Inheritance of tolerance to low soil pH in barley. *Hereditas* **88**, 101–105.

Takagi, H. (1984). Exploration of aluminium tolerance genes in wheat. *In* 'Proceedings of the 6th International Wheat Genetics Symposium' (S. Sakamoto, ed.), pp. 143–146. Kyoto University, Kyoto.

Temple-Smith, M. G. and Koen, T. B. (1982). Comparative response of poppy (*Papaver somniferum* L.) and eight crop and vegetable species to manganese excess in solution culture. *Journal of Plant Nutrition* **5**, 1153–1169.

Vlamis, J. and Williams, D. E. (1967). Manganese and silicon interactions in the Gramineae. *Plant and Soil* **28**, 131–140.

Vlamis, J. and Williams, D. E. (1973). Manganese toxicity and marginal chlorosis of lettuce. *Plant and Soil* **39**, 245–251.

Vose, P. N. (1981). Crops for all conditions. *New Scientist* **89**, 688–690.

Vose, P. R. (1983). Rationale of selection for specific nutritional characters in crop improvement with *Phaseolus vulgaris* L. as a case study. *Plant and Soil* **72**, 351–364.

Vose, P. B. and Randall, P. J. (1962). Resistance to aluminium and manganese toxicities in plants related to variety and cation exchange capacity. *Nature* **196**, 85–86.

Williams, D. E. and Vlamis, J. (1957). The effect of silicon on yield and manganese-54 uptake and distribution in the leaves of barley plants grown in solution culture. *Plant Physiology* **32**, 404–409.

Wright, M. J. and Ferrari, A. S. (1976). 'Plant Adaptation to Mineral Stress in Problem Soils'. Proceedings of a Workshop, Beltsville. Cornell University Press, Ithaca, New York.

6

Amelioration of Soil Acidity by Liming and other Amendments

P. D. CREGAN, J. R. HIRTH and M. K. CONYERS

SOIL ACIDITY AND PLANT GROWTH
ISBN 0 12 590655 2

I. USE OF LIME IN AUSTRALIAN AGRICULTURE

> 'Brought home a load of lime for use in the building of an extension to the mansion. It cost £20 a ton so that it is too expensive to use on the land.'
>
> Diary of a Welsh Swagman 1869–1894 (Evans, 1977)

The area of strongly acid soils (pH < 5.0 (1:5 soil:0.01 M $CaCl_2$) in Australia is estimated to be in excess of 15 m ha, although few objective estimates of the extent and severity of the problem are available on a State or even regional basis. The area affected comprises much of the productive agricultural land in southern Australia. In addition there are substantial areas that are acidifying under current agricultural practice and that will, in time, fall below pH 5.0 (1:5 soil:0.01 M $CaCl_2$).

Given the severity of the problem, the use of lime in Australia for agricultural purposes is low by world standards, being estimated at less than 200 000 $t.year^{-1}$. This contrasts with an annual lime use in New Zealand of 1.5×10^6 $t.year^{-1}$ (D. C. Edmeades pers. comm.) and in the USA of c.30×10^6 $t.year^{-1}$ (Barber, 1984). In New South Wales alone, the estimated 7.5 m ha of low pH soils (pH < 5.5 (1:5 soil:water) would require about 2 m $t.year^{-1}$ to maintain soil pH.

In contrast to Europe and North America, land prices and production per hectare in Australia are low and freight costs are high. In many areas of eastern Australia sufficient lime to raise pH to 5.5 (1:5 soil:0.01 M $CaCl_2$) would represent 30 to 50 per cent of the land value. Consequently the Australian approach has been to develop techniques to minimise the need for lime; early success with lime pelleting inoculated seed and application of molybdenum and lime/superphosphate banding mitigated against more widespread use of lime.

The extensive, low input nature of Australian agriculture, the accelerated acidification resulting from the widespread legume-ley farming systems practised in southern Australia (see Chapter 2) and the comparatively high cost of lime thus pose unique technical and economic challenges for Australia.

Liming materials are used widely to raise soil pH levels to those desirable for plant growth. However, a better understanding of the acidification processes and their relative importance may allow modifications of farming systems to reduce the rate of acidification. Such management options could include the manipulation of fertilizer type, crop and pasture species, crop sowing time, crop varieties and stock management. Their success will be determined by the importance of the acidification process being manipulated (see Chapter 2 for details), the extent to which it can be modified and the cost of these practices compared with liming. Eventually an integrated approach involving liming, management changes and plant tolerance will probably be necessary, particularly where the acidification potential is high and acidification has extended into the subsoil.

II. SUMMARY OF RESPONSES TO LIMING IN AUSTRALIA

A century ago lime was considered too expensive for use in Australian agriculture, costing the equivalent of seven months' wages for an itinerant labourer per imperial ton (Evans, 1977). Nevertheless, in the early part of this century, lime was included along with superphosphate in 'manurial' trials established by the various State Departments of Agriculture. Presumably by then, lime had become an affordable commodity for agricultural use. Also, its inclusion in these field trials is consistent with the traditions of early Australian settlers who had come from regions where regular liming was a long-established practice. However, much of the early evaluation of liming Australian soils lacked clear objectivity and critical evaluation. Lime was often applied only to pasture and crop plants tolerant of acid soils, such as oats, subterranean clover, ryegrass and potatoes. Few measurements of soil pH were made and many of the early responses were recorded visually, without supporting data on plant yield. Thus response to liming often varied from none to substantial, depending on species and soil pH, with established pastures of subterranean clover frequently giving no worthwhile result (Tiver, 1957; Leeper, 1964). Perhaps for these reasons, together with the discovery that small amounts of banded lime were very successful in establishing subterranean clover on newly-cleared acid soils, a body of opinion developed against the use of lime on Australian soils (Ellington, 1984). Now with the recognition of problems in plant production associated with accelerated soil acidification and legume-based agriculture (Coventry, 1985; Porter and Wilson, 1984), the use of lime on crops and pastures is being closely scrutinised again.

A. Improved plant performance with lime

In this section, we have endeavoured to summarise the many lime responses recorded on Australian soils. We have recorded the responses in terms of the dominant mechanism(s) thought to contribute to the improved plant performances (Tables I–IX). This approach leaves a number of reports that simply recorded an observed or measured response; these references have been included in a general section (Table X). For completeness, Tables XI and XII list the recorded instances where lime depressed or did not improve plant growth.

Table I. Lime responses in the field due to improved legume nodulation.

Legume	Type[a]	Lime Rate[b]	Method[c]	Comment	Reference
Subterranean clover (*Trifolium subterraneum*)	C	L,M	D	Some lime necessary for DM[d] response to Mo (Expt. 3) (= improved nodulation?)	Anderson and Moye (1952)
	C	?	R	5 sites: improved nodule number (all sites) in 2nd year	Anderson and Spencer (1948)
	C	L,M	D,B	7 sites: 2 sites DM responses to both broadcast and drilled lime (= nodulation response?)	Drake and Kehoe (1954a)
	C	L	D	Visual improvement in nodulation.	Drake and Kehoe (1954b)
	C	H	I (10)	Nodulated plants only in limed soil (no data).	Fitzpatrick (1958)
	C	L	D	Responses in establishment year only	Fitzpatrick (1958)
	C?	L,M,H	B or I (10)	19 sites on established pasture: DM responses at 8 sites to low lime rates (= nodulation response?)	Glencross and Clarke (1984)
	C	M,H	I (5)	% nodulated plants increased; improved nodulation and N_2-fixation.	Helyar and Anderson (1970a)
	C	M,H	I (10)	Nodule number increased with lime rate (no data).	Hirth *et al.* (1982)
	C	M	R	2 sites: marked improvement in nodule number.	McLachlan (1953)
	C	M,H	B (seeding)	1968 site: soil conditions more favourable for nodulation?	Ozanne and Howes (1974)
	C	H	B (seeding)	57 sites: nodulation failure without lime at 34 sites (mixed legume pasture).	Paton (1958)
	C?	L	D	Lime improved nodulation (no data).	Tiver (1958)
	C	L,M,H	B or I (10)	53 established pastures; most DM responses (28/53) due to improved nodulation.	Yeates *et al.* (1984)

Lucerne (*Medicago sativa*)	C	M,H	I (5)	Improved nodulation.	Helyar and Anderson (1970a)
	C	?	?	Improved N_2-fixation (= improved nodulation or Mo response?).	Helyar and Anderson (1970b)
	C?	L	D	General comment of improved nodulation.	Tiver (1958)
White clover (*T. repens*)	C	M,H	I (5)	Improved nodulation.	Helyar and Anderson (1970a)
	C	H	B (seeding)	57 sites: nodulation failure without lime at 34 sites (mixed legume pasture).	Paton (1958)
	C	H	B	Nodulation suggested to explain part of DM response as soil Al and Mn not high	White (1967)
Red clover (*T. pratense*)	C	M,H	?	DM yield markedly improved by 750 kg.ha^{-1} lime (= nodulation response?)	Anon. (1964)
Burr medic (*M. polymorpha*)	C	H	?	DM yield response due to improved nodulation in known sensitive spp.?	Glencross and Clarke (1984)
Glycine (*Neonotonia wightii*)	C	H	B	3 sites: % plants nodulated increased (note: DM response at 1/3 sites only).	Diatloff and Luck (1972)
Desmodium (*Desmodium uncinatum*)	C	L,M,H	I	Expt. 1: early nodulation improved. Differences had disappeared at second harvest.	Andrew and Bryan (1958)

[a] Type of lime: C = $CaCO_3$
[b] Rate of lime (kg.ha^{-1}): L = < 500, M = 500–2000, H > 2000
[c] Method of application: B = broadcast on surface, I = incorporated (depth in cm), D = drilled in with seed, R = raked in by hand
[d] DM = dry matter

Table II. Lime responses in pots due to improved legume nodulation.

Legume	Type[a]	Lime Rate[b]	Method[c]	Comment	Reference
Subterranean clover (*Trifolium subterraneum*)	C	pH 6.0 (1:5 water)	M	33 soils: marked improvement in nodulation (visual) on 15 soils (all pH < 5.2 1:5 water)	Bromfield *et al.* (1983)
	C	0.66	M	Improved number of non-crown and total nodules/plant.	Cordero and Blair (1978)
	C	0–5.4 (4)	S	Nodule number increased.	McLachlan (1980b)
	C	0–17.7 (5)	M (2.5)	Nodulation failure without lime.	Paton and Loneragan (1960)
	H	0–8.3 (8)	M	Nodule number increased.	Spencer (1950)
Lucerne	H	1.15	M (1.2)	Improved nodule number.	Anderson and Oertel (1946)
(*Medicago sativa*)	C	1.01	M	Lime increased herbage N% on 12/75 soils (= nodulation response?)	Haddad and Kaldor (1982)
	C	0–1.5 (4)	M	% plants nodulated improved on moderately acid soils (pH 5.5–6.0 1:5 water).	Munns (1965)
	C	0–3.2 (8)	M	Once Ca deficiency overcome, nodulation increased.	Andrew and Norris (1961)
White clover	C?	1.6	M	Nodulation enhaned on 3 soils.	Kerridge *et al.* (1972)
(*T. repens*)	C	0–3.2 (8)	M	Once Ca deficiency overcome, nodulation increased.	Andrew and Norris (1961)
Strawberry clover (*T. fragiferum*)	C	0–3.2 (8)	M	Once Ca deficiency overcome, nodulation increased.	Andrew and Norris (1961)
Barrel medic (*M. truncatula*)	C	0–3.2 (8)	M	Once Ca deficiency overcome, nodulation increased.	Andrew and Norris (1961)
Cluster clover (*T. glomeratum*)	C	0.66	M	Number of non-crown and total nodules/plant improved.	Cordero and Blair (1978)
Yellow serradella (*Ornithopus compressus*)	C	0.66	M		

Glycine (*Neonotonia wightii*)	C	5^d	M (?)	Increased nodule number dominant effect.	Lee and Wilson (1972)
Lotononis (*Lotononis bainesii*)	C	5^d	M (?)	Increased nodule number dominant effect.	Lee and Wilson (1972)
Desmodium (*Desmodium uncinatum*)	C	0–3.2 (8)	M	Nodule number and total nodule mass increased.	Andrew and Norris (1961)
Indigofera (*Indigofera spicata*)	C	0–3.2 (8)	M	Total nodule mass increased by first lime increment.	Andrew and Norris (1961)
Centro (*Centrosema pubescens*)	C	0–3.2 (8)	M	Nodule number and mass increased.	Andrew and Norris (1961)
Stylo (*Stylosanthes guianensis*)	C	0–3.2 (8)	M	Small increases in nodule mass with each lime increment.	Andrew and Norris (1961)
	C	0–1.4	M	Nodulation severely depressed in unlimed Coolum soil.	de Carvalho *et al.* (1980)
Caribbean stylo (*S. hamata*)	C	0–1.4	M	Nodulation poorer on unlimed, low phosphorus Kogan soil for *S. scabra*, *S. viscosa* and *S. fruiticosa*.	de Carvalho *et al.* (1980)
Townsville stylo (*S. humilis*)	C	0–1.4	M		
S. scabra	C	0–1.4	M		
S. viscosa	C	0–1.4	M		
S. fruiticosa	C	0–1.4	M		
Phasey Bean (*Macroptilium lathyroides*)	C	0–3.2 (8)	M	Large increase in nodule mass with first lime increment.	Andrew and Norris (1961)

[a] Type of lime: C = $CaCO_3$, H = $Ca(OH)_2$

[b] Rate of lime: Ranges of rates in $g.pot^{-1}$ (no. of rates)

[c] Method of application: M = mixed throughout (pots), M(x) = mixed to x cm, S = surface application

[d] Rates given in $kg.ha^{-1}$ equivalent on surface area basis.

Table III. Lime responses due to a release of plant available molybdenum.

Legume	Type[a]	Lime Rate[b]	Method[c]	Comment	Reference
Field responses					
Subterranean clover (*Trifolium subterraneum*)	C,H	L,M	B,D	Provided nodulation satisfactory, Mo substituted for heavy rates of lime.	Anderson and Moye (1952)
	C	M	B	Lime and Mo both improved establishment and growth at 6/15 sites.	Drake and Kehoe (1954a)
	–	–	–	General comment that heavy dressings of lime released previously unavailable Mo.	Simmons (1962)
Subterranean/white clover (*T. subterraneum* and *T. repens*)	C	H	B	Established pastures: main effect of heavy lime is to increase Mo supply.	Paton (1956, 1960)
	C	H	B	Lime and Mo produced equal growth responses at 7/57 sites on established pastures	Paton (1958)
White clover (*T. repens*)	C	H	B	65% of lime response due to Mo.	White (1967)
Lucerne (*Medicago sativa*)	C	?	?	Lime improved N_2-fixation; no basal Mo [= Mo response?]	Helyar and Anderson (1970b)
Pot responses					
White clover (*T. repens*)	C	1.6	M	Negative interaction between lime × Mo on 3/4 basalts (Expt. 2).	Kerridge *et al.* (1972)
	H	1.8	M	2/4 lime responses completely, and 1/4 partially, substituted by Mo.	Kerridge and Everett (1975)
	C	0–12.8	M	Mo completely substituted lime response (Expt. 3)	White (1967)
Subterranean clover (*T. subterraneum*)	H	1.15	M (1.0)	Negative interaction between lime × Mo (Expts. 4 and 6).	Anderson and Oertel (1946)
	C	?	M	Lime response replaced by Mo.	Oertel *et al.* (1946)
Desmodium (*Desmodium intortum*)	C,H	1.6,2,4	M	Negative interaction between lime × Mo on 9/16 soils (Expts. 2, 4, 6, 7b and 8).	Kerridge *et al.* (1972)

[a] Type of lime: C = $CaCO_3$, H = $Ca(OH)_2$
[b] Rate of lime: FIELD in $kg.ha^{-1}$ L = < 500, M = 500–2000, H = > 2000; POTS in $g.pot^{-1}$
[c] Method of application: B = broadcast, D = drilled in with seed, M = mixed throughout (pots), M(x) = mixed to x cm depth (pots)

Table IV. Lime responses due to overcoming a calcium deficiency of the plant.

Plant	Soil calcium meq.100g^{-1}	Soil calcium % TEC[a]	Type[b]	Lime Rate[c]	Method[d]	Comments	Reference
Field responses							
White clover (*Trifolium repens*)	1.6	21	C	L,M,H	I	$MgCO_3$ exacerbated Ca deficiency.	Andrew and Bryan (1958)
Subterranean clover (*T. subterraneum*)	1.0	?	C	M	B (seeding)	1967 site: increased seed yield by overcoming Ca deficiency.	Ozanne and Howes (1974)
Pot responses							
White clover (*T. repens*)	0.1	3	C	0–3.2 (8)	M	Ca deficiency symptoms evident on unlimed plants.	Andrew and Norris (1961)
Strawberry clover (*T. fragiferum*)	0.1	3	C	0–3.2 (8)	M		
Lucerne (*Medicago sativa*)	0.1	3	C	0–3.2 (8)	M		
Barrel medic (*M. truncatula*)	0.1	3	C	0–3.2 (8)	M		
Stylo (*Stylosanthes guianensis*)	0.09	?	C	0–0.17 (4)	M	DM[f] responses due to supply of Ca, not pH changes.	Probert (1980)

Cont'd on pg 214

Table IV. *cont'd*

Plant	Soil calcium meq.100g^{-1}	% TEC[a]	Type[b]	Lime Rate[c]	Method[d]	Comment	Reference
Stylo (*S. guianensis*)	0.19	5	C	0–0.67 (5)	M	DM increase to low rates of lime only (Expt. 1). Leaf Ca % suggests Ca deficiency.	Probert *et al.* (1979)
Townsville stylo (*S. humilis*)	0.06–0.19	1–5	C	0.2	M	Ca deficiency overcome by very low rate of lime (10/11 soils).	Isbell *et al.* (1972)
Glycine (*Neonotonia wightii*)	0.88	10	C	L,H[e]	M (?)	Ca effect on glycine DM suggested.	Lee and Wilson (1972)
Rye (*Cereale secale*)	?	?	C	3.0	M	Suggested Ca deficiency (soil 2)	Lipsett (1980)
Barley (*Hordeum vulgare*)	?	?	C	3.0	M	Suggested Ca deficiency (soil 2)	Lipsett (1980)

[a] % TEC = percentage of total acidic cation exchange capacity
[b] Type of lime: C = $CaCO_3$
[c] Rate of lime: FIELD in kg.ha^{-1} L = < 500, M = 500–2000, H = > 2000; POTS in g.pot^{-1} (no. of rates)
[d] Method of application: I = incorporated, B = broadcast, M = mixed throughout (pots)
[e] Rate expressed in kg.ha^{-1} equivalents only
[f] DM = dry matter

Table V. Lime responses and improved phosphorus nutrition.

Plant	Type[a]	Lime Rate[b]	Method[c]	Comments	Reference
Field response					
Subterranean clover (*Trifolium subterraneum*)	C	H	R	Lime reduced leaching of P added to an acid sand (Expt. 2).	Paton and Loneragan (1960)
Pot response					
White clover (*T. repens*)	C	0–10.0	M	4/6 responses in absence of added P attributed to increased availability of soil P.	Holford (1985)

[a] Type of lime: C = $CaCO_3$
[b] Rate of lime: FIELD H = > 2000 $kg.ha^{-1}$; POTS in $g.pot^{-1}$
[c] Method of application: R = raked in by hand, M = mixed throughout (pots)

Table VI. Lime responses due to overcoming multiple acidity factors.

Plant	Type[a]	Lime Rate[b]	Method[c]	Comments	Reference
Field responses					
Subterranean clover (*Trifolium subterraneum*)	C	L,M	B	3 sites that responded to lime but not Mo (= overcoming toxic acidity factors?)	Drake and Kehoe (1954a)
Lucerne (*Medicago sativa*)	C	?	I	Al and Mn toxicity of seedlings overcome.	Mahoney *et al.* (1981)
Glycine (*Neonotonia wightii*)	–	–	–	Lime responses often reported because of glycine's sensitivity to Al and Mn.	Anon (1967)
Wheat (*Triticum aestivum*)	C	M,H	I	Increased wheat grain yields associated with reduced Al and Mn levels	Reeves *et al.* (1982)
Annual grasses (*Hordeum leporinum*,) *Lolium rigidum*, *Vulpia* spp.)	C	H	B	Grass composition changed: barley grass encouraged, ryegrass and silver grass suppressed. Differential tolerance to high levels of Al and Mn?	McGowan (1967)
Pot responses					
Lucerne (*M. sativa*)	C	0–36.9	M	Of 31/40 DM[d] responses, maximum of 11 could be due to multiple acidity factors.	Jones (1975)
	C	1.01	M	DM responses on coarse-textured soils due in part to eliminating toxic Al/Mn.	Haddad and Kaldor (1982)
	C	pH 6.5 (1:5 water)	M	Both Al and Mn toxicities overcome to increase top and root growth in surface and subsoil	Pinkerton and Simpson (1982)
	C	3.0	M	Undefined toxicity factors involved in light-textured soil (site no. 3).	Lipsett (1980)

Subterranean clover (*T. subterraneum*)	C	0–36.9	M	Of 27/40 DM responses, 2 could be due to multiple acidity factors.	Jones (1975)
	C	0.66	M	Reduced acidity factors resulted in improved nodulation.	Cordero and Blair (1978)
Cluster clover (*T. glomeratum*)	C	0.66	M	Reduced acidity factors resulted in improved nodulation.	Cordero and Blair (1978)
Yellow serradella (*Ornithopus compressus*)	C	0.66	M	Reduced acidity factors resulted in improved nodulation.	Cordero and Blair (1978)
Safflower (*Carthamus tinctorius*)	C	3.0	M	Undefined toxicity factors involved on light-texture soil (site no. 3)	Lipsett (1980)
Linseed (*Linum usitatissimum*)	C	3.0	M	Undefined toxicity factors involved on light-texture soil (site no. 3)	Lipsett (1980)

[a] Type of lime: C = $CaCO_3$
[b] Rate of lime: FIELD in $kg.ha^{-1}$ L = < 500, M = 500–2000, H = > 2000; POTS in $g.pot^{-1}$
[c] Method of application: B = broadcast, I = incorporated, M = mixed throughout (pots)
[d] DM = dry matter

Table VII. Lime responses due to overcoming aluminium toxicity.

Plant	Type[a]	Lime Rate[b]	Method[c]	Comments	Reference
Field responses					
Subterranean clover (*Trifolium subterraneum*)	C	H	B	Root growth improved by heavy rates of lime (+ overcoming Al toxicity?)	Fitzpatrick (1958)
Lucerne (*Medicago sativa*)	C	H	B	Negative interaction between lime × superphosphate indicative of Al toxicity limiting growth.	Munns and Rudczuk (1966)
Perennial ryegrass (*Lolium perenne*)	C	M,H	I (5)	DM[d] increased in presence of supplied N (= overcoming Al toxicity?)	Helyar and Anderson (1970a)
Phalaris (*Phalaris aquatica*)	C	M,H	I (5)		
Wheat (*Triticum aestivum*)	C	?	I	Wheat grain yields improved on high Al soil.	Scott (1982)
Poppy (*Papaver somniferum*)	C	H	I	Poppy yield increased primarily by alleviating Al toxicity.	Temple-Smith *et al.* (1983)
Mixed pasture (*M. sativa*, *T. repens*, (*T. subterraneum*, *Lolium perenne*)	C	H	I (10)	Total DM increased 24/26 harvests (7 sites). Lucerne % in sward increased markedly and soil Al reduced to < 1% of CEC[d] at all sites.	Vimpany and Bradley (1979)
Pot Responses					
Lucerne (*M. sativa*)	C	0–4.34	M	Response to first increment of lime in presence of supplied N (= overcoming Al toxicity?)	Helyar and Anderson (1971)
	C	0–3.0	M	Negative interaction between lime × P on 3 soils suggests Al toxicity (Expt. 4).	Munns (1965)

	C	0–8.5	M	Toxic Al effects on root growth overcome in acidic subsoil.	Simpson *et al.* (1979)
	C	0–36.9	M	Of 31/40 responsive soils, 22 attributed to reducing Al_{ex}.	Jones (1975)
Glycine (*Neonotonia wightii*)	C	0.4,4.2	M	29% response in shoot growth to subsoil liming (Al sensitive spp.?)	Simpson and Pinkerton (1978)
Subterranean clover (*T. subterraneum*)	C	11.2	M	Root growth markedly increased and P deficiency symptoms overcome on 6/10 soils (Expt. 3) (= overcoming Al toxicity?)	Kehoe and Curnow (1963)
	C	0–36.9	M	Of 27/40 responsive soils, 19 attributed to reducing Al_{ex}	Jones (1975)
White clover (*T. repens*)	C	0–10.0	M	Lime overcame toxic effects on 1/15 soils.	Holford (1985)
Phalaris (*Phalaris aquatica*)	C	0–4.34	M	DM response to first level of lime in presence of supplied N (= overcoming Al toxicity?)	Helyar and Anderson (1971)
Stylos (*Stylosanthes humilis, S. guianensis, S. hamata, S. viscosa, S. fruiticosa*)	C	0–6.8	M	DM response of all 6 species due to overcoming Al toxicity. On Coolum and Kogan soils, lime overcame failed or restricted nodulation caused by Al toxicity.	de Carvalho *et al.* (1980)

[a] Type of lime: C = $CaCO_3$

[b] Rate of lime: FIELD in $kg.ha^{-1}$ M = 500–2000, H = > 2000; POTS in $g.pot^{-1}$

[c] Method of application: B = broadcast, I = incorporated (depth in cm), M = mixed throughout (pots)

[d] DM = dry matter, CEC = cation exchange capacity

[e] Al_{ex} = exchangeable aluminium

Table VIII. Lime responses and manganese toxicity.

Plant	Type[a]	Lime Rate[b]	Method[c]	Comments	Reference
Field responses					
Glycine (*Neonotonia wightii*)	C	H	B	Foliar chlorosis attributed to Mn excess, disappeared with liming (minor effect).	Diatloff and Luck (1972)
French bean (*Phaseolus vulgaris*)	C	H	I?	Severe Mn toxicity in winter crops eliminated by intermediate rates.	Siman *et al.* (1971)
Pot responses					
Subterranean clover (*Trifolium subterraneum*)	C	0–36.9	M	Of 27/40 responsive soils, 3 attributed to lowered Mn levels.	Jones (1975)
	C	pH 6.0 (1:5 water)	M	Mn toxicity symptoms on 5/20 unlimed soils. Growth response attributed to alleviation of Mn toxicity on only 1 of these soils.	Bromfield *et al.* (1983)
White clover (*T. repens*)	C	0–10.0	M	DM[e] response of 2/15 soils attributed to reduction in excess soluble Mn.	Holford (1985)
Lucerne (*Medicago sativa*)	C	H[d]	?	Mn toxicity suggested for large DM response (limited data).	Flemons and Siman (1970)
	C	3.0	M	DM response attributed to high soil Mn on 1/3 soils.	Lipsett (1980)
	C	0–36.9	M	Of 31/40 responsive soils, 12 attributed to lowered Mn levels.	Jones (1975)
Wheat (*Triticum aestivum*)	C	3.0	M	DM yield responses of 8 crop plants attributed to high availability of Mn	Lipsett (1980)
Oats (*Avena sativa*)	C	3.0	M		
Rye (*Cereale secale*)	C	3.0	M		

Barley (*Hordeum vulgare*)	C	3.0	M	DM yield responses of 8 crop plants attributed to high availability of Mn	Lipsett (1980)
Sunflower (*Helianthus annuus*)	C	3.0	M M		
Safflower (*Carthamus tinctorius*)	C	3.0	M		
Lupin (*Lupinus angustifolius*)	C	3.0	M		
Rapeseed (*Brassica napus*)	C	pH 6.0 (1:5 water)		Mn toxicity symptoms developed on an unspecified number out of 20 unlimed soils.	Bromfield *et al.* (1983)

[a] Type of lime: C = $CaCO_3$

[b] Rate of lime: FIELD H = 2000 $kg.ha^{-1}$; POTS in $g.pot^{-1}$

[c] Method of application: B = broadcast, I = incorporated, M = mixed throughout (pots)

[d] Rates give in $kg.ha^{-1}$ equivalent on surface area basis

[e] DM = dry matter

Table IX. Lime responses and cycling of organic matter.

Plant	Type[a]	Lime Rate[b]	Method[c]	Comments	Reference
Field responses					
Silver grass (*Vulpia* spp.)	LS	H	B	Stimulation of organic N mineralisation (only the grass component in 3/4 established pastures responded to lime).	Barrow (1964)
Subterranean clover (*Trifolium subterraneum*)	C	H	B	Stimulation of organic S mineralisation speculated (clover DM[d] increased on 2 clover-dominant, established pastures).	Barrow (1965)
Annual grasses	C	H	B	Stimulation of organic N mineralisation (grass DM increased on 1 grass-dominant, established pasture).	Barrow (1965)
	C?	L,M,H	B or I	Described improvement in grass colour and vigour in autumn and winter (= N mineralisation response?)	Glencross and Clarke (1984)
Pot response					
Annual ryegrass (*Lolium rigidum*)	C	0–8.0	M	Herbage N% increased linearly with lime rate (= increased organic N mineralisation?)	Barrow (1965)

[a] Type of lime: C = $CaCO_3$, LS = lime sand (30% acid insoluble)

[b] Rates of lime: FIELD in $kg.ha^{-1}$ L = < 500, M = 500–2000, H = > 2000; POTS in $g.pot^{-1}$

[c] Method of application: B = broadcast, I = incorporated, M = mixed throughout (pots)

[d] DM = dry matter

Table X. Lime responses not attributed to any particular response mechanism(s).

Plant	Type[a]	Lime Rate[b]	Method[c]	Comments	Reference
1. Occurred in the presence of added molybdenum					
Subterranean clover	C	0–8.0	M (pots)	DM[d] increased by first 2 increments.	Barrow (1965)
(*Trifolium subterraneum*)	C	L,M,H	B or I (10)	3/26 responses: > 15% increase in DM.	Glencross and Clarke (1984)
Perennial pasture (*T. repens, Lolium perenne*)	C	H	B	DM response occurred once pH > 6.2 (1:5 water) and lime exceeded 7000 $kg.ha^{-1}$	Rowe (1982)
Phalaris (*Phalaris aquatica*)	C	M	R	Overall yields extremely low but lime doubled DM at 1/2 sites.	McLachlan (1953)
2. Occurred in the absence of applied molybdenum					
Wheat (*Triticum aestivum*)	C	M,H	I	Average grain yield over 4 years increased by 32% at low rate (plant disease reduced yields at higher rates).	Anon. (1960)
	C	M,H	I	Cereal grain yields increased by intermediate rates.	Doyle and Bradley (1982)
Oats (*Avena sativa*)	C	M,H	I		
Triticale (*Triticosecale* spp.)	C	M,H	I		
Barley (*Hordeum vulgare*)	C	M,H	I		
3. No yield data given/responses visually assessed					
Perennial pasture (*T. repens, L. Perenne*)	C	M,H	B	Lime enabled establishment on very acid (pH <5.5 (1:5 water)) soils in 750–1000 mm	Hosking (1968)
Perennial pasture (*T. subterraneum*, (*L. perenne, P. aquatica*)	C	M,H	B	rainfall zone. 7/20 responsive sites in 650–950 mm rainfall zone.	Newman (1948)

(A brace groups the Doyle and Bradley (1982) row with Oats, Triticale and Barley.)

[a] Type of lime: C = $CaCO_3$
[b] Rate of lime: FIELD in $kg.ha^{-1}$ L = < 500, M = 500–2000, H = > 2000; POTS in $g.pot^{-1}$
[c] Method of application: B = broadcast, I = incorporated (depth in cm), R = raked in by hand, M = mixed throughout (pots)
[d] DM = dry matter

Table XI. Negative responses and nutrient deficiencies induced by liming

Plant	Type[a]	Lime Rate[b]	Method[c]	Comments	Reference
Phosphorus deficiency					
Subterranean clover (*Trifolium subterraneum*)	C	17.8	M (1.2)	High rates induced symptoms (Expt. A) and lowered P availability (Expt. B) (limed pH 6.6 (1:5 0.01 M $CaCl_2$).	Paton and Loneragan (1960)
	C	8.8	M	Herbage analysis suggest P deficiency for 1/4 DM[d] depressions (limed pH 6.4 (1:5 water)).	Jones (1975)
	C	5.4	S	Highest rate reduced yields; decreased P uptake and soil P availability suggested (no pH data)	McLachlan (1980b)
	C	4.3	M	Yields and herbage P% depressed by lime (Expt. 4) (limed pH 7.7 (1:5 water)).	Heylar and Anderson (1971)
	C	5.4	M	Lime depressed yield by reducing phosphate absorption (limed pH 6.3 (1:5 water)).	Robson and Loneragan (1970)
Barrel medic (*Medicago truncatula*)	C	5.4	M		
Disc medic (*M. tornata*)	C	5.4	M		
Woolly burr medic (*M. minima*)	C	5.4	M		
White clover (*T. repens*)	C	2.2,4.3	M	Yields and herbage P% depressed by lime (Expt. 4) (limed pH > 6.9 (1:5 water)).	Helyar and Anderson (1971)
Perennial ryegrass (*Lolium perenne*)	C	1.1–4.3	M	Yields and herbage P% depressed by lime (Expt. 4) (limed pH > 6.9 (1:5 water)).	Helyar and Anderson (1971)

Phalaris (*Phalaris aquatica*)	C	1.34	M	High lime rates induced P deficiency symptoms and depressed yield (limed pH 5.7 (1/5 water)).	Helyar and Anderson (1974)
Zinc (and copper) deficiency					
Subterranean clover (*T. subterraneum*)	C	L,H	B?	High rates induced Zn deficiencies on different soils (no yield or pH data).	Tiver (1957)
White clover (*T. repens*)	C	1.6	M	Lime induced yield depressions corrected by Zn on 1/20 soils (no pH data).	Kerridge *et al.* (1972)
Desmodium (*Desmodium intortum*)	C	1.6	M	Lime induced yield depression corrected by Zn on 5/20 soils and Zn + Cu on 2/20 soils (no pH data).	Kerridge *et al.* (1972)
Manganese (and zinc) deficiency					
Subterranean clover (*T. subterraneum*)	C?	L,M,H	B or I (10)	Mn deficiency (herbage analysis) induced by high rates at 1/26 sites (no pH data).	Glencross and Clarke (1984)
	C	L,H	B?	Low rate induced Zn and Mn deficiencies on Kangaroo Island (no yield or pH data).	Tiver (1957)
Boron (and zinc) deficiency					
Subterranean clover (*T. subterraneum*)	C	5.7,11.3	M?	Lime induced yield depression overcome by added B (Expts. 2b and 3). (Limed pH > 6.0 (1:5 water)).	Lipsett *et al.* (1979)
Lucerne (*M. sativa*)	C	1.0	M	Onset and severity of B deficiency exacerbated by lime, even though lime increased DM yield. Addition of B in presence of lime further increased DM yield on 3 soil types	Haddad and Kaldor (1982)
Oats (*Avena sativa*)	C	3.0	M	B and Zn deficiencies suggested to explain the lime induced yield depression on soil 2 (limed pH 7.0 (1:5 water)).	Lipsett (1980)

Cont'd on page 226

Table XI. *cont'd*

Plant	Type[a]	Lime Rate[b]	Method[c]	Comments	Reference
Sunflower (*Helianthus annuus*)	C	3.0	M	B and Zn deficiencies suggested to explain the lime induced yield depression on soil 2 (limed pH 7.0 (1:5 water)).	Lipsett (1980)
Linseed (*Linum usitatissimum*)	C	3.0	M		
Safflower (*Carthamus tinctorius*)	C	3.0	M		
Lupin (*Lupinus angustifolius*)	C	3.0	M		
Other deficiencies					
Stylo (*Stylosanthes guianensis*)	C	0.67	M	High rates induced Mg deficiency symptoms; applying $CaCO_3$ + $MgCO_3$ did not. (Limed pH 7.8 (1:5 water) at sowing, 6.7 at harvesting).	Probert *et al.* (1979)
Subterranean Clover (*T. subterraneum*)	C	7.1–11.3	M	Fe deficiency symptoms on 1 soil. Induced P and micronutrient deficiencies suggested for remaining 2/40 yield depressions (no data) (limed pH 6.4 (1:5 water)).	Jones (1975)
Unexplained yield depressions					
Subterranean clover (*T. subterraneum*)	C	8.0	M	High lime rates reduced DM yields to unlimed yields or less (limed pH 6.6 (1:5 0.01 M $CaCl_2$)).	Barrow (1965)

[a] Type of lime: C = $CaCO_3$

[b] Rate of lime: FIELD in $kg.ha^{-1}$ L = < 500, M = 500–2000, H = > 2000; POTS in $g.pot^{-1}$

[c] Method of application: B = broadcast, I = incorporated, M = mixed throughout (pots), M(x) = mixed to x cm depth (pots), S = surface applied (pots)

[d] DM = dry matter

B. Experiences with dolomite

Only a few instances of dolomite use on Australian soils have been recorded, most commonly topdressed onto existing pastures and often compared with an equivalent amount of limestone. In most cases, lime and dolomite gave similar responses (Newman, 1948; Hosking, 1972), suggesting that the main effect of the dolomite was to increase the soil pH. Many subsoils in south-eastern Australia are high in magnesium (Williams and Raupach, 1983), and raising the pH of the surface soil enables the plant roots to grow into the subsoil and access the subsoil magnesium. Thus plant responses to dolomite are unlikely to be due to magnesium deficiency even though the magnesium level of the surface soil is low.

C. Depressed plant performance

Yield depressions following the application of lime have been reported for a wide range of crop and pasture species on Australian soils (Table XI). In almost every case, single or multiple nutrient deficiencies induced by the lime were suggested or shown to be the reason for the negative growth response. The pH of limed soils (1:5 soil:water) are also given in Table XI; these show that most of the yield depressions in pots occurred at pH levels lower than neutrality, illustrating that Australian surface soils can be readily overlimed because of their highly-leached and weakly-buffered nature. Also, very few yield depressions occurred under field conditions, suggesting that, unlike in pots, plant roots are able to access plant nutrients from either incompletely limed surface soils, or from unlimed subsurface soils.

D. Lack of plant growth responses to lime

In discussions about the usefulness of lime in Australian agriculture, the point is invariably made about the many non-responses to lime, as an indication of the limited value of lime on many acid soils. In Table XII, the non-responses recorded in the available literature for Australian soils have been listed. In doing so, we recognise that this list is incomplete, as probably many experiments where no response to lime was observed were not recorded at the time and are now long forgotten. Also, many of the early reports of non-responsiveness should be treated with caution, given that another two to three decades of soil acidification has occurred on these soils. Thus, the many Victorian pastures reported as giving 'disappointing' responses to topdressed lime in the 1930s and 1940's

Table XII. Lack of plant growth responses to lime.

Plant	Type[a]	Lime Rate[b]	Method[c]	Comments	Reference
Field experiments					
Subterranean clover (*Trifolium subterraneum*)	C	H	B	No responses on 4/7 established annual pastures.	Barow (1965)
	C	L,H	B,D	4/15 sites non-responsive to heavy rate, although 1 responded to drilled lime.	Drake and Kehoe (1954a)
	C	M,H	B or I (10)	No responses in pasture growth at 11/19 sites.	Glencross and Clarke (1984)
	C	?	?	No response in either winter or spring.	Helyar and Andeson (1970b)
	C	H	B	17/74 sites non-responsive to lime.	Paton (1958)
	C	L	D	Lime drilled with seed did not improve nodulation at 4 sites in the first year.	Simmons (1962)
	C,	L,M,H	B or I	No responses on 25/53 established annual pastures.	Yeates *et al.* (1984)
White clover (*T. repens*)	C	?	?	No response in either winter or spring.	Helyar and Anderson (1970b)
Desmodium	C	L,M,H	I	Did not respond to lime over 4 harvests (Expt. 1).	Andrew and Bryan (1958)
(*Desmodium intortum*)	C	H	B	No yield increase at 2 sites, despite improved nodulation.	Diatloff and Luck (1972)
Glycine (*Neonotonia wightii*)	C	H	B		
Phalaris (*Phalaris aquatica*)	C	M	R	No response at 1/2 sites.	McLachlan (1953)
	C	M,H	I (5)	Little or no DM[d] increase in the absence of supplied N.	Helyar and Anderson (1970a)

Perennial ryegrass (*Lolium perenne*)	C	M,H	I (5)	Little or no DM[d] increase in the absence of supplied N.	Helyar and Anderson (1970a)
Perennial pasture (*L. perenne*, *T. repens*)	C	H	B	No effect on pasture yield and composition, nor on animal production and health (not acid enough?)	Savage and Hosking (1982)
Wheat (*Triticum aestivum*)	C	H	I	No grain yield responses at 18/20 sites because of subsoil acidity.	Porter and Wilson (1984)
Barley (*Hordeum vulgare*)	C	M,H	I (10)	No response to lime in first or second year on 3 major soil types.	Glencross and Clarke (1984)
Rapeseed (*Brassica napus*)	C	M,H	I (10)		
Lupin (*Lupinus* spp.)	C	M,H	I (10)		
Field pea (*Pisum sativum*)	C	M,H	I (10)		
Pot experiments					
Subterranean clover (*T. subterraneum*)	C	?	?	Yield of unlimed plants 101% of limed plants.	Helyar and Anderson (1970b)
	C	0.7–12.1	M	No responses on 9/40 soils.	Jones (1975)
	C	2.0	M	Non-responsive with or without added N on 7 soils (Expt. 1).	Munns (1965)
White clover (*T. repens*)	C	pH 6.0	M	No DM responses on 18/33 soils.	Bromfield *et al.* (1983)
	C	1.6	M	1/4 soils non-responsive.	Kerridge *et al.* (1972)
	C	?	?	DM of unlimed plants 114% of limed plants.	Helyar and Anderson (1970b)
	C	0.35,4.2	M	No top response to liming acid subsoil.	Simpson and Pinkerton (1978)
Kenya white clover (*T. semipilosum*)	C	0.35,4.2	M		

Cont'd on page 230

Table XII. *cont'd*

Plant	Type[a]	Lime Rate[b]	Method[c]	Comments	Reference
Phasey bean (*Macroptilium atropurpurem*)	C	0.35,4.2	M	No top response to liming acid subsoil.	Simpson and Pinkerton (1978)
Desmodium	C	0.35,4.2	M		
(*D. intortum*)	C	1.6	M	Non-responsive on 2/14 soils.	Kerridge *et al.* (1972)
Lucerne	C	0.7–7.1	M	No responses on 9/40 soils.	Jones (1975)
(*M. sativa*)	C	1.0	M	No responses on 3 (alluvial)/15 soils.	Haddad and Kaldor (1982)
Perennial ryegrass (*Lolium perenne*)	C	?	?	DM yield of unlimed plants 114% of limed plants.	Helyar and Anderson (1970b)
	H	1.15	M (1.0)	No response to lime, although added N increased yield.	Anderson and Oertel (1946)

[a] Type of lime: C = $CaCO_3$, H = $Ca(OH)_2$

[b] Rate of lime: FIELD in kg.ha^{-1} L = < 500, M = 500–2000, H = > 2000; POTS in g.pot^{-1}

[c] Method of application: B = broadcast, I = incorporated (depth in cm), D = drilled with the seed, R = raked in by hand, M = mixed throughout (pots), M(x) = mixed to x cm depth (pots)

[d] DM = dry matter

(Twentyman, 1937; 1938; Andrew, 1947; Newman, 1948) may respond differently today.

In all, some 81 field sites did not respond to lime out of a total of 183 sites, most of which (61) were pasture sites (Table XII) where the lime was broadcast onto the surface.

E. Low input treatment of acid soil problems

Molybdenum deficiency has been widely overcome by the addition of small amounts of molybdenum to the superphosphate applied to pastures. Other methods of ensuring adequate molybdenum supply in acid soils include spraying molybdenum directly onto the crop, or onto seed crops to ensure high molybdenum seed content, seed dusting and the addition of molybdenum to the lime used for lime-pelleting legume seed (Gupta and Lipsett, 1981). The dramatic and frequent responses achieved from application of molybdenum to legume-based pastures on acid soils during the 1950s and 1960s led to the common opinion that molybdenum deficiency was the only important nutritional problem of acid soils.

Many problems with legume nodulation are related to soil acidity and have also been overcome without resorting to broadcasting lime at heavy rates. Nodulation in many cases has been successfully achieved through either the use of lime pelleting or lime/superphosphate banding. Success for these methods depends on soil acidity limiting nodulation during the establishment phase only. With subterranean clover this practice appears successful, providing soil pH (1 : 5 soil : water) is above 5.0 (Roughley and Walker, 1973). At lower pH values, nodulation, nodule function and/or plant growth may be reduced (see Chapter 3). For more acid-tolerant species and rhizobia, e.g. some tropical legumes, pelleting may not be beneficial. On the other hand, for more sensitive species, e.g. lucerne (*Medicago sativa*), pH (1 : 5 soil : water) may need to be higher than 5.5 before lime pelleting or banding is satisfactory. These techniques, all using small amounts of lime, are most successful where the lime facilitates acid-sensitive processes or where near-surface (7–10 cm) banding enables plant roots to grow through acid surface horizons into more favourable subsoil environments. However, both lime pelleting and banding are only potentially successful where soil acidity is mild.

III. FACTORS MODIFYING PLANT RESPONSE TO LIME

A. Infertile acid soils

Although the alleviation of toxic effects of soil acidity is a prime consideration in liming acid soils, the simultaneous changes in nutrient availability should not be overlooked as potential growth-limiting factors. On weakly-buffered, highly-leached acid soils this is particularly so. If liming is necessary for these infertile soils, it needs to be done judiciously to overcome the toxic levels of aluminium and manganese but not to induce nutrient deficiencies (Bouton *et al.*, 1981). Otherwise, liming highly-leached acid soils to near neutrality can depress the yields of many plant species (Kerridge *et al.*, 1972; Munns and Fox, 1976; Sumner, 1979). Also, many acid soils are so grossly infertile that appropriate fertilization is needed together with lime to achieve any growth responses (McLachlan, 1980a). Phosphorus deficiencies in particular occur on many strongly acid soils and additions of lime and phosphorus, the latter often as superphosphate, are necessary for successful plant establishment and growth (Kunishi, 1982; McLachlan, 1980b; Helyar and Anderson, 1970a; Paton and Loneragan, 1960; Brooks *et al.*, 1982; Munns and Rudczuk, 1966; Drake and Kehoe, 1954a). Symptoms of phosphorus deficiency also occur on plants growing in acid soils high in available aluminium because of inhibition of root growth by aluminium. However, this apparently large requirement for phosphorus on these unlimed soils is not to be confused with a need for heavy phosphorus fertilization, as the symptoms can be readily remedied by liming alone.

Liming can improve phosphorus uptake through improved root growth (Friesen *et al.*, 1980) and by reducing the solubility of aluminium (Haynes, 1983).

B. Subsurface soil acidity

In some acid soils, the acid toxicity problem extends down the soil profile (Williams, 1980). In these situations, applying lime to amend the acidic topsoil gives only small yield improvements because the zone of ameliorated soil does not extend far enough down the profile (Doyle and Bradley, 1982; Pinkerton and Simpson, 1982; Temple-Smith *et al.*, 1983; Porter and Wilson, 1984). Root growth is frequently confined to the amended topsoil because the toxicities of the soil below, particularly the excess aluminium, strongly restrict root extension and root growth in this zone (Doss and Lund, 1975; Gonzalez-Erico *et al.*, 1979).

C. Plant sensitivity

The responsiveness of crop and fodder plants to liming is determined in part by their sensitivity to the toxicity or deficiency factors and the infertility of acid soils. Legumes in particular can be sensitive to both factors. In both tropical and temperate plants there are species and, within a particular species, cultivars that are less tolerant of soil acidity factors (Munns and Fox, 1976; Edwards *et al.*, 1981; Cregan *et al.*, 1984). Sensitive legume species and cultivars often respond markedly to low rates of lime, simply because small amounts of lime greatly improve their ability to nodulate and fix nitrogen. The larger responses to liming in the grain yield of tolerant cereal crops, as reported by Scott (1982) and Doyle and Bradley (1982), were due to the presence of acidic subsurface soils below the limed soil. These reports illustrate the advantages of using lime, together with plant tolerance, to overcome productivity problems associated with subsurface soil acidity (discussed in greater detail in Chapter 5).

D. Changes to the physical properties of soils

Limestone is primarily applied to soils to amend the toxicities associated with acid soils. It can also be used to supply calcium for the improvement of soil structure through the displacement of excess sodium and magnesium although the improvement in soil structure following the use of limestone is uncommon, even on acid sodic soils. However, De Boodt *et al.* (1961) reported that soil treated with limestone was more stable structurally than unlimed soil, when compared by wet and dry sieving, and Van Wambeke (1976) claimed improvement in the structure of alfisols under cropping following liming. During *et al.* (1984a,b) attributed growth responses in pasture during late summer to increased moisture contents of limed soil. They measured improvements in the stability of the limed soil aggregates and hypothesised improved infiltration during heavy rain. However, Shannon *et al.* (1984) working on similar soils, considered that it was not possible to determine the cause of late summer and early autumn responses to liming. They postulated that several factors, such as changed pasture composition, lesser water stress due to improved rooting volume in the absence of aluminium, more rapid rewetting of dry soils which were limed, and seasonal improvement in phosphorus status due to mineralisation of organic phosphorus, were all possible causes. Hoyt (1981) found that soils limed six years previously had a higher proportion of coarse dry aggregates, even when initially low in exchangeable sodium. It was considered that the limestone had increased the resistance of the soil to pulverising when cultivated. In the same experiment, calcium hydroxide also improved soil structure but decreased the level of organic matter. He concluded that the improved soil structure and associated improvement in rapeseed emergence was a direct effect of liming. While calcium is known to flocculate soil colloids, neither Hoyt (1981) nor other authors have suggested or demonstrated a cementing agent effect. No data on the beneficial effect of lime on the physical characteristics of Australian soils and any subsequent effects on plant growth are known by the authors. In contrast, the application of limestone to the exposed subsoil of a krasnozem, which may possess a net positive charge, can move the soil to zero charge and deflocculation, hence increasing erodibility (Peele, 1936).

The main limitation of liming to improve soil structure is its low solubility, even at low soil pH. For acid soils that are also sodic, Abbott and McKenzie (1986) have suggested a mixture of limestone and gypsum.

IV. LIMING MATERIALS AND THEIR APPLICATION

A. Materials and manufacture

Naturally occurring minerals which are commonly used to raise soil pH are $CaCO_3$ (limestone) and $CaMg(CO_3)_2$ (dolomite). Magnesite, $MgCO_3$, is sometimes used in conjunction with $CaCO_3$ as a magnesium source in magnesium-deficient acid soils and crushed serpentine ($Mg_3Si_2O_5(OH)_4$) is also used as a magnesium supplement. Other less-commonly used liming materials are quick or burnt lime, CaO; hydrated or slaked lime, $Ca(OH)_2$ and magnesia, MgO.

Waste products from manufacturing processes using limestone can be used for neutralising soil acidity. The production of CaO produces a dust composed of varying proportions of $CaCO_3$ and CaO, and precipitator dusts from cement manufacture additionally contain about 5% K_2O. Steel production also produces alkaline slags of variable composition (Barber, 1967); an analysis of some New South Wales alkaline waste materials is provided by Thompson *et al.* (1984). Experiments in southern New South Wales (R. Cumming, pers. comm.; E. Havilah, pers. comm.) indicate that the slag materials are useful provided they are crushed finely. Human wastes can also be effective liming materials; sewage ash from Canberra has been evaluated, with results superior to limestone, principally due to its higher phosphorus content (Willett *et al.*, 1984; Dear *et al.*, 1985). However, the restricted production and the conditions imposed on its handling have hindered its agricultural use.

The most common liming material is $CaCO_3$. In eastern Australia it commonly occurs as calcite, which may be metamorphosed to marble. The process of calcite extraction includes quarrying, jaw crushing to 7–15 cm diameter, hammer milling and finally ball milling to fine particle sizes. Stone and particle size fractions are separated by mechanical screens and finally by air centrifugation. This can be contrasted with the situation in Western Australia where limestone sands, of variable purity, are simply excavated and sold unprocessed. Additional detail on liming materials and production technology can be found in Boynton (1966) and the distribution of limestones in Australia is described by Knight (1976).

B. Quality

The quality of a liming material is determined by two characteristics: its neutralising value (NV) and its fineness. These characteristics influence the rate of dissolution and the extent and efficiency of soil neutralisation.

The NV of a liming material is its ability to neutralise acidity, relative to the mass of a standard material, generally pure $CaCO_3$. With pure minerals and compounds, the NV is the percentage ratio of the molecular weight of $CaCO_3$ to that of the material under question. For example, Mt Gambier dolomite is about 63% $CaCO_3$ and 32% $MgCO_3$, with 5% impurities and the mole weights of $CaCO_3$ and $MgCO_3$ are 100.09 and 84.32 respectively.

Therefore,

$$\text{NV of } CaCO_3 = 63\% \times \frac{100.09}{100.09} = 63$$

$$\text{NV of } MgCO_3 = 32\% \times \frac{100.09}{84.32} = 38$$

$$\text{NV of impurities} = 5\% \times 0 = \frac{0}{101}$$

With waste materials or materials of complex composition, the relative NV can be determined by laboratory titrations or by field trials. Unfortunately, chemical assessments of NV can be misleading. The NV of a precipitator dust which is predominantly CaO at the time of production can approach 178%. After 30 days of open-air storage, the NV of this hygroscopic material declines to approximately 135%. The hydrated lime then continues to form re-carbonated external surfaces and, in time, the NV stabilises at about 120%. Purchase prices based on NV at point-of-sale are therefore necessary for quality control, but would involve additional costs. In field experiments, however, precipitator dust materials (NV approximately 120%) have only been as effective in changing soil pH as superfine limestone (99.5 per cent of this material is finer than 25O μm) with an NV of 98%. This might be explained by some of the CaO fraction of the precipitator dust reacting directly with soil organic matter. Products containing CaO have two additional problems;

first, they are caustic and are dangerous to handle and second, a saturated $Ca(OH)_2$ solution has a pH of 12.5, which may cause temporary soil sterilisation (Hutchinson and MacLennan, 1914) and possibly induce short-term micronutrient deficiencies.

The fineness of limestone determines the efficiency of the soil-limestone reaction and to a large extent the rate of reaction. Efficiency is measured by the maximum change in soil pH per mass of liming material of a given NV. Ground limestone (NV 95 to 99%) reaches 100 per cent efficiency at a fineness of about BSS or USS 60 mesh size, i.e. 150 μm (Barber, 1967; Tisdale and Nelson, 1975; Adams, 1978).

Attempts to explain the efficiency and rates of reaction for calcium salts have tended to be particle- and laboratory-orientated (Barber, 1967; Keren and Shainberg, 1981). In an effort to better understand field results (such as presented in Tisdale and Nelson, 1975), the efficiency and rates of reaction of limestone are now examined in terms of field geometry, and the probable solubility of $CaCO_3$ in soils.

Using the maximum planar cross-section of spherical particles, the surface area covered by various particle sizes can be calculated. Smaller particle diameter means more particles and a greater area covered. Whilst 2.72 t (or 1 m^3) of 100 mesh material will cover 1 ha, the same quantity of 16 mesh material would only cover 0.15 ha. Total coverage with 16 mesh material would require 18.13 $t.ha^{-1}$ (Table XIII). However, the reality of this mathematical approach is confounded because lime-

Table XIII. Efficiency of two-dimensional distribution of limestone of varying particle sizes.

Individual particle sizes				
Mesh (BSS)	16	30	60	100
Particle diameter (μm)	1000	500	250	150
Maximum surface area covered by 1 m^3 or 2.72 t (ha)	0.15	0.30	0.61	1.00
Amount to cover 1 ha (t)	18.13	9.07	4.36	2.72
Commercial product particle size ranges				
Products	Agricultural[a]		Superfine[b]	
Maximum surface area covered by 1 m^3 or 2.72 t (ha)				
Monolayer[c]	11.26		15.5	
Median particle[a, b]	1.44		3.32	

[a] Agricultural 70% < 250 μm, median diameter = 70 μm
[b] Superfine 100% < 250 μm, median diameter = 30 μm
[c] Using total particle size analysis over a planar area

stone products are not homogeneous and cover a range of particle sizes from 1 mm to 1 μm; also small particles tend to aggregate.

The surfaces covered by 1 m^3 of two liming products have been calculated using both the total particle size distribution and the median particle size methods (Table XIII). The relative differences in coverage between products is greater (130 per cent) with the latter method, which better reflects the actual coverage efficiency.

The above, two-dimensional approach is valid for topdressed lime. Incorporated lime requires consideration of the third dimension (depth) together with the diffusion of Ca^{2+} and HCO_3^-, when assessing efficiency of lime reaction (Table XIV).

Experimentally, 60 to 100 mesh material (250–150 μm) is the largest particle size that is 100 per cent efficient agriculturally (in terms of changes in soil pH). Using 60 mesh material (100 per cent efficient), and assuming a distance of ion diffusion independent of the solid particle diameter, proportional decreases in the efficiency of coarser particles can be calculated. The results from calculations using this approach are consistent with the data of Adams (1978) and Tisdale and Nelson (1975) (Fig. 1). At application rates greater than about 2.5 $t.ha^{-1}$ the relative efficiency of coarse particles increases because of the greater total density of particles.

Although the rate of reaction of limestone depends (in part) on its fineness, the pH values of the suspensions of limed soils used in laboratory determinations of soil pH can fluctuate significantly (0.3 pH units) over relatively short periods (30 min). This is due to the saturated

Table XIV. Efficiency of limestone particle sizes with three-dimensional distribution of 1 $t.ha^{-1}$.

Mesh (BSS)	Particle diameter (μm)	Particles[a] per cm^3	Percentage of soil volume potentially[b] covered after different distances of diffusion[c]			
			5000	2000	1000	500 μm
16	1000	0.7	50	4.6	1	0.3
60	250	51	2500	182	27	5
100	150	209	11 450	785	109	17

[a] An application rate of 1 $t.ha^{-1}$ mixed to a 10 cm depth.

[b] Percentages greater than 100 indicate that the potential volume of influence exceeds the soil volume.

[c] Ca^{2+} and HCO_3^- are assumed to have similar diffusion distances for a first approximation. The calculated percentage of soil volume covered includes the solid particle radius as well as the diffusion distance.

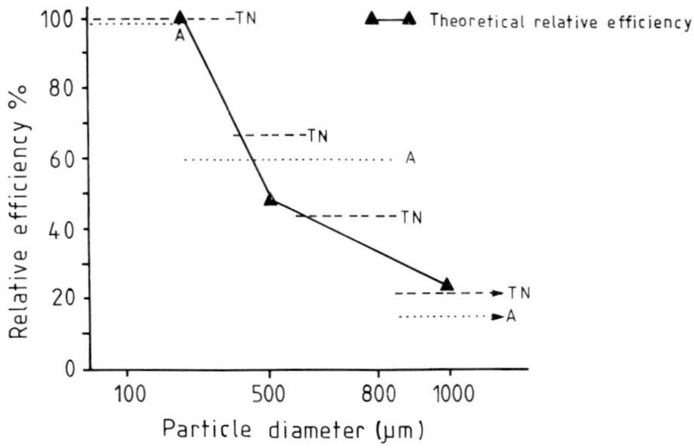

Fig. 1. Relative efficiencies of limestones of various particle sizes in changing soil pH for an application rate of approximately 2.5 t.ha^{-1}: calculated efficiency using the three-dimensional ion diffusion model; and the efficiencies from the experimental data cited by Tisdale and Nelson (1975) -------TN and of Adams (1978) .... A. Length of bars represent the range in particle size of each liming material.

condition of the solution in the test tube creating a reaction which has still to happen in the field. In the test tube the rate of dissolution of limestone is competing with the rate of soil titration. The closer the former reaction gets to depleting the undissolved limestone, the smaller the variation in test tube estimates of soil pH become. Therefore, the extent or apparent rate of limestone reaction needs to be evaluated by considering the solubility of calcitic limestone in water and the total amount of water passing through the soil horizon.

The solubility of calcite in pure water at 20°C is 0.013 g.L^{-1}. Therefore, to dissolve 2.5 t.ha^{-1} of lime incorporated to a depth of 10 cm would require 19.23 mm of rain. However, rain water is never pure (atmospheric CO_2 forms carbonic acid in solution) and the soil contains many acids, and has a negative charge which absorbs Ca^{2+} from solution. Assuming that limestone dissolution is dominated by the acid-base reaction rather than by Ca^{2+} absorption, and that the initial pH of water is governed by the first dissociation constant of carbonic acid, the amount of rain to dissolve a dressing of limestone can be estimated from the pH-rainfall relationship (Table XV) provided the slope of the soil's titration curve is known. Nevertheless, the rainfall figures calculated only represent 'middle of the road' estimates of the moisture requirement for the calcite-soil reaction. For hard limestones and marble, for imperfect mixing and distribution of the limestone into the soil, and where there is a proportion of material coarser than 60 mesh, more than the calculated

Table XV. The solubility of calcite at various soil solution pH values and the rainfall required for dissolution of 2.5 $t.ha^{-1}$ in the 10 cm surface layer.

pH 20°C	Solubility of[a] $CaCO_3$ $g.L^{-1}$	Rainfall to dissolve 2.5 $t.ha^{-1}$ (mm)
7.00	0.013	19.2
5.70	0.052	4.8
5.20	0.117	2.4
4.95	0.170	1.5
4.45	0.390	0.6
3.95	0.891	0.3

[a] Fine calcite $CaCO_3$, based on Bradfield (1941).

rainfall is required for a full reaction. For limestones more soluble than the calcite specified, less rain will be needed. Where soil is saturated for long periods, and therefore allows greater diffusion of Ca^{2+} and HCO_3^- and increased dissolution of the solid phase $CaCO_3$ without a flux of water, the above calculation will be an overestimate of the rainfall required for full reaction.

Over the range of soil pH values in which limestone is likely to be applied, some 200 to 3000 mm of rain will be required to pass through the soil to dissolve 2.5 $t.ha^{-1}$. An approximate time scale can therefore be estimated on the basis of average, annual rainfall.

C. Evaluation of liming materials

An evaluation of liming materials involves consideration of the NV, fineness and the cost of cartage and spreading. Once the agricultural efficiency of a product has been calculated (see Tisdale and Nelson (1975) for the procedure) and the cost per tonne spread on the paddock is known, the relative values of alternative liming products can be assessed from Figure 2. The evaluation of plant nutrient constituents such as potassium, magnesium or phosphorus may also be necessary.

Generally, the benefits of waste products include cheapness and 'bonus' nutrients such as phosphorus (sewage sludge) and potassium (cement kiln dust). The disadvantages of waste materials include:

(1) variable analysis. Precipitator dusts can fluctuate from 5 to 28% CaO over a 30-day period;

(2) higher application rates. Poorer quality lime-sands, basic slags and materials of low NV have to be applied at higher rates to achieve the

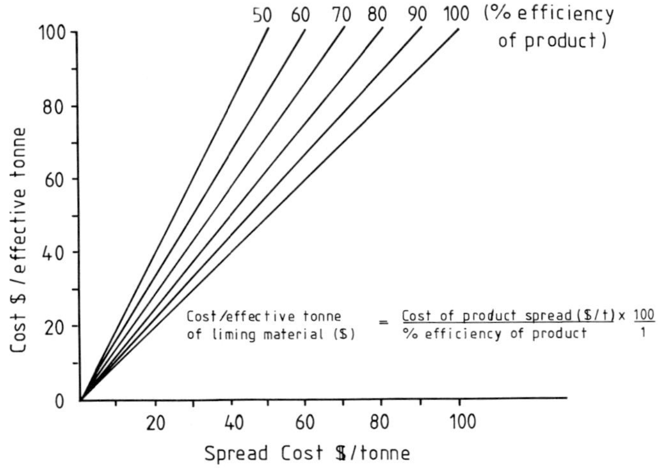

Fig. 2. The relationship between the cost of spreading liming materials and cost per tonne of effective liming product for liming materials of various agricultural efficiencies. Spread cost includes purchase, cartage and spreading.

same effect on soil pH as good quality limestones. Thus, cartage and spreading costs are higher and, despite lower ex work prices, their radial distribution from their sources can be limited;

(3) availability. Changes in the mode of production, the amount of material produced and the seasonality of production affect the ease of obtaining alternative materials;

(4) handling problems. Products containing CaO are dangerous to the skin and eyes, while sewage ash often has to be carted in fully-enclosed trucks.

D. Chemical effects of liming materials

The beneficial effects of liming on the soil chemistry are largely a reversal of the problems associated with acid soils. Negative effects of liming can include reduced phosphate availability, suppressed availability of magnesium and potassium, and induced deficiencies of micronutrients such as boron. Reviews detailing these effects include Kamprath and Foy (1971), Adams (1978), Pearson and Adams (1967), Foy (1984) and Chapter 1 of this monograph.

E. Lime requirement

A limiting factor in acid soils is aluminium toxicity and this has led to the development of methods to assess lime requirement based on exchangeable aluminium (Al_{ex}). Commonly, the Al_{ex} value is multiplied by a factor to determine a lime rate sufficient to neutralise most of the Al_{ex} (Kamprath, 1970). Such an approach is limited by the soil-specific relationship between Al_{ex} and the non-aluminium acidity. Soils A and B in Figure 3 are used to illustrate how such relationships can vary. The exchangeable calcium (Ca_{ex}) is used as an index of lime titration and shows the relative neutralisation and removal rates of Al_{ex} for these soils. If the lime requirement is calculated as equal to $Al_{ex} \times 1.5$ (Kamprath, 1970) this would result in soil A, with a mean initial Al_{ex} of 1.48 cmol(+).kg^{-1}, having no Al_{ex} after 2.22 cmol(+).kg^{-1} of calcium was applied as $CaCO_3$. At site B however, with a mean initial Al_{ex} of 1.6 cmol(+).kg^{-1}, the addition of 2.4 cmol(+).kg^{-1} of calcium as $CaCO_3$ would leave nearly 1 cmol(+).kg^{-1} of Al_{ex} in the soil. Although this level of Al_{ex} represents less than 15 per cent of the CEC, as predicted by Kamprath (1970), it is too much aluminium for the healthy growth of acid-sensitive species. Additionally it indicates that soil responses to the application of liming

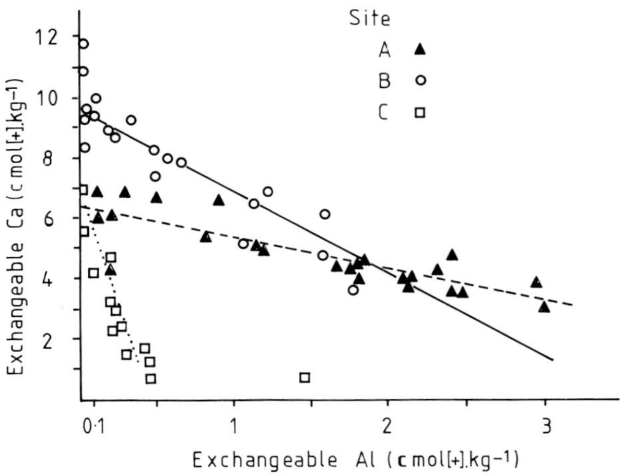

Fig. 3. The relationship between exchangeable aluminium Al_{ex} and exchangeable calcium Ca_{ex} for two sites on the southern tablelands of New South Wales (A and B) and for a composite of three sites (C) on the south-western slopes of New South Wales. Site A soil was developed on basaltic colluvium, site B on interbedded sandstone and shale, and sites marked C were developed on granodiorite or sedimentary material with substantial windblown deposition.

rates based on this formula diverge significantly. A composite of three sites from the southern wheatbelt of New South Wales (C) is shown as a further contrast.

The aluminium-based approach of Kamprath (1970) is likely to work satisfactorily where the proportions of aluminium and non-aluminium acidity are similar to those in North Carolina and Panama soils (Kamprath, 1978) , where there is 1:1 calcium:aluminium replacement on a cmol(+) basis, and where aluminium toxicity is the dominant problem to be corrected. Just as pH is inadequate for determining lime requirement without a knowledge of the soil's buffering capacity or titration curves, so the static measurement of Al_{ex} is inadequate to indicate the changes which will occur with increases in pH and Ca_{ex}.

When aluminium is expressed as a percentage of the effective cation exchange capacity (ECEC) or extracted with $CaCl_2$, it generally reaches insignificant levels at pH 5.0–5.2 (1:5 soil:0.01 M $CaCl_2$) for many Australian soils (Bromfield *et al.*, 1983; Conyers 1983 and Chapter 7). This finding is consistent with a range of published data for North American soils (Kamprath, 1978) and indicates that a target pH approach to lime requirement is more useful universally, although the target may vary with soil type and the acid sensitivity of the species to be grown.

The use of the target pH approach to lime requirement is supported indirectly by the survey of Follett and Follett (1983) which shows that methods for estimating lime requirement based on pH and target pH levels of 5.5 to 7 (1:5 soil:water) have remained the approach of State advisory services in the USA. The most common test for lime requirement is that proposed by Shoemaker *et al.* (1961), and comparative studies have tended to favour this method (Fox, 1980; Tran and van Lierop, 1981; 1982) though not always (Loynachan, 1981). The most recent study (Brown and Cisco, 1984) claims that an improved Woodruff buffer is superior to the method proposed by Shoemaker *et al.* (1961) as it predicts both high and low lime requirements with more accuracy. However, Follett and Follett (1983) demonstrated that the selection of a soil test is often a parochial process, presumably because no method has a monopoly on accuracy or precision on all soil types. Despite the concern about the precision of achieving target pH or aluminium levels, the depth of incorporation of the limestone, relative to the soil acidity profile, is likely to be critical to both economics and plant performance.

F. Distribution of lime

The uneven distribution of lime over an area will reduce the potential yield response and this will be most significant in very infertile, acid soils

(Stewart and Bandel, undated). Thus, where lime is a high cost input, machinery that ensures an even distribution should be used.

G. Spatial variation and lime application

1. Lateral variation

Lateral variation in soil properties has been long recognised (Robinson and Lloyd, 1915) and the practical implications with respect to nutrient availability and physical factors have been explored (Fairfield-Smith, 1938; Burridge and Cunningham, 1960; Webster and Butler, 1976). These lateral variations are of particular importance with regard to soil acidity. Liming can increase soil heterogeneity and small shifts in pH values of composite samples (approximately 0.5 pH units) can give an inadequate description of the effect of liming. This heterogeneity is illustrated for site A in Figure 4.

At another site (initial composite pH = 4.3, 1:5 soil:0.01 M $CaCl_2$), the pH ranged from 4.3 to 6.1 (n = 25) 18 months after 7.5 $t.ha^{-1}$ of limestone was applied. Field determinations of pH using Raupach indicator solution gave pH values ranging from 4.5 to 8 across individual peds. Even with precise spreading there will always be areas that are overlimed, whilst others will receive insufficient lime.

Often the heterogeneity is not clear-cut but a mosaic of less acid and more acid areas. Such spatial distribution of pH is common and poses a major problem for measuring the growth responses to lime treatments in small plot research, and for determining a suitable target pH (see page

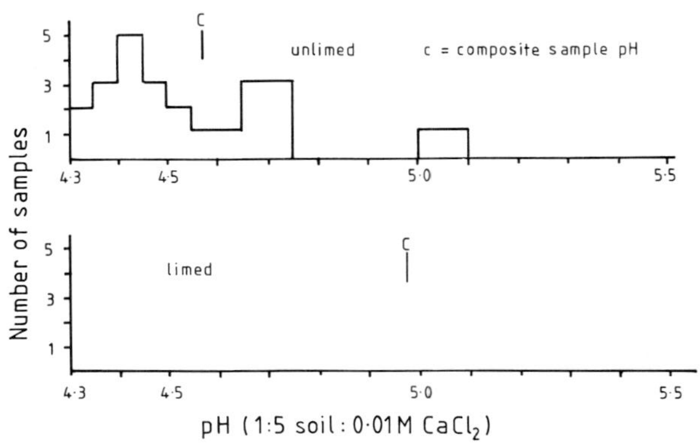

Fig. 4. Frequency distribution of pH in a limed and unlimed basaltic soil on the southern tablelands of New South Wales. The limestone was applied at 5 $t.ha^{-1}$ by a direct drop machine and incorporated with a disc plough 18 months prior to sampling.

253). For example, a mean paddock pH = 5.0 (1:5 soil:0.01 M $CaCl_2$) leaves 50% or so of the paddock at pH < 5.0. The target pH should therefore be greater than pH 5.0 (1:5 soil:0.01 M $CaCl_2$) so as to correct the acidity over most of the paddock (Russell, 1984; Draycott and Messem, 1979). Where changes in pH and/or CEC are known, each area (soil) can be treated separately, even within the one paddock.

2. *Profile variation*

Very few soils have a uniform pH throughout the profile. The deep acid sands of north-western New South Wales are a notable exception (Doyle and Bradley, 1982). Most soils that have acidified due to agricultural practice would be described as having three acidification horizons. The immediate soil surface (1–3 cm) is commonly less acidic than the soil below. Beyond this second, more acid horizon the soil becomes less acidic with depth (e.g. Williams, 1980). The depth to which acidification has occurred is influenced by the period of agriculture, the degree of the acid imbalance of the system, the bulk density of the soil and the buffer capacity of the soil layers (see Chapter 2). Where the problem acid layer is confined to near the surface, lime mixed through this layer can be totally effective in ameliorating soil acidity. Therefore, responses to liming are common where the species grown are sensitive to the acid soil factors present (Coventry, 1985). Some soils, e.g. many of the soils of the eastern wheat belt of Western Australia, have naturally more acidic subsoils and the use of tolerant species or varieties may be the only practical short-term solution, as the responses to liming are infrequent (Porter and Wilson, 1984) The importance of subsoils to plant nutrition and the plant's response to liming is frequently overlooked, even though the impact of acid subsoils in depressing yield and modifying responses to liming are widely reported (Pinkerton and Simpson, 1982; Adams *et al.*, 1967; Scott, 1982; Mahoney *et al.*, 1981; Sanchez and Salinas, 1981).

H. Lime placement

1. *Surface application*

In Australia many areas with problem acid soils are not arable. Here, top-dressing lime is the only possible method of application. Data on plant responses to changes in soil pH over time from surface applications of lime are limited. Woodhouse (1956) showed slower response to surface application, compared with thorough incorporation. In north-west Tasmania (approximately 900 mm average annual rainfall), Rowe (1982) reported responses to surface-applied lime in the second year of two experiments for rates in excess of 7.5 $t.ha^{-1}$. With lower application rates and a lower rainfall a longer lag phase would be expected. In the same experiments, pH changes down the profile were significant to a depth of approximately 15 cm, eighteen months after the last lime application and following a total application of 12.5 $t.ha^{-1}$ of topdressed lime over five years. These results indicate that, although topdressing may be less efficient in terms of the rate of change in soil pH (due to the surface layering of the lime hindering the three dimensional dissolution of CO_3^{2-} and HCO_3^-), the potential for surface applications to increase pH at depth does exist with high lime rates.

2. Mixing lime with soil

Little information is available on the effect of tillage practices, and the uniformity and thoroughness of lime incorporation, on plant response, although Barber (1967; 1984) describes the soil mixing characteristics of some tillage practices. Following two years of experimentation at Wagga Wagga, B. J. Scott (pers. comm.) reports that two passes with a tined implement gave 60–70 per cent of the yield response of rotary-hoed treatments. Responses to the extent of lime incorportion could be expected to be rate-dependent with the need for thorough incorporation diminishing as the rate increases. This is supported by Kauffman and Gardner (1978) who showed that maximum wheat yields could be achieved when only 30 per cent of the soil was adequately limed and that wheat roots were chemotrophic towards limed segments. Conventional cultivation will only incorporate lime to the depth of cultivation. Where soils are very highly acid below the depth of incorporation, adequate amelioration is currently impractical. To overcome this limitation, attempts have been made to supply liming materials to the subsoil (Kauffman and Gardner, 1978; Anderson and Henrick, 1983).

I. Maximising the benefit of lime within a farming system

Given the generally beneficial effects of liming, consideration needs to be given to how lime can be used and incorporated to maximise its benefit within different farming systems. In a rotation of four years crop: four years pasture, where cropping was the more profitable component, lime is most efficient when applied just prior to the start of the cropping phase as it can be incorporated as part of the conventional seedbed preparation.

Table XVI. The changes in pH (0.01 M $CaCl_2$) and exchangeable calcium (Ca_{ex}) (cmol(+).kg^{-1}) at two depths in a krasnozem after 7.5 t.ha^{-1} of limestone was applied over a 10-year period.

	Depth (cm)	Unlimed	Limed	Difference
pH	0–10	4.57	5.27	+ 0.70
	15–25	4.39	4.48	+ 0.09
Ca_{ex}	0–10	3.45	6.23	+ 2.78
	15–25	1.91	2.50	+ 0.59

Subsequent crops can then be established using reduced tillage techniques (Cregan *et al.*, 1984). In this system the period for repaying the capital invested in the lime is short, sometimes requiring only one crop to pay for the lime (May and Godyn, 1984). This system also ensures the adequate incorporation of lime. At the end of the pasture phase, soil tests can monitor soil pH changes and determine the need for additional liming. As a general guide, lime should be applied prior to the growing of the most acid-sensitive crop in the rotation.

J. Movement of pH change through the profile following liming

Except by mechanical means, the movement of alkali (HCO_3^- and CO_3^{2-} down the profile requires the addition of lime far in excess of the lime requirement (lime to reach pH 5.5 (1:5 soil:0.01 M $CaCl_2$) in top 10 cm). In some highly-acidifying systems, the application of limestone, even when incorporated in the surface 10 cm, may be ineffective in preventing subsoil acidification (Jones, 1976) or have only minimal effect (Friesen *et al.*, 1982). Data from the southern highlands of New South Wales (1500 mm average annual rainfall) (Table XVI) show that despite a total application of 7.5 t.ha^{-1} of limestone over the decade preceding sampling, and a pH difference of 0.7 in the surface 10 cm (relative to the unlimed soil), the limestone has had a minimal effect on pH ($\Delta pH = 0.09$) at 15–25 cm in this krasnozem. The $\Delta Ca/\Delta pH$ ratio for the top 10 cm indicates a relationship of 3.97 cmol(+).kg^{-1} of calcium per pH unit. At 15–25 cm the ratio is 6.55 indicating that calcium has moved down the soil profile more rapidly than either the CO_3^{2-} or HCO_3^-. The same effects were reported by Friesen *et al.* (1982) for tropical conditions and the data of Adams *et al.* (1967) also support this conclusion.

Although pH changes at depth may be insignificant, the increase in Ca_{ex} may be beneficial to plant nutrition.

K. Longevity of liming effects

In this chapter the longevity of lime application is considered in terms of its effects on soil pH (where the net change in pH returns to zero) rather than on changes in productivity. Inputs of acid after liming are considered to come from the same processes that resulted in acidification prior to liming. These processes are detailed in Chapter 2.

Constant and regular inputs of acidity into a soil could be expected to result in a logarithmic decline in soil pH because of the logarithmic nature of the pH scale. However, as most acid agricultural soils have near-linear titration curves over the pH range of 4 to 6 (Loynachan, 1981; Tisdale and Nelson, 1975; Shoemaker *et al.*, 1961), the pH decline would be expected to be linear over time if the inputs of acidity are uniform. Available observations for unlimed (Williams, 1980) and limed soils (Mahoney and Uren, 1982) suggest that this is not the case. The greater annual rate of pH decline for higher pH limed soils (Fig. 5) implies that there is a greater rate of acidification in higher pH soils (there was little movement of limestone down the profile in the soils tested). This would then also explain the non-linearity of the rate of pH decline for unlimed

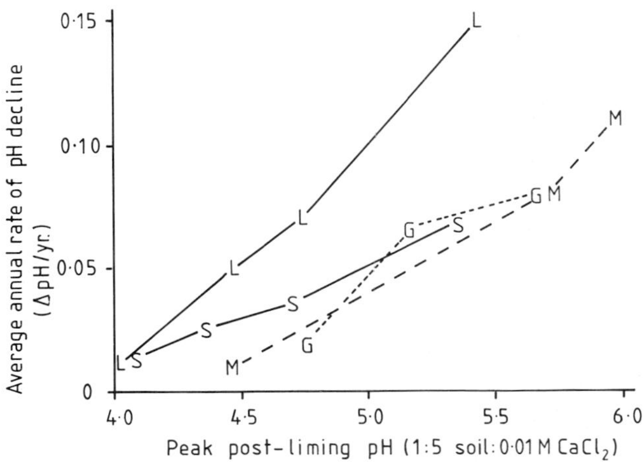

Fig. 5. The average annual rate of pH decline as a function of peak pH after liming (1:5 soil:0.01 M $CaCl_2$) for sites: Lima (L), Gooram (G) and Strathbogie (S) (Mahoney and Uren, 1982) and Mardin (M) (Southern Limestone, unpublished data).

soils (Williams, 1980). For limed soils there is often a steeper rate of pH decline in the first one or two years after liming than over the subsequent six to eight years of observation (see Chapter 7). However, this may be due to unreacted limestone dissolving in the test tube during pH determinations of samples taken soon after liming. Presumably the acid inputs decline with the pH, and levels of productivity. This observation is supported by New Zealand data where more limestone is required to retain pH > 6 than pH 5.6–5.8 (Edmeades *et al.*, 1984). In the high rainfall, west coast region of the South Island of New Zealand, 2.5 $t.ha^{-1}$ lime is needed every three to four years to maintain a pH of 5.8–6.0 (1:5 soil:water). In Canterbury, a pH in excess of 6.0 (1:5 soil:water) is maintained by 3 $t.ha^{-1}$ every six to eight years and a pH of 5.6–5.8 (1/5 water) by 1 $t.ha^{-1}$ every six to eight years (Edmeades, 1981). The average annual losses of limestone calculated from Mahoney and Uren (1982) for south-eastern Australia are in agreement with this.

The rate of pH decline therefore depends on the peak pH after liming, the annual input of acid and the slope of the soil's titration curve.

V. AMELIORATION OF SOIL ACIDITY WITHOUT LIME

Acidification of the soil occurs slowly (see Chapter 2). Our concern is to retard the acidification that results from agricultural practices and minimise the pH decline. For such an approach to succeed the sources of acid, the importance of additional acid, and the cause of alkali loss need to be identified (Cregan and Helyar, 1986; also see Chapter 2).

A. Nitrogen management

The chemical forms in which nitrogen enters and leaves the soil strongly influence the acidification rates of soils (Tisdale and Nelson, 1975; Helyar, 1976). For example, the use of less-acidifying fertilizers is a powerful tool in lessening acidification, and cost differences between nitrogen forms will need to be considered in the light of their acidication potential. Alternatively, practices that increase the plant's use of nitrate per unit input, and/or minimise nitrate leaching, decrease acidification (Table XVII). As nitrate in solution moves down through the soil profile, active plant growth is required to follow and use this mobile nitrogen. Ideally the plant's sink for nitrogen and the supply of nitrogen should be balanced to eliminate any leaching of nitrate and associated cations from the root zone. Management options available to improve the efficiency of use of nitrogen include:

1. reducing unnecessary (beyond the needs of the seedling plant) nitrogen mineralisation by eliminating or modifying traditional fallowing;
2. using split applications of fertilizer to ensure that nitrogen needs of crop and pasture are not exceeded. The presence of high soil levels of nitrogen increases the potential for nitrate leaching and acidification;
3. using deep-rooted grass, pasture and crop species that can use the available nitrate (Hood, 1976) and which may increase the rhizosphere pH by way of excess anion absorption (Israel and Jackson, 1978);
4. using perennial pastures that are more effective year-round sinks for nitrate. This may be most beneficial where regular wetting and drying cycles result in fluxes of nitrate down through the profile (Simpson, 1962) and where heavy stocking may result in nitrogen leaching from urine patches (Ball and Ryden, 1984). Perennial grasses and/or grass/clover mixtures, compared with legume-only pastures, also reduce the potential for nitrate leaching (Kilmer, 1974);
5. using crops which have appropriate root distribution patterns and early sowing times to minimise wastage of stored soil moisture and loss of nitrate. For crops and pastures grown in environments where regular cycles of wetting and drying can occur during the growing season, minimising the loss of nitrate can be equated with maximising the use of available soil moisture, as the nitrate moves through the profile with the wetting front (Olsen *et al.*, 1964);
6. improving crop and pasture growth to enable the plant to become a larger sink for nitrogen should also minimise nitrate leaching. This option involves nutritional and soil physical factors and the choice of varieties and species;
7. incorporating crop residues low in nitrogen. Cereal straw, for example, may potentially delay the leaching of nitrate mineralised during a period of fallowing. This nitrogen is used by soil microorganisms maintaining a constant C:N ratio (Russell, 1961) and may subsequently become available at a time of plant demand.

B. Product removal

Under Australian conditions little can be done to reduce the accumulation of alkalinity by grazing stock camping and depositing their dung in localised campsites. However, these patterns of fertility, including pH, should be recognised and considered when spreading lime or fertilizer (Hilder, 1966).

Table XVII. An example of a highly acidifying system: the annual pasture ley – cropping rotation

Seasonal events	Chemical, biological and physical processes that influence acidification	Reaction + = acid added − = alkali added o = balanced	Possible management modifications
A Crop			
Fallowing	Nitrification of organic N	+	Reduce or eliminate cultivated fallows
	Accumulation of NO_3^-	o	
	Oxidation of organic anions (surface-dominant reaction)	−	Earlier sowing to give more opportunity to use NO_3^-
Sowing	Nitrification of NH_4^+ fertilizer	++	Use NO_3^- fertilizers
	Nitrification of organic N	+	Use urea and NH_4NO_3 in preference to DAP, MAP and $(NH_4)_2SO_4$
	Leaching NO_3^- from surface		Use DAP in preference to MAP
Crop growth	NO_3^- uptake (and reduction to protein)	−	Improve crop growth to provide a larger NO_3^- sink
	Further leaching NO_3^- (deeper than roots)	o	
	Grain legumes — excess cation uptake → H+ excretion (dominant in the root zone)	+	Use cereals instead of grain legumes. Use deep rooted species for NO_3^- recovery
Harvest	Product removal		
	legume grain	++	Use cereals
	cereals	+ (very small)	
	Stubble returned to surface	−	

B Pasture			
Dead and decaying pasture	Organic matter accumulation	+	
	Oxidation of organic anions	−	
	Nitrification of organic N	+	Use perennial pastures with summer growth potential to use accumulating NO_3^-.
	Feeding conserved forage	−	Eliminate large stock camps and night paddock to reduce intrafarm transfers
Seasonal break	Further nitrification	+	Use permanent autumn-growing pasture
	Leaching of accumulated NO_3^-	o	
	Limited NO_3^- uptake by pasture	−	
	Feeding conserved forages	−	
Pasture growth	Urea — N → NO_3^-[a]	+	Avoid heavy stock concentrations
	NO_3^- leaching to subsoil[b]		Use vigorous deep-rooting species
	NO_3^- uptake	−	
	Excess cation uptake by legume → H^+ excretion	+	Use grass/legume pastures
	Fodder conservation:		
	grass hay	+	
	legume hay	++	Use mixed pastures
Senescence	Animal products removed[a]	+	

[a] Can occur in any season

[b] Acidification can be associated with cation losses and alkali addition with cation gains

[c] DAP = di-ammonium phosphate; MAP = mono-ammonium phosphate

The complete removal of plant material from the pasture in the form of conserved fodder, particularly of plants with a high residual alkalinity, such as lucerne (see Pierre and Banwart, 1973; Jarvis and Robson, 1983), will increase acidity. Feeding animals repeatedly in the same paddock with conserved fodder or holding dairy cows overnight in the same paddock often results in a substantial accumulation of alkali in these paddocks, with equivalent acidification of donor paddocks. Returning conserved fodder to the paddocks from which it was conserved, and rotation of night and/or holding paddocks, will reduce such trends.

Where the crop product is removed from the site of production, growing crops with low ash alkalinity, e.g. cereal grain crops, will minimise acidification. In pastures, a mixture of grasses (low ash alkalinity) and legumes (generally high ash alkalinity) is preferable to legume monocultures. Grain legume crops and hay legume crops are the most acidifying because they fix nitrogen, which is potentially acidifying (Nyatsanga and Pierre, 1973; Haynes, 1983). On an equivalence basis, the excess cation uptake of crops with high ash alkalinity is equal to the increase in soil acidity when crop products are removed (Israel and Jackson, 1978). Conversely, for plants with low ash alkalinity, the uptake of nitrate results in an addition of alkali to the soil potentially equal to the excess anion uptake.

VI. A PHILOSOPHY FOR THE TREATMENT OF ACID SOIL INFERTILITY

Under many farming and pastoral systems, soil pH will decline at an accelerated rate until it stabilises at a low terminal steady-state value (discussed in Chapter 2). To allow this process to proceed knowingly is extreme resource exploitation; restoration of such soils to former levels of productivity is unlikely to be totally successful. The greatest risks of such exploitation, besides the obvious effects of low yields and restricted choice of species, are the reduction in the fertility of subsoils and the high erosion risk following the failure of plants to provide adequate ground cover and soil stabilisation (Cumming, 1980).

A. Opportunity cost of soil acidity

As soils become more acidic the number of species that can be successfully grown becomes fewer until at pH levels where exchangeable Al exceeds 30 per cent of ECEC, only the most tolerant grasses and a few tropical legumes can grow satisfactorily. A farmer's choice thus becomes very narrow. If, because of this limitation, the most profitable enterprises cannot be undertaken, there will be significant opportunity costs or, as is more likely, a mix of direct losses (due to acid soil factors reducing yield) and opportunity costs. Detailed economic models are thus necessary to evaluate optimum strategies for liming over time. An example is given in Chapter 7. In addition, the economic implications of differences in the tolerance of crop and pasture species, and the effect of farming practices on acidification rate (in both the surface and the subsoils), need to be examined.

B. Lime use management

Since liming to maintain a near-neutral, arbitrary pH, e.g. pH 6.5 (1:5 soil:water) cannot be justified in terms of plant response across a range of agricultural species, criteria need to be developed to answer the following, much-asked questions: When should a soil be limed and how much lime should be applied?

Both biological and economic issues are involved and the conclusions will be determined by the following factors: (1) the inputs and outputs of acid and alkali over time; (2) the shape of the response surface for each particular situation; (3) the farmer's knowledge; (4) the farmer' s attitude to risk; and (5) government policy. The starting point for any well-informed decision making is a base of sound, biological response data. Methods for determining fertilizer requirements (including lime) for low-input agriculture have been reviewed by Sanchez and Salinas (1981). They conclude that, where the available capital is scarce and yields are less certain, the use of a linear plateau response function is most appropriate.

For the farmer to establish if soil acidity is likely to affect the yield of his pastures/crops he needs to know: (1) the level of soil acidity factors in his soil; and (2) the tolerance of the plants he wishes to grow. Once the soil acidity factors exceed the tolerance of the desired crops/pastures, yields will decline. This approach requires regular monitoring of soil acidity factors (soil test, tissue tests and biological assays can all be used), as well as the calibration of test values with plant growth responses and the establishment of critical values. The advantage of this approach is that the problem can be identified and remedied long before substantial losses occur, so preventing the ecosystem from being irretrievably damaged. Therefore, from a biological viewpoint, liming should start when the most acid-sensitive agricultural species will respond to lime. Any economic analyses used to assist decision making on lime use also need to consider the potentially debilitating, long-term effects of acidification if lime is withheld.

A similar system, based solely on Al_{ex}, was proposed by Cochrane *et al.* (1980) and Cregan (1980). They suggested that the objective of liming should be to reduce Al_{ex} to below the toxic level for the species/variety in question and, in addition, the lime rate should be calculated from the Al_{ex}. Although appealing due to its simplicity and use of existing soil test information, this method faced the same limitations as Kamprath's (1970) lime requirement estimation, on which it is based (see page 241). Furthermore, the amount of lime required for the satisfactory growth of a tolerant species is inadequate to restore flexibility to the farming system, because other problems of acid soils (manganese toxicity and *Rhizobium* infection) may remain. Also there is no residual alkali to ameliorate either infertile, acid subsoils or correct acid hot spots (see page x). As relatively low rates of lime are applied frequently, cartage costs may make lime more expensive on a unit-applied basis. This approach is thus not universally applicable, but does have a role in low-input agriculture where Al_{ex} dominates the ECEC.

Despite the above limitations, liming Australian soils to a pH at which most of the aluminium is precipitated from the soil solution (pH 5.0–5.5 (1:5 soil:0.01 M $CaCl_2$)) is a practical and sound recommendation. The practice of liming soils to near neutrality, as suggested in some European (e.g. Russell, 1961; Draycott and Messem, 1979) and North American literature (e.g. McLean, 1970) is excessive and expensive, and for less fertile Australian soils introduces the possibility of lime-induced nutritional imbalances leading to yield depressions (Kamprath and Foy, 1971; Sumner, 1982). Lime application rates can be determined readily by buffer pH methods calibrated to the lower target pH values as suggested by Tran and van Lierop (1979). This approach maintains an objective similar to that of Kamprath (1970) but is more widely applicable, as the observations of Woodruff (1967) in the midwestern USA and data from Australia (see page 242) show.

C. Government planning

Governments and their advisers, as well as all field-orientated advisers and researchers need to develop strategies for the future preservation of the fertility of agricultural soils. A *laissez-faire* non-interventionist approach may be inappropriate when the consequences of individual actions are the permanent degradation of a finite resource, our soil. A number of management strategies are presented in Table XVIII. The implementation of these strategies presents a significant socio-economic and environmental challenge for individuals and governments alike.

Table XVIII. Selected management strategies for different acid soil problems.

Problem and profile description[a]	Soil constitutes management problem (yes/no)	Management options
Very slightly acid to alkaline soils	no	Monitor pH at long intervals
Slight/moderate acidity; potential problem soils	no	1. Monitor pH at frequent intervals 2. prophylactic lime once soils reach "critical" pH
Only surface moderately acid	Yes (problems induced & natural)	1. Use tolerant species + monitoring 2. Lime banding + prophylactic; use all species 3. Lime + prophylactic or monitoring; use all species
All profile strongly acid	Yes (problems induced & natural)	1. Tolerant species + deep-placed lime and prophylactic 2. Non-agricultural use, e.g. national park, urban landuse
Only surface strongly acid	Yes (problems induced & natural)	1. Lime + prophylactic; use all species 2. Lime banding + prophylactic; tolerant species
Only subsoil strongly acid	Yes (problems induced & natural)	1. Use tolerant species + prophylactic 2. Use tolerant species + deep-placed lime + prophylactic

[a] Slightly acid: no effect on plant yield
Moderately acid: affects yield of sensitive and moderately sensitive species
Strongly acid: affects yield of tolerant species
Lime: Lime rate sufficient to raise pH to eliminate exchangeable Al (c. pH 5.2; 1:5 0.01M $CaCl_2$)
Prophylactic: applying sufficient lime at regular intervals to balance acidification
Critical pH: soil pH at which yield of sensitive species is affected (pH < 5.5; 1:5 0.01 M $CaCl_2$)

REFERENCES

Abbott, T. S. and McKenzie, D. C. (1986) Improving soil structure with gypsum. Agfact A. C. 10. New South Wales Department of Agriculture.

Adams, F. (1978). Liming and fertilization of Ultisols and Oxisols. *In* 'Mineral Nutrition of Legumes in Tropical and Subtropical Soils'. (C. S. Andrew and E. J. Kamprath, eds.), pp. 377–394. CSIRO, Melbourne.

Adams, W. E., Pearson, R. W., Jackson, W. A. and McCreery, R. A. (1967). Influence of limestone and nitrogen on soil pH and coastal bermuda grass. *Agronomy Journal* **59**, 450–453.

Anderson, A. J. and Moye, D. V. (1952). Lime and molybdenum in clover development on acid soils. *Australian Journal of Agricultural Research* **3**, 95–110.

Anderson, A. J. and Oertel , A. C. (1946). Plant responses to molybdenum as a fertilizer. 2. Factors affecting the response of plants to molybdenum. *Council for Scientific and Industrial Research (Australia) Bulletin* **198**, 25–44.

Anderson, A. J. and Spencer, D. (1948). Lime in relation to clover nodulation at sites on the southern tablelands of New South Wales. *Journal of the Australian lnsitute of Agricultural Science* **14**, 39–41.

Anderson, D. L. and Hendrick, J. G. (1983). Subsoil lime injector. *Soil Science Society of America Journal* **47**, 337–339.

Andrew, C. S. and Bryan, W. W. (1958). Pasture studies on the coastal lowlands of subtropical Queensland. III. The nutrient requirements and potentialities of *Desmodium uncinatum* and white clover on a lateritic podzolic soil. *Australian Journal of Agricultural Research* **9**, 267–286.

Andrew, C. S. and Norris, D. O. (1961). Comparative responses to calcium of five tropical and four temperate pasture legume species. *Australian Journal of Agricultural Research* **12**, 40–55.

Andrew, W. D. (1947). Pasture investigations. *Journal of the Department of Agriculture, Victoria* **45**, 489–500.

Anon. (1960). Guide to the Rutherglen Research Station 1959–60. Department of Agriculture, Victoria.

Anon. (1964). Soil and Pasture Research on the Northern Tablelands of New South Wales. CSIRO, Melbourne.

Anon. (1967). *Glycine javanica* — its strengths and weaknesses. *Rural Research in CSIRO* **59**, 11–17.

Ball, P. R. and Ryden, J. C. (1984). Nitrogen relationships in intensively managed temperate grasslands. *Plant and Soil* **76**, 23–33.

Barber, S. A. (1967). Liming materials and practices. *In* 'Soil Acidity and Liming' (R. W. Pearson and F. Adams, eds.), pp. 125–160. *Agronomy Monograph* **12**. American Society of Agronomy, Madison, Wisconsin.

Barber, S. A. (1984). Liming materials and practices *In* 'Soil Acidity and Liming' (F. Adams, ed.) pp. 171–209. Agronomy Monograph 12 Second edition American Society of Agronomy, Madison, Wisconsin.

Barrow, N. J. (1964). Some responses to lime on established pasture. *Australian Journal of Experimental Agriculture and Animal Husbandry* **4**, 30–33.

Barrow, N. J. (1965). Further investigations of the use of lime on established pastures. *Australian Journal of Experimental Agriculture and Animal Husbandry* **5**, 442–449.

Bouton, J. H., Syers, J. K. and Sumner, M. E. (1981). Alfalfa, *Medicago sativa* L., in highly weathered, acid soils. I. Effect of lime and P application on yield and acetylene reduction (N_2-fixation) of young plants. *Plant and Soil* **59**, 455–463.

Boynton, R. S. (1966). 'Chemistry and Technology of Lime and Limestone'. Interscience (John Wiley and Sons) New York, London, Sydney.

Bradfield, R. (1941). Calcium in the soil. 1. Physico-chemical relations. *Soil Science Society of America Proceedings* **6**, 8–15.

Bromfield, S. M., Cummings, R. W., David, D. J. and Williams, C. H. (1983). The assessment of available manganese and aluminium status in acid soils under subterranean clover pastures of various ages. *Australian Journal of Experimental Agriculture and Animal Husbandry* **23**, 192–200.

Brooks, C. O., Bouton, J. H. and Sumner, M. E. (1982). Alfalfa, *Medicago sativa* L., in highly weathered, acid soil. III. The effects of seedling selection in an acid soil on alfalfa growth at varying levels of phosphorus and lime. *Plant and Soil* **65**, 27–33.

Brown, J. R. and Cisco, J. R. (1984). An improved Woodruff buffer for estimation of lime requirement. *Soil Science Society of America Journal* **48**, 587–592.

Burridge, J. C. and Cunningham, R. K., (1960). Cacao yield maps and soil fertility in Ghana. *Empire Journal of Experimental Agriculture* **28**, 327–334.

Cochrane, T. T., Salinas, J. G. and Sanchez, P. A. (1980). An equation for liming acid mineral soils to compensate crop aluminium tolerance. *Tropical Agriculture* **57**, 133–140.

Conyers, M. K. (1983). Soil acidity and liming in the Southern Highlands of N. S. W. M. Sc. Agric. thesis, University of Sydney, Australia.

Cordero, S. J. and Blair, G. J. (1978). The effects of lime-pelleting and lime-superphosphate fertilizer on the growth of three annual legumes in an acid sandy soil. *Plant and Soil* **50**, 257–268.

Coventry, D. R. (1985). Changes in agricultural systems on acid soils in Southern Australia. Proceedings Third Australian Agronomy Conference, Hobart, Jan, 1985, 126–145.

Cregan, P. D. (1980). Soil acidity and associated problems. Guidelines for farmer recommendations. AG bulletin No. 7, New South Wales Department of Agriculture, Sydney.

Cregan, P. D. and Heylar, K. R. (1986). Non-acidifying farming systems. Proceedings 15th Riverina Outlook Conference, Wagga Wagga 49–62.

Cregan, P. D., Scott, B. J. and Cumming, R. W. (1984). Liming problem acid soils. Agfact Pl.4.1. New South Wales Department of Agriculture.

Cumming, R. W. (1980). Problems of fertility related to long periods of pasture improvement and management in a small urban water supply catchment. Proceedings First Australian Agronomy Conference, Gatton, 1980, 308–309.

De Boodt , M., De Leenheer, L. and Kirkham, D. (1961). Soil aggregate stability indexes and crop yields. *Soil Science* **91**, 138–146.

de Carvalho, M. M., Andrew, C. S., Edwards, D. G. and Asher, C. J. (1980). Comparative performance of six *Stylosanthes* species in three acid soils. *Australian Journal of Agricultural Research* **31**, 67–76.

Dear, B. S., Dann, P. and Loveland, B. (1985). A comparison of lime and sewage ash to correct soil acidity. Proceedings Third Australian Agronomy Conference, Hobart, January, 1985, 230.

Diatloff, A. and Luck, P. E. (1972). The effects of the interactions between seed inoculation, pelleting and fertiliser on the growth and nodulation of desmodium and glycine on two soils in S. E. Queensland. *Tropical Grasslands* **6**, 33–38.

Doss, B. D. and Lund, Z. F. (1975). Subsoil pH effects on growth and yield of cotton. *Agronomy Journal* **67**, 193–196.

Doyle A. D. and Bradley, J. (1982). Lime for cereals on soils in northern New South Wales. Proceedings Second Australian Agronomy Conference, Wagga Wagga, July 1982, 259.

Drake, F. R. and Kehoe, J. K. (1954a). Pasture and soil fertility investigations in East Gippsland. *Journal of the Department of Agriculture, Victoria.* **52**, 337–347.

Drake, F. R. and Kehoe, J. K. (1954b). Subterranean clover establishment on acid sandy soil in East Gippsland. *Journal of the Department of Agriculture, Victoria.* **52**, 385–388.

Draycott, P. and Messem, A. (1979). Soil acidity — the need for a systematic approach to liming. *British Sugar Beet Review* **47**, 21–23.

During, C., Jackson, B. L. J. and Dyson, C. B. (1984a). Lime effects on hill country. I. Effect of lime and monocalcium phosphate on soil moisture. *New Zealand Journal of Agricultural Research* **27**, 383–387.

During, C., Jackson, B. L. J. and Dyson, C. B. (1984b). Lime effects on hill country. 2. Relation of soil moisture, pH and aluminium values with pasture yield. *New Zealand Journal of Agricultural Research* **27**, 531–535.

Edmeades, D. C. (1981). Lime research by the Ministry of Agriculture and Fisheries. *New Zealand Agricultural Science* **15**, 192–199.

Edmeades, D. C., Pringle, R. M., Mansell, G. P. and Shannon, P. W. (1984). Effects of lime on pasture production on soils in the north island of New Zealand. I. Introduction and description of data. *New Zealand Journal of Agricultural Research* **27**, 349–356.

Edwards, D. G., Kang, B. T. and Danso, S. K. A. (1981). Differential response of six cowpea (*Vigna unguiculata* (L.) Walp.) cultivars to liming in an Ultisol. *Plant and Soil* **59**, 61–73.

Ellington, A. (1984). Soil acidity — effects and remedies: A review. Department of Agriculture, Victoria. Technical Report Series No. **102**.

Evans, W. (1977). 'Diary of a Welsh Swagman, 1869–1894'. Sun Books, Melbourne.

Fairfield-Smith, H. (1938). An empirical law describing heterogeneity in the yields of agricultural crops. *Journal of Agricultural Science*, (Cambridge) **28**, 1–23.

Fitzpatrick, E. N. (1958). Pasture development on peaty sands. *Journal of the Department of Agriculture, Western Australia (3rd Series)* **7**, 505–510.

Flemons, K. and Siman, A. (1970). Goulburn lucerne failure linked with induced manganese toxicity. *Agricultural Gazette of New South Wales* **81**, 662–663.

Follett, R. H. and Follett, R. F. (1983). Soil and lime requirement tests for the 50 states and Puerto Rico. *Journal of Agronomic Education* **12**, 9–17.

Fox, R. H. (1980). Comparison of several lime requirement methods for agricultural soils in Pennsylvania. *Communications in Soil Science and Plant Analysis* **11**, 57–69.

Foy, C. D. (1984). Physiological effects of hydrogen, aluminium and manganese toxicities in acid soils. *In* 'Soil Acidity and Liming'. Second Edition (F. Adams, ed.), pp. 57–97. American Society of Agronomy, Madison Wisconsin.

Friesen, D. K., Miller, M. H. and Juo, A. S. R. (1980). Liming and lime-phosphorus-zinc interactions in two Nigerian Utisols: II. Effects on maize root and shoot growth. *Soil Science Society of America Journal* **44**, 1227–1232.

Friesen, D. K., Juo, A. S. R. and Miller, M. H. (1982). Residual value of lime and leaching of calcium in a kaolinitic ultisol in the high rainfall tropics. *Soil Science Society of America Journal* **46**, 1184–1189.

Glencross, R. N. and Clarke, M. G. (1984). Soil acidity in the lower Great Southern. *Journal of Agriculture, Western Australia (4th series)* **25**, 142–145.

Gonzalez-Erico, E., Kamprath, E. J., Naderman, G. C. and Soares, W. V. (1979). Effect of depth of lime incorporation on the growth of corn on an oxisol of central Brazil. *Soil Science Society of America Journal* **43**, 1155–1158.

Gupta, U. C. and Lipsett, J. (1981). Molybdenum in soils, plants and animals. *Advances in Agronomy* **34**, 73–115.

Haddad, K. S. and Kaldor, C. J. (1982). Effect of parent material, natural available soil boron, and applied boron and lime on the growth and chemical composition of lucerne on some acidic soils of the Central Tablelands of New South Wales. *Australian Journal of Experimental Agriculture and Animal Husbandry* **22**, 317–323.

Haynes, R. J. (1983). Soil acidification induced by leguminous crops. *Grass and Forage Science* **38**, 1–11.

Helyar, K. R. (1976). Nitrogen cycling and soil acidification. *Journal of the Australian Institute of Agricultural Science* **42**, 217–221.

Helyar, K. R. and Anderson, A. J. (1970a). Responses of five pasture species to phosphorus, lime and nitrogen on an infertile acid soil with a high phosphate sorption capacity. *Australian Journal of Agricultural Research* **21**, 677–692.

Helyar, K. R. and Anderson, A. J. (1970b). Some effects of the soil pH on different species and on the soil solution for a soil high in exchangeable aluminium. Proceedings of the 11th International Grasslands Congress, Surfers Paradise, Australia, April, 1970. pp. 431–434. University of Queensland Press, St. Lucia.

Helyar, K. R. and Anderson, A. J. (1971). Effects of lime on the growth of five species, on aluminium toxicity and on phosphorus availability. *Australian Journal of Agricultural Research* **22**, 707–721.

Helyar, K. R. and Anderson, A. J. (1974). Effects of calcium carbonate on the availability of nutrients in an acid soil. *Soil Science Society America Proceedings* **38**, 341–346.

Hilder, E. J. (1966). Distribution of excreta by sheep at pasture. Proceedings Tenth International Grasslands Congress, Helsinki, 1966, 977–981.

Hirth, J. R., Coventry, D. R. and Reeves, T. G. (1982). Soil management studies and growth of subterranean clover in North-East Victoria. Proceedings Second Australian Agronomy Conference, Wagga Wagga, July, 1982, 263.

Holford, I. C. R (1985). Effects of lime on yields and phosphate uptake by clover in relation to changes in soil phosphate and related characteristics. *Australian Journal of Soil Research* **23**, 75–83.

Hood, A. F. M. (1976). The leaching of nitrates from intensively managed grassland at Jealott's Hill. *In* 'Agriculture and Water Quality'. pp. 201–221. Ministry of Agriculture, Fisheries and Food, Technical Bulletin No. 32, London.

Hosking, J. W. (1968). The development of the South Gippsland plains. Natural Resources Conservation League of Victoria Convention, Leongatha, October, 1968. pp. 33–43.

Hosking, J. W. (1972). The effect of lime and dolomite on pasture growth. Pasture Refresher Course, Glenormiston, Victoria, September, 1972.

Hoyt, P. B. (1981). Improvements in soil tilth and rapeseed emergence by lime applications on acid soils in the Peace River region. *Canadian Journal of Soil Science* **61**, 91–98.

Hutchinson, H. B. and MacLennan, K. (1914) The relative effect of lime as oxide and carbonate on certain soils. *Journal of Agricultural Science* **6**, 302–322.

Isbell, R. F. , Jones, R. K. and Gillman, G. P. (1976). Plant nutrition studies on some yellow and red earth soils in northern Cape York Peninsula. I. Soils and their nutrient status. *Australian Journal of Experimental Agriculture and Animal Husbandry* **16**, 532–541.

Israel, D. W. and Jackson, W. A. (1978). The influence of nitrogen nutrition on ion uptake and translocation by leguminous plants. *In* 'Mineral Nutrition of Legumes in Tropical and Subtropical Soils'. (C. S. Andrew and E. J. Kamprath, eds.), pp. 113–129. CSIRO, Melbourne.

Jarvis, S. C. and Robson, A. D. (1983). The effect of nitrogen nutrition of plants on the development of acidity in Western Australian soils. II. Effects of differences in cation/anion balance between plant species grown under non-leaching conditions. *Australian Journal of Agricultural Research* **34**, 355–365.

Jones, H. R. (1975). Factors affecting plant responses to lime on acid soils. M. Agr. Sc. Thesis, La Trobe University.

Jones, M. J. (1976). Effects of three nitrogen fertilizers on exchangeable cation content at different depths in cropped soils at two sites in the Nigerian savanna. *Tropical Agriculture* **53**, 243–254.

Kamprath, E. J. (1970). Exchangeable aluminium as a criteria for liming leached mineral soils. *Soil Science Society of America Proceedings* **34**, 252–254.

Kamprath, E. J. (1978). Lime in relation to Al toxicity in tropical soils. *In* 'Mineral Nutrition of Legumes in Tropical and Subtropical Soils'. (C. S. Andrew and E. J. Kamprath, eds.), pp. 233–245. CSIRO, Melbourne.

Kamprath, E. J. and Foy, C. D. (1971). Lime-fertilizer-plant interaction in acid soils. *In* 'Fertilizer Technology and Use'. (R. A. Olson, ed.) pp. 105–151. Soil Science Society of America, Madison, Wisconsin.

Kauffman, M. D. and Gardner, C. H. (1978). Segmented liming of soil and its effects on the growth of wheat. *Agronomy Journal* **70**, 331–336.

Kehoe, J. K. and Curnow, B. (1963). Root growth of subterranean clover on some acid sandy soils in Victoria. *Australian Journal of Experimental Agriculture and Animal Husbandry* **3**, 11–16.

Keren, R. and Shainberg, I. (1981). Effect of dissolution rate on the efficiency of industrial and mined gypsum in improving infiltration of a sodic soil. *Soil Science Society of America Journal* **45**, 103–107.

Kerridge, P. C. and Everett, M. L. (1975). Nutrient studies on some soils from Eungella and East Funnel Creek, Mackay hinterland, Queensland. *Tropical Grasslands* **9**, 219–228.

Kerridge, P. C., Andrew, C. S. and Murtha, G. G. (1972). Plant nutrient status of soils of the Atherton Tableland, North Queensland. *Australian Journal of Experimental Agriculture and Animal Husbandry* **12**, 618–627.

Kilmer, V. J. (1974). Nutrient losses from grassland through leaching and runoff. *In* 'Forage Fertilization'. (D. A. May, ed.), pp. 341–362. American Society of Agronomy, Madison, Wisconsin.

Knight, C. L. (1976). Limestone. *In* 'Economic Geology of Australia and Papua New Guinea, 4, Industrial Minerals and Rocks, Monograph 8'. (C. L. Knight, ed.), pp. 173–212. The Australasian Institute of Mining and Metallurgy.

Kunishi, H. M. (1982). Combined effect of lime, phosphate fertiliser, and aluminium on plant yield from an acid soil of the southeastern United States. *Soil Science* **134**, 233–238.

Lee, M. T. and Wilson, G. L. (1972). The calcium and pH components of lime responses in tropical legumes. *Australian Journal of Agricultural Research* **23**, 257–265.

Leeper, G. W. (1964). 'Introduction to Soil Science'. Melbourne University Press, Carlton.

Lipsett, J. (1980). Effect on growth of several crop species after liming light-textured acid soil in southern New South Wales. *Journal of the Australian Institute of Agricultural Science*, **46**, 245–247.

Lipsett, J., Pinkerton, A. and David, D. J. (1979). Boron deficiency as a factor in the reclamation by liming of a soil contaminated by mine waste. *Environmental Pollution* **20**, 231–240.

Loynachan, T. E. (1981). Lime requirement methods for cold region soils. *Soil Science Society of America Journal* **45**, 75–80.

McGowan, A. A. (1967). Ecological studies of Wimmera ryegrass and associated annual grasses in the clover ley rotation in North Eastern Victoria. M. Agr. Sc. Thesis, University of Melbourne.

McLachlan, K. D. (1953). Effects of lime and copper on subterranean clover grown on acid soil. *Australian Journal of Agricultural Research* **4**, 151–159.

McLachlan, K. D. (1980a). Nutrient problems in sown pasture on an acid soil. I. Survey of the problems and suggested corrective treatments. *Australian Journal of Experimental Agriculture and Animal Husbandry* **20**, 319–326.

McLachlan, K. D. (1980b). Nutrient problems in sown pasture on an acid soil. II. Role of lime and superphosphate. *Australian Journal of Experimental Agriculture and Animal Husbandry* **20**, 568–575.

McLean, E. O. (1970). Lime requirements of soils — Inactive toxic substances or favorable pH range. *Soil Science Society of America Proceedings* **34**, 363–364.

Mahoney, G. P. and Uren, N. (1982). The long-term effect of lime on soil pH. Proceedings Second Australian Agronomy Conference. Wagga Wagga, July 1982, p. 258.

Mahoney, G. P. , Jones, H. R. and Hunter, J. M. (1981). Effect of lime on lucerne in relation to soil acidity factors. Proceedings 14th International Grasslands Congress, Lexington, Kentucky, p. 299.

May, M. G. and Godyn, D. L. (1984). Test your lime spreading rates. *Farm Magazine* **5**, 68–69.

Munns, D. N (1965). Soil acidity and the growth of a legume. I. Interactions of lime with nitrogen and phosphate on growth of *Medicago sativa* L. and *Trifolium subterraneum* L. *Australian Journal of Agricultural Research* **16**, 773–741.

Munns, D. N. and Fox, R. L. (1976). Depression of legume growth by liming. *Plant and Soil* **45**, 701–705.

Munns, D. N. and Rudczuk, E. (1966). Lime and superphosphate factorial field trial with lucerne on an acid soil. CSIRO Plant Industry Division, Field Station Record **5**, 63–66.

Newman, R. J. (1948). Pasture investigations in the Western District. *Journal of the Department of Agriculture, Victoria* **46**, 49–57.

Nyatsanga, T. and Pierre, W. H. (1973). Effect of nitrogen fixation by legumes on soil acidity. *Agronomy Journal* **65**, 936–940.

Oertel, A. C., Prescott, J. A. and Stephens, C. G. (1946). The influence of soil reaction on the availability of molybdenum to subterranean clover. *Australia Journal of Science* **9**, 27–28.

Olsen, R. A., Frank, K. D. and Dreier, A. F. (1964). Controlling losses of fertilizer nitrogen from soils. Proceedings Eighth International Congress of Soil Science, Bucharest, Romania p. 1023.

Ozanne, P. G. and Howes, K. M. W. (1974). Increased seed production of subterranean clover pastures in response to fertilizers supplying calcium. *Australian Journal of Experimental Agriculture and Animal Husbandry* **14**, 749–757.

Paton, D. F. (1956). Investigations on trace element deficiencies of pasture in Tasmania. *Journal of the Australian Institute of Agricultural Science* **22**, 33–36.

Paton, D. F. (1958). The use of lime in clover establishment on acid soils in Tasmania. Australian Agrostology Conference, Armidale. New South Wales, December, 1958 **1**(1), 25–1 to 25–7.

Paton, D. F. (1960). Problems in legume establishment with special reference to nodulation and nutrition. *Journal of the Australian Institute of Agricultural Science*, **26**, 32–37.

Paton, D. F. and Loneragan, J. F. (1960). An effect of lime on residual phosphorus in soil. *Australian Journal of Agricultural Research* **11**, 524–529.

Pearson, R. W. and Adams, F. (1967). 'Soil Acidity and Liming'. American Society of Agronomy Monograph 12, Madison, Wisconsin.

Peele, T. C. (1936). The effect of calcium on the erodibility of soils. *Soil Science Society of America Proceedings* **1**, 47–59.

Pierre, W. H. and Banwart, W. L. (1973). The excess-base and excess-base/nitrogen ratio of various crop species and of plant parts. *Agronomy Journal* **64**, 91–96.

Pinkerton, A. and Simpson, J. R. (1982). Root growth in soils rendered acid by improved pastures. Proceedings Second Australian Agronomy Conference, Wagga Wagga, July, 1982, p. 262.

Porter, W. M. and Wilson, I. R. (1984). Soil acidity in the eastern wheatbelt. *Journal of Agriculture, Western Australia* **25**(4), 132–135.

Probert, M. E. (1980). Growth responses to various calcium sources in a yellow earth soil with low calcium status. *Australian Journal of Experimental Agriculture and Animal Husbandry* **20**, 240–246.

Probert, M. E. Winter, W. H. and Jones, R. K. (1979). Plant nutrition studies on some yellow and red earth soils in northern Cape York Peninsula. 3. Effects of lime and placement on responses to applied phosphorus. *Australian Journal of Experimental Agriculture and Animal Husbandry* **19**, 583–589.

Reeves, T. G., Ellington, A., Coventry, D. R., Hirth, J. R. and Steed, G. R. (1982). Improved soil management in N. E. Victoria. Proceedings Second Australian Agronomy Conference, Wagga Wagga; July 1982, p. 326.

Robinson, G. W. and Lloyd, W. E. (1915). On the probable error of sampling in soil surveys. *Journal of Agricultural Science, Cambridge* **7**, 144–153.

Robson, A. D. and Loneragan, J. R. (1970). Sensitivity of annual *Medicago* species to manganese toxicity as affected by calcium and pH. *Australian Journal of Agricultural Research* **21**, 223–232.

Roughley, R. J. and Walker, H. M. (1973). A study of inoculation and sowing methods of *Trifolium subterraneum* in New South Wales. *Australian Journal of Experimental Agriculture and Animal Husbandry* **13**, 284–291.

Rowe, B. A. (1982). Effects of limestone on pasture yields and the pH of two krasnozems in north-west Tasmania. *Australian Journal of Experimental Agriculture and Animal Husbandry* **22**, 100–105.

Russell, E. W. (1961). 'Soil Conditions and Plant Growth' (9th edition), Longmans, London.

Russell, E. W. (1984). 'Soil Conditions and Plant Growth' (10th edition), Longmans, London.

Sanchez, A. and Salinas, G. (1981). Low input technology for managing oxisols and ultisols in tropical America. *Advances in Agronomy* **34**, 279–406.

Savage, G. F. J. and Hosking, W. J. (1982). Effect of liming pasture on milk production and animal health and on growth and composition of pastures. Proceedings 14th International Grasslands Congress (J. A. Smith, ed.). June, 1981. Lexington, Kentucky. p. 128.

Scott, B. J. (1982). Responses to lime by cereals. Proceedings Second Australian Agronomy Conference. Wagga Wagga, July 1982, p. 260.

Shannon, P. W., Pringle, R. M. Mansell, G. P. and Edmeades, D. C. (1984). Effects of lime on pasture production on soils in the North Island of New Zealand. 2. The nattire of seasonal pasture growth responses to lime application. *New Zealand Journal of Agricultural Research* **27**, 357–361.

Shoemaker, H. E., McLean, E. O. and Pratt, P. F. (1961). Buffer methods for determining lime requirement of soils with appreciable amounts of extractable aluminum. *Soil Science Society of America Proceedings* **25**, 274–277.

Siman, A., Craddock, F. W. Nicholls, P. J. and Kirton, H. C. (1971). Effects of calcium carbonate and ammonium sulphate on manganese toxicity in an acid soil. *Australian Journal of Agricultural Research* **22**, 201–214.

Simmons, K. V. (1962). Establishing subterranean clover on acid soils. *Agricultural Gazette of New South Wales* **73**, 183–188.

Simpson, J. R. (1962). Mineral nitrogen fluctuations in soils under improved pastures in southern New South Wales. *Australian Journal of Agricultural Research* **13**, 1059–1072.

Simpson, J. R. and Pinkerton, A. (1978). Effects of subsoil acidity and liming on the root behaviour and shoot growth of some leguminous plants. *In* 'Plant Nutrition 1978'. Proceedings Eighth International Colloquium Plant Analysis Fertiliser Problems. Auckland, New Zealand, September, 1978. pp. 449–455. Information Series No. 134. New Zealand Department of Scientific and Industrial Research, Wellington.

Simpson, J. R., Pinkerton, A. and Lazdovskis, J. (1979). Interacting effects of subsoil acidity and water on the root behaviour and shoot growth of some genotypes of lucerne (*Medicago sativa* L.). *Australian Journal of Agricultural Research* **30**, 609–619.

Spencer, D. (1950). The effect of calcium and soil pH on nodulation of *T. subterraneum* L. clover on a yellow podsol. *Australian Journal of Agricultural Research* **1**, 374–381.

Stewart, and Bandel, V. (undated). Uniform lime and fertilizer spreading. Bulletin 254. Co-operative Extension Service, University of Maryland, USA.

Sumner, M. E. (1979). Response of alfalfa and sorghum to lime and P on highly weathered soils. *Agronomy Journal* **71**, 763–766.

Temple-Smith, M. G., Wright, D. N., Laughlin, J. C. and Hoare, B. J. (1983). Field response of poppies (*Papaver somniferum* L.) to lime application on acid krasnozems in Tasmania. *Journal of Agricultural Science, Cambridge* **100**, 485–492.

Thompson, S. C., Gemell, G. and Roughley, R. J. (1984). Host specificity for nodulation among Australian Acacia. Proceedings Seventh Australian Legume Nodulation Conference. Australian Institute of Agricultural Science Occasional Publication No. 12. 27–28.

Tisdale, S. L. and Nelson, W. L. (1975). 'Soil Fertility and Fertilizers' (3rd edition) MacMillan, New York.

Tiver, N. S. (1957). Lime — a fresh look at an old fertilizer. *Journal of Agriculture, South Australia* **61**, 359–363.

Tiver, N. S. (1958). Fertilizers on deep sands — new facts about applying lime, trace elements. *Journal of Agriculture, South Australia* **62**, 141–143; 153.

Tran, T. S and van Lierop, W. (1981). Evaluation and improvement of buffer–pH lime requirement methods. *Soil Science* **131**, 178–188.

Tran, T. S. and van Lierop, W. (1982). Lime requirement determination for attaining pH 5.5 and 6.0 of coarse-textured soils using buffer pH methods. *Soil Science Society of America Journal* **46**, 1008–1014.

Twentyman, R. L. (1937). Pasture improvement work. Results of 1936 tests. *Journal of the Department of Agriculture, Victoria* **35**, 173–194.

Twentyman, R. L. (1938). Pasture improvement. Results of 1937 tests. *Journal of the Department of Agriculture, Victoria* **36**, 157–170; 197–204.

Van Wambeke, A. (1976). Formation, distribution and consequences of acid soils in agricultural development. *In* 'Plant Adaptations to Mineral Stress in Problem Soils. Proceedings of a Workshop'. (M. J. Wright, ed.), pp. 15–24. Cornell University, New York.

Vimpany, I. A. and Bradley, J. (1979). Development of field and glasshouse tests for potassium and phosphorus. Final Report to the Australian Wool Corporation, New South Wales Department of Agriculture.

Webster, R. and Butler, B. E. (1976). Soil classification and survey studies at Ginninderra. *Australian Journal of Soil Research*, **14**, 1–24.

White, R. E. (1967). Nutrient requirements of white clover based pastures on a Krasnozem basalt at Maleny, south eastern Queensland. *Australian Journal of Experimental Agriculture and Animal Husbandry* **7**, 509–514.

Willett, I. R., Jakobsen, P., Cunningham, R. B. and Gunthorpe, J. R. (1984). Effects of lime-treated sewage sludge on soil properties and plant growth. CSIRO Division of Soils, Divisional Report No. 67, CSIRO, Australia.

Williams, C. H. (1980). Soil acidification under clover pasture *Australian Journal of Experimental Agriculture and Animal Husbandry* **20**, 561–567.

Williams, C. H. and Raupach, M. (1983). Plant nutrients in Australian soils. *In* 'Soils: An Australian Viewpoint'. Division of Soils, CSIRO, pp. 777–793. CSIRO: Melbourne/ Academic Press: London.

Woodhouse, W. W., Jnr. (1956). Effect of placement and rate of phosphate, potash and limestone on the growth of alfalfa and Lespedeza. *Soil Science Society of America Proceedings* **20**, 15–18.

Woodruff, C. M. (1967). Crop responses to lime in the midwestern United States. *In* 'Soil Acidity and Liming'. (R. W. Pearson and F. Adams, ed.), pp. 207–227. American Society of Agronomy, Madison, Wisconsin.

Yeates, J. S., McGhie, D. A. and Wilson, I. R. (1984). Soil acidity on high rainfall pastures. *Journal of Agriculture, Western Australia (4th series).* **25**, 138–141.

7

The Integration of Data on Lime Use by Modelling

Z. HOCHMAN, D. L. GODYN and B. J. SCOTT

I. INTRODUCTION

There is a wealth of detailed experimental data regarding many aspects of the soil acidity problem. This information is generally compartmentalised into the areas of expertise indicated by chapter titles in this monograph. As a result, specialists tend to delve in ever greater detail into various aspects of their disciplines. By contrast, practical advice to farmers relies on rules of thumb.

Some degree of synthesis is attempted by all people concerned with the problem of soil acidity. However it is impossible for the human brain to consider and properly weigh the importance of the many interrelated

SOIL ACIDITY AND PLANT GROWTH
ISBN 0 12 590655 2

factors. Therefore researchers are often biased towards the importance of their own area of expertise, while farmers and advisory workers tend to rely on guesses, intuition and limited experience. There is a need for information from the relevant disciplines to be synthesised to a level appropriate to rational decision making on practical questions such as: 'How much lime should I use on this paddock to grow that crop?'

Modelling can provide the tools needed to integrate and quantify our knowledge and understanding of the relevant chemical and biological processes. This aspect of modelling is explored briefly (Section II) with an outline of a conceptual model for crop growth on acid soils.

Another important role for modelling is illustrated with a simple decision-oriented model — 'Lime-it'. The micro-computer provides the medium with which the best currently available information on production and economics is integrated at a level appropriate for on-farm decisions.

II. THE CONCEPTUAL BASIS OF A CROP GROWTH MODEL FOR ACID SOILS

A framework for a detailed mechanistic model, describing crop growth on acid soils, is outlined in this section. The purpose of such a model is not to predict yields but to give an overall view and summary of how yields may be affected by soil acidity in terms of the interacting underlying processes. In outlining this model, we hope to encourage an integrated approach to research so that specialists in any component of the system become more aware of, and pay more attention to, interactions with other components. Such an approach would eventually provide modellers with the quantitative relationships required to describe crop growth on acid soils. The model is represented by a flow diagram (Fig. 1) in which material and information flow, state, rate, driving and auxiliary variables are identified for a single integration interval. At this stage there is no attempt to quantify these relationships. In Figure 1, crop growth is described as a state variable which is dependent on the other rate, auxiliary and exogenous variables. The potential rate of crop growth is calculated in the 'net assimilation' variable. This rate variable is in turn calculated from the exogenous variables 'light' and 'temperature' and from 'soil moisture' (which is a state variable of the soil submodel). Actual crop growth also depends on the crop nutrient status, through an auxiliary function which is calculated from the state variable 'nutrients in crops'.

The degree of crop injury due to aluminium or manganese toxicity affects crop growth directly through chemical damage of roots. In the

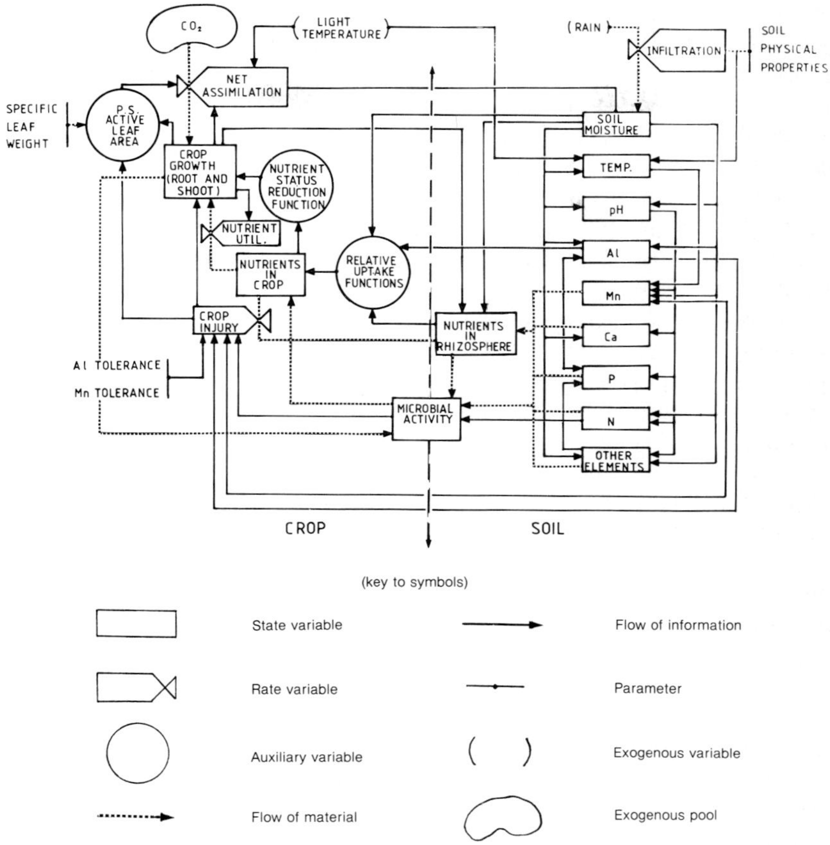

Fig. 1. A flow diagram of a crop growth model for acid soils.

case of aluminium toxicity, crop injury also affects flow of nutrients from rhizosphere to crop. Accumulation of manganese in leaf tissue reduces the photosynthetic efficiency of leaves, thus reducing the rate of assimilation. The 'crop injury' function is derived from information on soil levels of aluminium and manganese together with specific crop (or cultivar) tolerance factors. The effects of soil acidity (both harmful and beneficial) on crop injury through soil-borne disease (microbial sub-routine) must also be considered.

The material flows considered for 'crop growth' are atmospheric CO_2 (an exogenous pool) and nutrients. These flows are regulated by the rate variables 'net assimilation' and 'nutrient utilisation (demand)' respectively. The nutrient utilisation function reflects an assumption that nutrient flow to growing organs is determined by 'demand' as well as supply. Flow of material away from crop growth is also indicated. This is the flow of assimilates to the 'microbial activity' state variable. This flow represents the cost side of the symbiotic relationship of crops with rhizobia and mycorrhizas.

The state variable 'nutrients in crop' provides an important link between the crop and soil subroutines. Values are obtained from the state variables 'nutrients in rhizosphere' and 'microbial activity'. The functions necessary to derive values for 'nutrients in crop' from 'nutrients in rhizosphere' must be provided in the auxiliary variable 'relative uptake functions'. These functions also require information on aluminium in the soil and on soil moisture which differentially affects the uptake of various nutrients. 'Microbial activity' contributes to 'nutrients in crop', especially nitrogen which is fixed by rhizobia in the roots of leguminous crops and pastures.

The soil submodel contains physical and chemical elements. Soil moisture and soil temperature are determined by physical parameters. Soil pH, aluminium, manganese and the various nutrients are the interrelated chemical state variables. The soil section of Figure 1 represents one of an unspecified number of soil compartments in a vertical profile.

'Soil moisture' is a function of 'rain'. The flow of rain water through soil is regulated by 'infiltration' which can be determined from physical constants specific to the soil in question. Soil temperature is either measured or determined from air temperature and soil physical properties which determine conductivity of temperature through the soil.

The soil chemical parameters are derived initially from soil analysis. Their dynamic states are further defined by soil moisture and temperature as well as the complex interrelationships between these state variables. Presentation of the soil chemistry section of the relational diagram (Fig. 1) is simplified by the omission of all rate, and auxiliary variables. Soil pH, which is a measured parameter, affects all the other soil chemical state variables. The state variables designated 'Al', 'Mn', 'Ca', 'P', etc. refer (notionally at least) to concentrations of these elements available to the plant. Aluminium is a function of pH, soil moisture, calcium and phosphorus. Manganese is a function of pH, soil moisture and soil temperature. Calcium is a function of pH. Phosphorus is a function of pH, aluminium, calcium and 'other elements'. Nitrogen and 'other elements' are a function of pH and soil moisture.

The values of the state variables 'Mn', 'Ca', 'P', 'N', 'trace and other elements' are used to calculate the state variables 'nutrients in rhizosphere' and 'microbial activity' as indicated by the 'flow of material' lines.

'Microbial activity', which could encompass rhizobia, mycorrhizas and soilborne diseases, is also a function of the state variables 'soil moisture', 'soil temperature', 'pH', 'Al', 'Ca', 'N' and 'trace and other elements'.

This flow diagram represents a single time interval of integration (Δt) which may, be in the order of one day to one week in a model of this level of abstraction. Once these relations are quantified, the actual simulation procedure simply requires the calculation of all rate variables so that they may be used to update the state variables as follows: state variable at time $t + \Delta t$ is equal to state variable at time t plus its net rate of change at time t multiplied by the interval of integration Δt.

The development of a major model of the type outlined above would provide a useful framework within which to integrate and evaluate research in a number of disciplines. It would also serve to identify where information is most deficient and to provide a guide for future research and a means of co-ordinating it. However such a model would be inappropriate as a tool to assess on-farm decisions regarding use of lime. A simpler, more empirically-based model is required for that purpose.

III. 'LIME-IT' — A SIMPLE MODEL Of LIME USE IN LONG-TERM SUBTERRANEAN CLOVER PASTURES.

A. Frame of reference for 'Lime-it'

Increasing numbers of primary producers are being forced to face up to a relatively new technology in Australian agriculture — use of lime. In agriculture, most decisions are risky because of the complexity of agricultural systems and the unpredictable nature of the environment and commodity prices. Four additional factors make decisions on lime use more difficult than most:

1. Because lime modifies many parameters which affect plant performance (not all favourably), it is difficult to predict crop responses.
2. The long residual effect of lime needs to be taken into account. This is particularly relevant in long-term pastures where there may be long intervals between crops. Given the greater efficiency of incorporated lime, compared with top-dressed lime, cropping provides an opportunity do integrate the liming operations with the existing production system. Therefore the residual value of liming must be considered over the whole crop-pasture phase.
3. There is very little direct experience among farmers in the technology of lime use. The latest statistics on production of agricultural crushed

lime in Australia are for 1983–84 when 171 104 tonnes of lime were crushed. This contrasts with 82 136 tonnes in 1972–73. Much of the increased production occurred after 1979–80 when 101 392 tonnes were crushed for agricultural use (Australian Bureau of Statistics Catalogue 8303.0). The Australian experience contrasts with that of the USA where 32.9 million tons of agricultural limestone were applied in 1979 (Hall, 1983).

4. Amelioration of soil acidity often requires a high cost input in an extensive system of agricultural production. This is further complicated by the fact that the cost of lime is a function of its source, reflecting cost of production as well as cost of transport from source to farm. The cost of an arbitrary rate of lime as a fraction of total income for production for various crops and locations (Table I) also serves to illustrate the economic folly of 'importing' recommendations from

Table I. Cost of lime as a fraction of product value for various crops and locations.

Crop	Location	Value of product ($[a]/t.ha^{-1}year^{-1})	Cost of lime @ 2.5 t.ha^{-1}	Cost/value
Lucerne[b] (for hay)	Mayfield, Kentucky, USA	260	33	12.7
Soybean[b]	Crossville, Alabama, USA	595	33	6.5
Corn[b]	Southern Coastal Plain, Virginia USA	740	33	4.5
Winter wheat[c]	Central Clay area, Netherlands.	3551	380	10.7
Potatoes[d]	Central Clay area, Netherlands.	9450	380	4.0
Dairy pasture[e]	Volcanic Ash Soils, New Zealand.	460	42	9.1
Sheep pasture[f]	'Easy Hill Country, King Country, New Zealand.	300	71	23.7
Pasture for sheep[g]	Southern Slopes, New South Wales, Australia.	211	150	71.1
Wheat[g]	Wagga Wagga, New South Wales, Australia.	264	150	56.8

[a] $ symbol used to represent local currency
[b] (Hall, 1983)
[c] (Proef Station voor de Akkerbouw en Groenteteeld in Voilegrond, 1982)
[d] (Landbouw Economisch Instituut, 1986)
[e] (Edmeades *et al.*, 1981)
[f] (Edmeades *et al.*, 1985)
[g] (Authors' estimates for 1985)

other countries or of 'rule of thumb' or 'blanket' recommendations. Long-term pastures based on subterranean clover present a particularly challenging area to tackle as we must deal with a species which is relatively tolerant of soil acidity (Helyar, 1978) in an extensive production system where many producers might dismiss liming as too expensive.

The model 'Lime-it' was developed to facilitate rational decisions on the use of lime. Specifically the objectives were: to develop a model as a tool for integrating currently available data; and, using reasonable assumptions, to predict soil and production parameters required to assess the benefits of liming and the effects of liming on returns to investment and cash flow.

B. A descriptive outline of 'Lime-it'

The effect of liming is calculated in four submodels representing the soil, pasture, livestock and economic factors, using an annual time step and a 10-year calculation period. The calculation sequence for a single year is described below and illustrated in a flow diagram (Fig. 2).

1. Soil submodel

pH and total exchangeable cations (TEC) are predicted from initial values and lime rate. Exchangeable aluminium (Al) is predicted from its initial value, soil pH and TEC. The decline of pH with time is predicted from empirical functions and adjusted for the pH buffer capacity of the soil which is calculated in the soil submodel.

2. Pasture submodel

A Mitscherlich function is used to generate 10 optional response curves for calculating the percentage of optimum pasture production as a function of soil pH. This is calculated for each year the model is run.

3. Livestock production submodel

Pasture production as a percentage of potential pasture production, current stocking rate, and gross margin per dry sheep equivalent (DSE) is used to calculate potential livestock production and the stocking rate required to utilise additional pasture production.

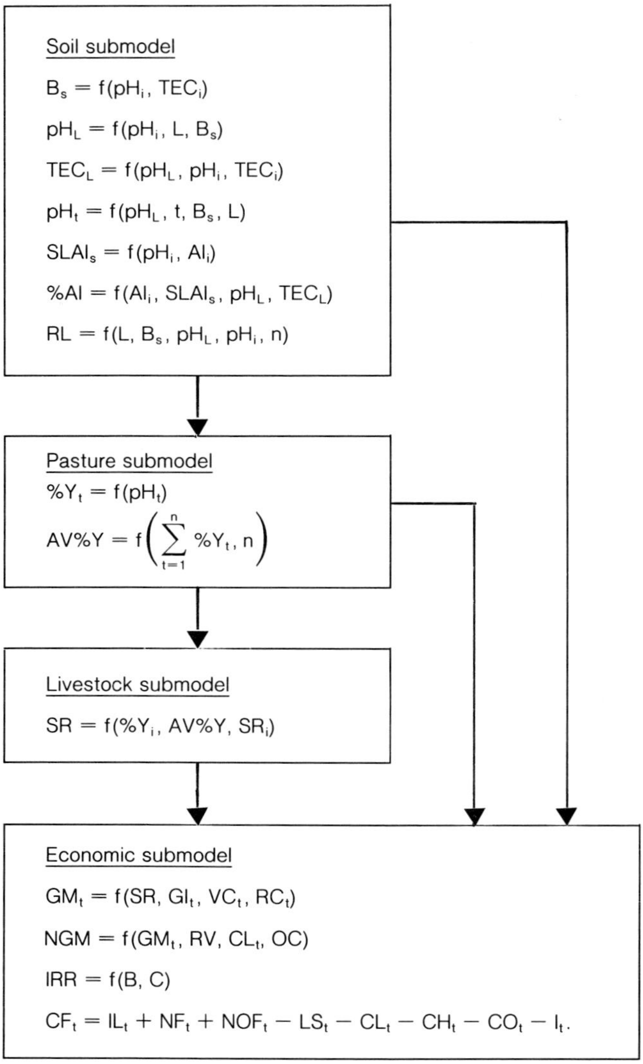

Fig. 2. The calculation sequence involved in running 'Lime-it' for a single year.

FIGURE 2 (LEGEND)

AV%Y	=	Percent pasture production averaged over n years
B	=	Benefit due to liming ($)
B_s	=	Buffering capacity of soil s
C	=	Cost due to liming ($)
CF_t	=	Cash flow in year t ($)
CH_t	=	Household cost in year t ($)
CL_t	=	Cost of lime in year t ($)
CO_t	=	Overhead costs in year t ($)
GI_t	=	Gross income per dry sheep equivalent (DSE) per hectare ($)
GM_t	=	Gross margin in year t ($)
I_t	=	Interest rate in year t (%)
IL_t	=	Additional income from livestock in year t due to liming ($)
IRR	=	Internal rate of return (%)
L	=	Liming rate (t/ha)
LS_t	=	Value of additional livestock in year t ($)
n	=	Number of years for which the model is to be run
NF_t	=	Net farm income after tax in year t ($)
NGM	=	Net gross margin ($)
NOF_t	=	Net off farm income after tax in year t ($)
OC	=	Opportunity costs ($)
pH_i	=	Soil pH before liming
pH_L	=	Soil pH one year after liming
pH_t	=	Soil pH in year t
RC_t	=	Replacement cost of livestock per DSE in year t ($)
RL	=	Residual effect of lime on soil pH compared with unlimed soil in year n.
RV	=	Residual value of lime in year n ($)
SR	=	The average stocking rate which can be maintained over n years (DSE)
SR_i	=	The stocking rate which applies before liming (DSE)
t	=	Time step for model run (years)
TEC_i	=	Total exchangeable cations before liming (me%)
TEC_L	=	Total exchangeable cations one year after liming (me%)
VC_t	=	Variable case costs per DSE in year t ($)
$\%Y_i$	=	Relative yield of unlimed pasture compared with yield at the biologically optimal lime rate
$\%Y_t$	=	Relative pasture yield in year t.
Al_i	=	Exchangeable aluminium before liming
$SLAl_s$	=	Slope of pH, aluminium relationship of exchangeable cations
%Al	=	Percentage aluminium saturation of exchangeable cations [Al (me%) divided by sum (me%) of exchangeable cations (Ca, Mg, K, Na, Mn, Al) expressed as a percentage]

4. Economic submodel

Using production data from the livestock production submodel, an investment in lime is considered from 3 approaches to economic analysis. Marginal analysis is used to determine the optimal rate of lime. For each rate of lime the internal rate of return (IRR) is calculated to enable a comparison with alternative investments. Once these two criteria have been used to select the optimal rate, a cash flow analysis is invoked to determine the effect of an investment in lime on the cash flow of the farm.

5. Input variables

The following input variables are required to run 'Lime-it':

soil parameters: pH (1:5 soil:0.01 M $CaC1_2$), Unless stated otherwise all pH values are measured using this technique.

TEC = Ca + Mg + K + Na + Mn + Al (me%) determined by leaching 4 g of soil with 200 mL of unbuffered 0.0125 M $BaC1_2$ using equipment described by Metson (1956). All cations were measured in the leachate by atomic absorption.

Al (exchangeable, as above).

production parameters: current stocking rate ($DSE.ha^{-1}$)

economic parameters: gross margin ($\$.DSE5^{-1}$)
cost of lime (spread) ($\$.t^{-1}$)
cost of additional livetock ($\$.DSE^{-1}$)
cash flow inputs (as in Table III)

C. The soil submodel

In this submodel we are concerned with prediction of pH, TEC and exchangeable aluminium in the top 10 cm of soil as functions of rate of lime applied and time after liming.

The concepts of lime requirement tests, either for a given soil pH or for a level of exchangeable aluminium, are discussed in Chapter 6. Lime requirement tests are a crude guide to decision making as they provide only one option (e.g. 4 $t.ha^{-1}$ to achieve a pH of 5.5). Producers will need to know the consequences of a range of lime rates. In the absence of a routine locally calibrated test for soil buffering, we developed empirically based predictive formulae for pH, TEC and exchangeable aluminium as a function of lime rate.

1. The data base

Data for the soil submodel were derived from 16 field experiments covering a range of soil types (yellow and red podsolic soils, red earth, krasnozem, black colluvial basalt) in New South Wales. Soils were initially acid with values ranging from pH 3.9 to 4.8. Fine lime (95 per cent passing 150 microns, neutralising value = 98%) was incorporated (at various rates) to a depth of approximately 10 cm and soils were sampled about one year after liming. Data values were derived from composite samples for each of the 448 individual plots.

2. The pH buffering capacity of soils

The rate of change in soil pH per unit of lime applied can be regarded as an index of the pH/lime buffering capacity of a soil. Such an index can be defined from the slope of the curve relating pH to the rate of lime application. For our 16 soils, equation (1) was used to describe the rate of change in soil pH due to liming.

$$\Delta \text{ pH} = B_s \text{ x } (\sqrt{(L + 1)} - 1) \qquad (1)$$

where Δ pH is pH increase due to liming, B_s is buffering capacity of the soils and L is lime (t.ha^{-1}).

This equation accounts for 93% of the total variance (Fig. 3). The B_s values derived for the 16 soils ranged from 0.34 to 1.55. That is, the addition of 1 t.ha^{-1} of lime increased soil pH by 0.14 to 0.64 units, depending on the soils' pH/lime buffering capacities. The problem now is to predict B_s from routinely measured soil properties.

3. The relationship between TEC and pH

The relationship between degree of change in soil pH and the initial cation exchange capacity of the soil has long been recognised (Lucas, 1942). Such a correlation was also obtained by Ross *et al.* (1964) who considered that the correlation should be expected because: 'at a given pH of an acid soil, cation exchange capacity may be related to reserve acidity. Because the major portion of lime reacts with reserve acidity, lime requirement increases with increased reserve acidity and increased CEC'.

The relationship between pH and TEC of our 16 soils was therefore investigated as a possible predictive index for pH buffering capacity. The

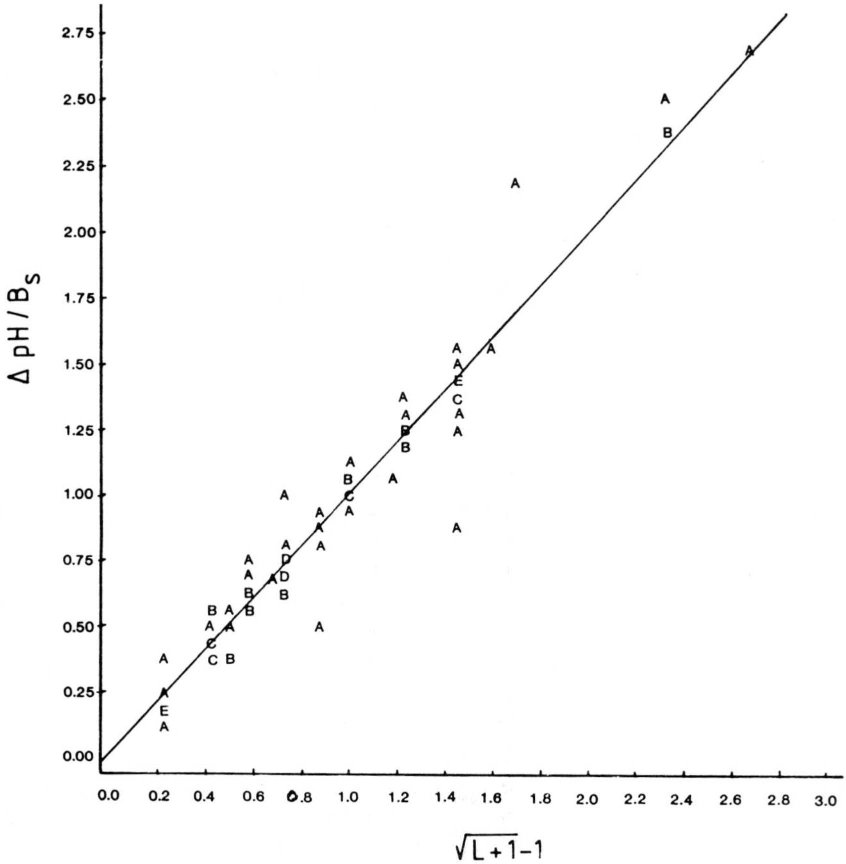

Fig. 3. The effect of lime application on the increase in soil pH (ΔpH) for 16 soils from New South Wales. The solid line represents equation (1). The letter A represents a value for a single site, the letter B represents a mean of values from two sites and so on. (L is lime rate in $t.ha^{-1}$; B_s is the pH buffering capacity of soil s). The equation accounts for 93% of the variance in the data.

regression equation (Equation (2) and Fig. 4) shows a linear correlation between pH and TEC for our data with SL_s as the soil specific variable which determines the slope of the line.

$$TEC = SL_s\,(pH - 2.043) - 0.957 \tag{2}$$

This equation accounted for 96% of the total variance. The value of

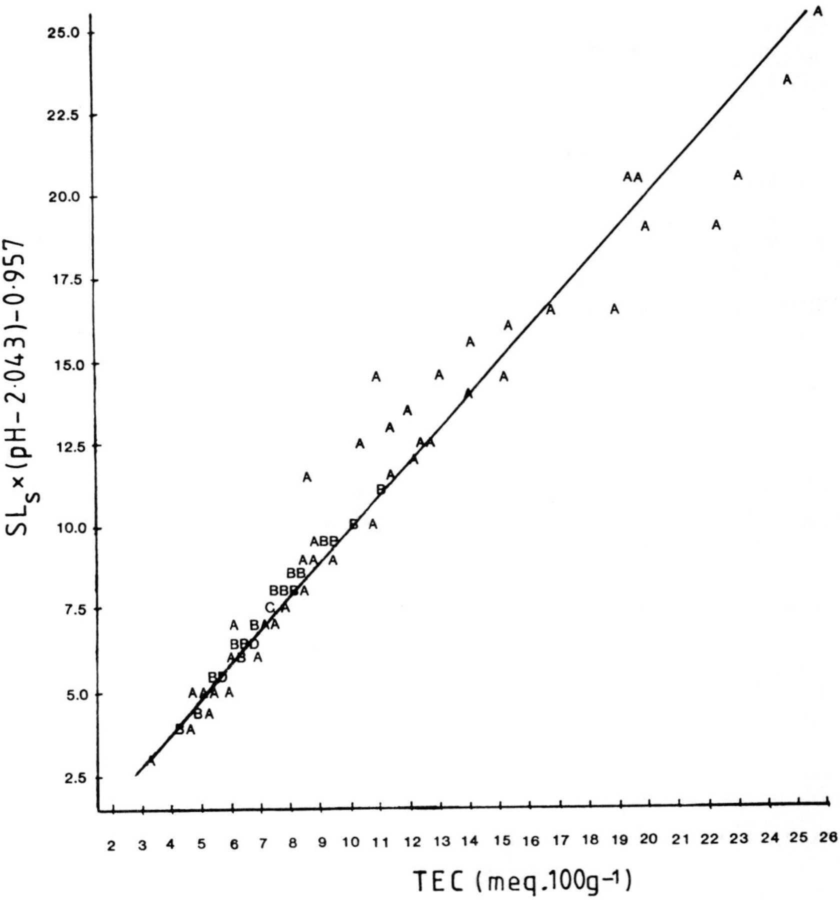

Fig. 4. The relationship between soil pH and total exchangeable cations (TEC) for 16 soils in New South Wales. The solid line represents equation (2). Letters as in Figure 3 (SL_s is the slope of the regression of pH and TEC for soil s). The equation accounts for 96% of variance in the data.

SL_s can now be predicted from measured pH and TEC values for unlimed soils by using equation (2 a) which is derived from equation (2).

$$SL_s = (TEC_i + O.957)/(pH_i - 2.043) \tag{2a}$$

The application of this formula is restricted to the soil types from which it was derived. Work by Helling *et al.* (1964) and Gessa *et al.*

(1978) may be used as a basis for a more universil model for which additional data on soil organic matter and clay contents would be required.

4. *Correlation between pH buffering capacity (B_s) and the slope of the regression of pH and TEC (SL_s)*

The potential use of SL_s as an index of pH buffering capacity was investigated by its correlation with B_s. The correlation obtained (equation (3) and Fig. 5), provides a basis for prediction of pH change for any rate of lime added.

$$B_s = (34.24/(SL_s + 9.5)) - 1.741 \qquad (3)$$

This equation accounted for 88% of the total variance. The calculation routine for predicting a change in soil pH in response to lime rates is as follows:
(a) derive a value for SL_s using equation (2 a)
(b) use the SL_s value in equation (3) to predict B_s
(c) use the B_s value in equation (1) to predict the change in soil pH.

5. *pH and exchangeable aluminium*

The relationship between soil pH and exchangeable aluminium has been

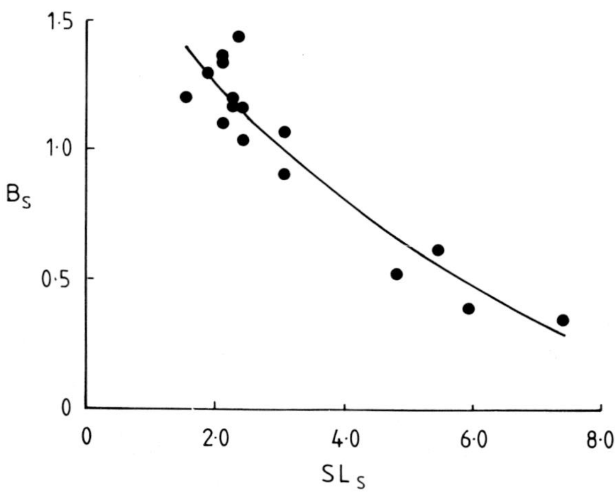

Fig. 5. Relationship between pH buffering capacity (B_s) and the slope (SL_s) of the regression of pH and total exchangeable cations for the 16 soils in New South Wales. The solid line represents equation (3). Each solid circle represents one of the 16 soils. The equation accounts for 88% of variance in the data.

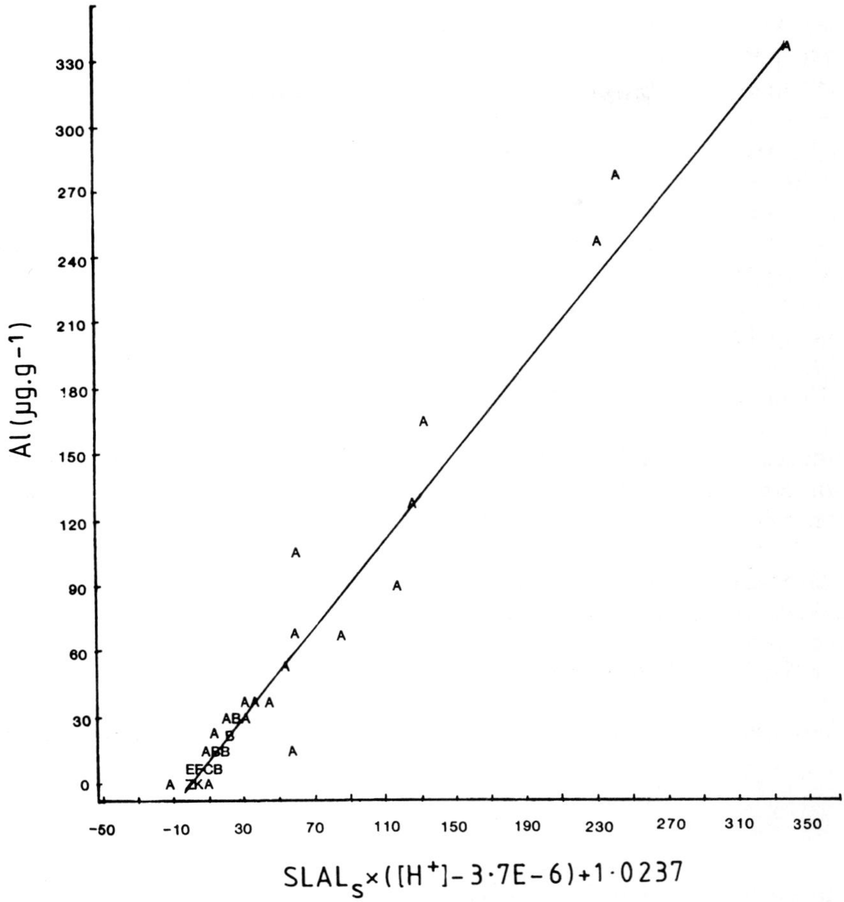

Fig. 6. The relationship between the concentration of exchangeable aluminium and the concentration of hydrogen ions for 16 soils in New South Wales. The solid line represents equation (4). Letters as in Figure 3. $SLAL_s$ is the slope of the regression of Al on H^+ for soil s. The equation accounts for 94% of variance in the data.

described for many soils and soil types (Coleman and Thomas, 1967; Kamprath, 1978; Bromfield *et al.*, 1983). Percentage aluminium saturation is considered as a useful indicator of lost yield potential (Kamprath, 1978) and should also be predicted as a function of lime added.

The relationship between pH (expressed as H^+ concentration) and exchangeable aluminium for the 16 soils is represented in Figure 6 and by equation (4).

$$SLAl_s = (Al.\ 90 - 1.0273)/([H^+] - 3.7.10^{-6}) \tag{4}$$

where $SLAl_s$ is the slope of the line relating exchangeable aluminium (Al) to $[H^+]$ for soil s.

Equation (4) accounted for 94% of the total variance. Therefore, pH and exchangeable aluminium of an unlimed soil are sufficient data to predict the slope of the $[H^+]$/Al line for that soil. Since both pH and TEC can be predicted from equations (1) and (2), percentage aluminium saturation (%Al) can now be predicted for any lime rate (equation (5))

$$\%Al = 100.(-1.233 + SLAl_s/10pH_L)/(TEC_L.90) \qquad (5)$$

where $pH_L = pH + \Delta pH$; and from equation (2): $TEC_L = SL_s.(pH_L - 2.043) - 0.957$

Thus far we have developed a predictive subroutine for changes in soil pH, exchangeable aluminium and TEC in response to liming. The advantages of this method are: (1) that it requires no laboratory tests other than those required for routine diagnosis; and (2) that it is not limited by preconceived 'target' levels of either pH or exchangeable aluminium.

D. Soil acidification and residual value of liming

To estimate how the benefits from liming are related to costs, the period over which lime is effective must be taken into account. After liming, soil will reacidify (see Chapter 2). A curve describing soil acidification as a function of time (t) under subterranean clover pasture (Bromfield *et al.*, 1983) is used as the basis for predicting acidification rate:

$$pH = 4.90 - 2.48\,(1 - e^{-0.0075.t}).$$

In this formula 4.90 is the native pH of the soil while 2.42 (i.e. 4.90– 2.48) is the implied asymptotic pH level. The constant, 0.0075, reflects the processes of acidification (nitrate leaching, product removal and organic matter build-up), characteristic of a subterranean clover pasture and the pH buffering capacity of the granitic soils used in that study.

In order to generalise the above formula to a range of soil types we replace the constant, 4.90, by a variable pH_v which is the 'virgin' pH of that soil. This value can generally be determined from unimproved land which is not in agricultural use. The constant, 0.0075, is multiplied by a buffering index. $B_s/1.3$ is used because the B_s value of the granitic soil for which data were presented by Bromfield *et al.* (1983) was calculated to be 1.3 using equations (2) and (3). The Bromfield equation can now be expressed as:

$$pH_t = 2.42 + (pH_v - 2.42)\, e^{(-0.0058.t.B_s)} \tag{6}$$

Equation (6) is still specific to acidification with a starting point of pH_v. It is necessary to add a value t_i to t in order to adjust for the current soil pH. From equation (6), t_i can be calculated as:

$$t_i = -\ln((pH_i - 2.42)/(pH_v - 2.42))/(0.0058.B_s) \tag{7}$$

and

$$pH_t = 2.42 + (pH_v - 2.42)\, e^{(-0.0058.(t+t_i).B_s)} \tag{8}$$

N.B. pH_v need not be specified accurately as $t + t_i$ will compensate for an inaccurate value.

In reviewing experiments measuring losses of lime from soils receiving increasing amounts of liming materials, Gasser (1973; 1985) observed that losses increase with increasing rate of application and that losses are greater in the years immediately following liming. Gasser proposed an empirical model:

$$\text{Lime loss } (t.ha^{-1}) = 0.168 \times 2^{(pH-5.5)}$$

The observed acidification of two long-term liming experiments was compared with the Gasser model (Bolton, 1977). Observed losses of lime were larger than those predicted by about 150 $kg.ha^{-1}.year^{-1}$ though the form of the curve was reflected in the data. The difference between observed and predicted values appears to be related mainly to the rate of addition of acids. In 'Lime-it' we used equation (8) to predict the rate of acid addition and modify the Gasser equation so that it predicts only the loss of added lime.

$$\text{Lime loss} = 0.168.(2^{(pH_{tLr}-5)} - 2^{(pH_{tL0}-5)}) \tag{9}$$

where pH_{tLr} is the soil pH predicted for the previous year for any lime rate; and pH_{tL0} is the soil pH predicted for the current year when no lime is added. Also note that 5 was used instead of the 5.5 of the Gasser model. This is to account for the difference in methods used to determine soil pH.

The lime loss from equation (9) can now be converted to soil pH by using equation (1). This pH drop is then subtracted from the soil pH calculated in equation (8).

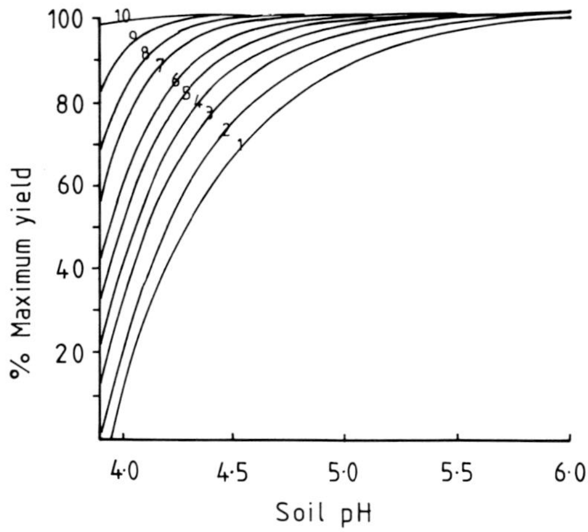

Fig. 7. A range of relative pasture responses to soil pH. The curves 1 to 10 each have a different value for the variable c in equation (10).

With more data, the constants used in the acidification functions could be calibrated and the functions used may be improved by accounting for additional factors such as rainfall (through drainage) and nitrate leaching.

E. Pasture response

Liming has increased the production of dry matter in pastures based on subterranean clover (Barrow, 1965; Helyar and Anderson, 1970; Munns and Fox, 1977; McLachlan, 1980) However it is not possible to predict responses to liming from soil measurements. Currently we must rely on local expertise based on field experimentation, strip trials and demonstrations as well as often limited on-farm experience to select a 'likely' production response curve. A generalised production function (equation (10)) is used to generate 10 optional response curves (Fig. 7). The 'local expert' can select any one of the curves to predict the response of pasture to a change in soil pH. The choice of curve determines the value of c in equation (10). Values of c range from 2.0 for curve 1 to 16.0 for curve 10.

$$\%Y = 100\,(1 - 2\,e^{-c(pH-3.6)}) \qquad (10)$$

where %Y = % maximum yield of pasture.

An interesting comparison may be made between the curves of Figure 7 and the relationships obtained by Edmeades *et al.* (1984) on pastures in the North Island of New Zealand. Their responses to liming

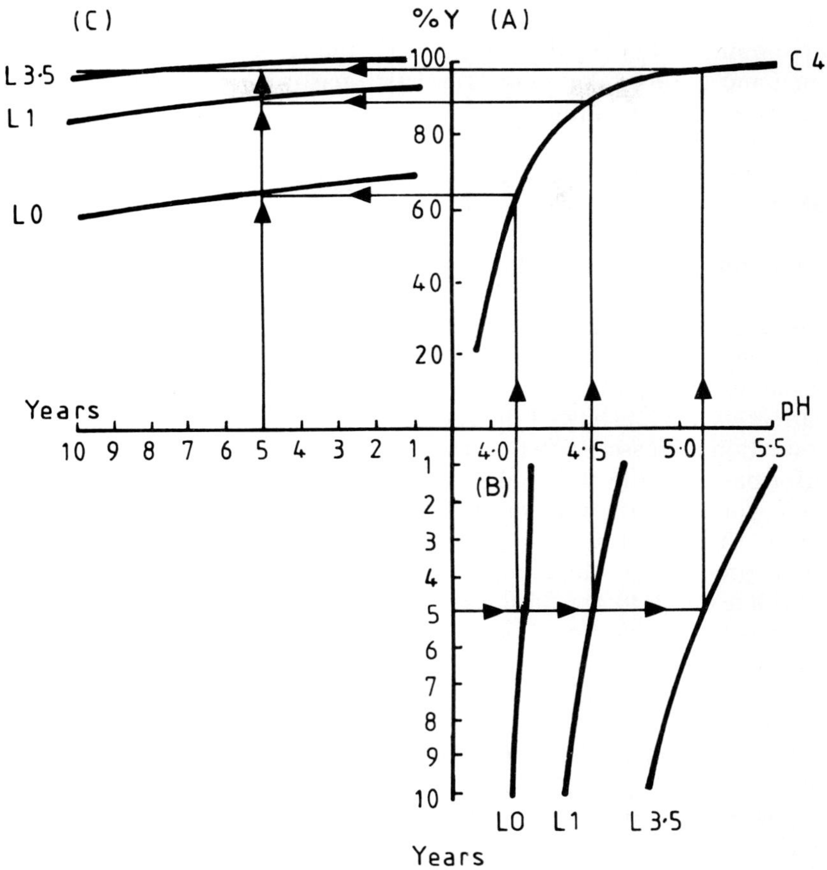

Fig. 8. The interrelationship between soil pH, time (years) and relative pasture yield (%Y) for soils with no lime (L_0), 1 t.ha^{-1} (L_1), and 3.5 t.ha^{-1} ($L_{3.5}$).
A The relationship between soil pH and relative pasture yield (%Y), using curve number 4 from Figure 7.
B The decline in soil pH over a 10-year period.
C Relative pasture production as a function of time.

on yellow-grey earths correspond with curves 5 to 6, yellow-brown loams correspond with curve 4 and sedimentary soils with curves 2 to 3. (An adjustment of 0.5 pH units was made to allow for differences in methods of pH measurement). Experiments on the southern slopes of New South Wales often range between curves 3 and 5 (unpublished data).

There is no attempt in this analysis to predict seasonal or year-to-year fluctuations in pasture responses to liming. The curve selected is assumed to be an 'average' response.

The model so far can be summarised by Figure 8, which synthesises:

(1) the relation between soil pH and relative pasture yield (in this case curve number 4 of Figure 7 is used); (2) the effect of time on soil pH of limed and unlimed soil; and (3) relative pasture yields as a function of time in limed and unlimed soils. As yet there is a lack of data to compare with this simple synthesis. Some support can be implied from results on a white clover (*Trifolium repens* L.) based pasture in Tasmania (Rowe, 1982).

F. Effects of lime on stocking rate and animal production

Most of the research examining responses in animal production to liming has been done in New Zealand with pastures based on white clover (O'Connor and Edmeades, 1984). A five-year grazing experiment (Bircham *et al.*, 1977) showed that ewe liveweight as well as wool and lamb production increased with liming. The authors attributed effects on stock performance and health to increased availability of pasture dry matter. Their work also showed, through positive lime by stocking rate interactions, that full utilisation of pasture is necessary to obtain responses in animal production. These findings have been supported recently by Rhys and Edmeades (1984) who showed that under common grazing, animals grazed limed and unlimed plots uniformly. There was no preference for herbage on limed pasture.

In the model, these results are interpreted by assuming that the percentage increase in stocking rate due to liming is linearly related (1:1) to the percentage increase in pasture production (Moyle, 1961). For any soil, current stocking rate must be supplied as input data. For any lime rate, the pasture model is then used to calculate the average stocking rate required to utilise the additional pasture production at a level which is similar to pasture utilization before liming.

This simple approach has the advantage of eliminating the need to use climatic and soil physical data to predict actual pasture yields. In addition, stocking rates should reflect the long-term average production as well as the farmer's preferances. Users of the model should readily relate to predicted increases in stock carrying capacity.

G. Economic submodel

1. Introduction

In the preceding sections the effects of liming were defined in chemical and biological terms. A farm may be considered as a bio-economic complex controlled to achieve certain economic objectives (Dent and

Anderson, 1971). Production is therefore linked with the overall objectives of farm managers. Depending on managerial preferences and financial situation, an optimal liming strategy is selected. Given reduced marginal returns and limited finance, the economic optimum is unlikely to be at the same input level as the biological maximum. In 'Lime-it', three economic criteria are used to assist managers in selecting a lime rate. These are marginal analysis, internal rate of return and cash flow analysis. Not included in this analysis are some second order benefits which may follow liming. For example, if a cereal crop were grown after the period of analysis, it could well benefit from increased nitrogen fixed in the pasture phase.

2. *Marginal analysis*

Marginal analysis is used to determine the lime rate that maximises returns. The marginal cost associated with liming consists of the marginal cost of lime (including application costs) and the marginal costs of additional livestock due to higher stocking rates per hectare being required to utilise increased pasture production. The marginal costs of livestock are the additional costs as defined in gross margin calculations, as well as the opportunity costs of additional livestock. The marginal benefits from liming include an increase in gross margin, as well as the residual value of lime and the value of additional livestock at the end of the analysis period. Calculations are done in constant or real dollars. Because money in year '1' is not valued the same as money in year 'n' (e.g. Chisholm and Dillon, 1966), income and costs are compounded, reflecting differences in investment opportunity over time. The model uses a real interest rate of 5%. Net returns are estimated at the end of the investment period (year n) and hence compounding rather than discounting is used in this analysis.

The gross margin of a livestock enterprise based on pasture is defined as the difference between gross income (including sale of culled animals) and variable costs (including replacement costs). The gross margin per hectare in year t, GM_t, is defined as:

$$GM_t = (GI_t - VC_t - RC_t).SR \tag{11}$$

where GI_t is the gross income per DSE from the pasture based enterprise in year t, VC_t is the variable cash costs per DSE in year t, RC_t is the replacement costs of livestock per DSE in year t and SR is the average stocking rate in DSE that can be maintained over that time period of the analysis.

The compounded total gross margins (TGM) over a period of n years from an investment in lime is:-

$$TGM = \sum_{t=1}^{n} (GM_t) \qquad (12)$$

The compounded total opportunity cost of livestock (OCL) is:

$$OCL = \sum_{t=1}^{n} (O_t . LS_t) \qquad (13)$$

where O_t is the opportunity cost of capital and LS_t the value of additional livestock required as a result of liming.

The net gross margin (NGM) for each rate of lime application at the end of n years is:

$$NGM = TGM + RV - CL - OC \qquad (14)$$

where RV is the residual value of lime at the end of the calculation period. The residual value of lime is calculated from the cost of lime required to equalise the soil pH of limed and unlimed soils at the end of the analysis period. CL is the cost of lime; OC is opportunity cost.

NGM is calculated for each incremental lime application. Optimal investment strategy is to apply the rate of lime required to maximise NGM. In practice however, because of diminishing marginal NGM, alternative investment options, risk and uncertainty, livestock producers will tend to select a sub-optimal rate of lime.

3. *Internal rate of return*

To enable a comparison between an investment in lime and alternative investment opportunities, an internal rate of return (IRR) for each lime rate is calculated. The IRR is the rate of discount which makes the present value of benefits equal to the present value of costs. That is,

$$\sum_{t=1}^{n} (B_t - C_t)/(1 + \lambda)^t = 0 \qquad (15)$$

where λ is the internal rate of return, B_t is the benefit of lime in dollars in year t and C_t is the cost of lime in year t.

The benefits, B_t, are:

$$B_t = TG_t + R_n \quad (16)$$

where TG_t is the total gross margin in year t and R_n is the residual value of lime in the final year of analysis.
The costs, C_t, are:

$$C_t = CL_1 + LS_t \quad (17)$$

where CL_1 is the cost of lime in year 1 and LS_t is the cash cost of livestock in year t.

4 Cash flow budget

Before a farmer invests in lime, the effects on cash flow should be considered. This becomes particularly important where liming is to be financed by a lending agency. Taxation may be important in a cash flow analysis as it will influence cash surplus and the repayment period. A decision to invest in lime can be further influenced by the overall farm income and expenditure pattern. Taxation and a financial overview of other income and expenses are therefore included in the analysis.

The cash flow effects of an application of lime for a given period can be expressed as:

$$CF_t = IL_t + NF_t + NOF_t - CL_t - LS_t - CH_t - CO_t - I_t \quad (18)$$

where CF_t is the cash flow in year t, IL_t is the additional income of livestock derived from liming in that year. NF_t is the net farm income after tax and NOF_t is the net off-farm income in year t after tax. CH_t is the household cost and CO_t is the overhead cost in year t. I_t is the interest rate in year t.

The effects of inflation and net cost/price squeeze on the cash flow can be accounted for by nominating a rate of inflation and a rate of change for the value of farm products.

The cash flow takes account of taxation by multiplying IL_t, CL_t, LS_t, CO_t and I_t by a tax factor TF_t.

$$TF_t = (100 - TR_t)/100 \quad (19)$$

where TR_t is the tax rate in year t. Interest paid or earned is calculated over the debt or surplus in each year. The debt or surplus is:

$$DS_t = DS_{t-1} + CF_t \tag{20}$$

where DS_t is the debt or surplus in year t, DS_{t-1} is the debt or surplus in the previous year and CF_t the cash flow in year t.

Finally the cash flow 'profit or loss' is calculated for year n.

$$PL_n = DS_n + LS_n + RV_n \tag{21}$$

where PL_n is the 'profit or loss' in year n, and LS_n the value of the additional livestock in that year and RV_n the value of residual lime in year n. The 'profit or loss' is expressed in nominal and in real or constant dollars.

IV. CASE STUDY

One of the benefits of model building is the feasibility and ease of 'experimentation' with a model compared with experimentation on the real system. To illustrate some of the possibilities with regard to 'Lime-it', we first present a model run using input data typical of a long-term pasture on a problem acid soil on the southern slopes of New South Wales. Subsequently values of economic inputs such as the cost of lime, rate of lime applied and gross margin per unit of livestock are varied.

Table II. An example of a model run: the effect of lime on the net gross margin (NGM) and the internal rate of return to investment. Stocking rate is 7.5 DSE.ha^{-1}; gross margin as \$12 per DSE; cost of lime spread is \$52.t^{-1}; cost of additional livestock is \$20 per DSE. Soil pH is 4.2 and TEC is 4.43 me%.

Lime Rate (t.ha^{-1})	Total gross margin (\$.ha^{-1})	Residual value of lime (\$.ha^{-1})	Total cost of lime (\$.ha^{-1})	Opportunity cost of livestock (\$.ha^{-1})	Net gross margin (\$.ha^{-1})	Internal rate of return (%)
0.0	1053	0	0	81	973	0
0.5	1331	17	52	101	1195	43
1.0	1457	32	94	111	1284	35
1.5	1524	46	137	116	1316	29
2.0	1562	58	179	120	1321	25
2.5	1586	70	222	121	1313	22
3.0	1601	82	264	122	1296	19
3.5	1611	92	306	124	1273	17
4.0	1618	102	349	124	1248	15
4.5	1624	111	391	125	1219	13

Table III. Information required for cash flow: an example

Selected curve no.	4[a]
Hectares to be limed	100
Lime applied	2 $t.ha^{-1}$
Interest rate payable on loan	15%
Interest rate earned on investment	12%
Marginal tax bracket	30%
Stocking rate without lime	7.5 $DSE.ha^{-1}$
Expected rate of inflation	10%
Annual price change for livestock	7%
Current net farm income	0 $\$.year^{-1}$
Current net off-farm income	0 $\$.year^{-1}$
Annual overheads	0 $\$.year^{-1}$
Annual household expenditure	0 $\$.year^{-1}$

[a] See Fig. 7 for characteristics of curve no. 4

Table II gives an example of a model run. The stocking rate is 7.5 $DSE.ha^{-1}$ and the gross margin $\$12.DSE^{-1}$. The cost of lime spread on the property is $\$52.t^{-1}$ and the cost of additional livestock $\$20.DSE^{-1}$. Soil pH in this example is 4.2 and total exchangeable cations 4.43 me%.

The maximum net gross margin (NGM) is obtained at a lime rate of 2.0 $t.ha^{-1}$. The average internal rate of return (IRR) at that lime rate is 25. The highest internal rate of return is 43% at a lime rate of 0.5 $t.ha^{-1}$. Beyond this application rate the increase in NGM (or marginal NGM) declines steadily. The marginal NGM from 1.5 to 2.0 t of lime is only $5. Given risk and uncertainty and alternative investment opportunities, farmers may opt for an investment strategy between 0.5 and 2.0 $t.ha^{-1}$ (Table II).

To complete the cash flow component of the model, further information is required. Table III gives details of an example.

Specific details are required on area to be limed, lime rate, expected interest rates, inflation and taxation. Further details may be added on other farm income, off-farm income, overhead and household expenditures to provide a whole-farm overview.

Table IV provides details on the cash flow defined in Tables II and III.

At the selected lime rate, it takes six years to repay the loan. Including the value of residual lime and the value of the additional livestock on the property, the total cash surplus or 'profit' at the end of 10 years in inflated dollars is $44 616. In constant dollars this is $18 922.

Table IV. 'Lime-it' 10-year cash flow (pH 4.2, TEC 4.43 me%). Data from Tables II and III apply. All values are dollars

Year	1	2	3	4	5	6	7	8	9	10
Income										
Added income due to liming:	0	2864	3127	3410	3715	4043	4395	4772	5178	5613
Other income										
net farm	0	0	0	0	0	0	0	0	0	0
net off-farm	0	0	0	0	0	0	0	0	0	0
Total income	0	2864	3127	3410	3715	4043	4395	4772	5178	5613
Costs										
Lime	7770									
Extra livestock		6091								
Household	0	0	0	0	0	0	0	0	0	0
Other overheads	0	0	0	0	0	0	0	0	0	0
Interest		809	1232	1033	784	476	101	−280	−704	−1198
Total costs	7700	6899	1232	1033	784	476	101	−280	−704	−1198
Cash flow	−7700	−4035	1895	2377	2931	3567	4293	5052	5882	6811
Debt or surplus	−7700	−11735	−9840	−7463	−4532	−965	3328	8381	14262	21073
Residual value of lime										13762
Value of added livestock										9781
Profit or Loss										44616
Profit or Loss in constant dollars										18922

V. SENSITIVITY OF 'LIME-IT' PREDICTIONS TO VARYING INPUT VALUES

A model run requires 18 variables to be specified. The relative influence of each of these variables on predicted values, such as period of loan repayment or optimal lime rate, can serve as a guide to their influence on rational lime use on the farm. In this sensitivity study all variables are initially set to the values used in the 'case study'. Input variables are investigated one at a time for a range of values while others are kept constant.

A. The effects of varying economic parameters

1. The effects of varying the cost of lime and the soil's pH buffering capacity

The model proved only moderately sensitive to changes in the cost of lime (Table V). Significant variation in the cost of lime from \$60 to \$20 per tonne resulted in relatively small changes in the predicted repayment period for an investment in 2 $t.ha^{-1}$ of lime; six to four years in the lightly buffered soil (TEC = 4.43me%) and seven to four years in the heavily buffered soil (TEC = 10.0me%). Similarly the predictions of optimal lime rate were only doubled for a three-fold difference in price. The changes in IRR for optimal lime rate are also relatively small. The well buffered soils tend to have longer repayment periods, higher optimal lime rates and lower values of IRR than the lightly buffered soils.

2. The effects of varying the rate of lime

The repayment period is affected by the rate of lime (Table VI). An application of one tonne of lime per hectare meant that a loan could be

Table V. The effect of lime cost changes and pH buffering capacity (TEC) on the repayment period, optimal lime rate and internal rate of return for a soil with an initial pH of 4.2. All other variables are constant at values listed in Table III.

Cost of lime ($S.t^{-1}$)	Repayment period on 2 $t.ha^{-1}$ (Years)		Optimal lime rate ($t.ha^{-1}$)		Internal rate of (%)	
TEC (me%)	4.43	10.00	4.43	10.00	4.43	10.00
20	4	4	3.0	4.0	39	30
40	5	6	2.0	3.0	31	20
52	6	7	2.0	3.0	25	16
60	6	7	1.5	2.0	26	18

Table VI. The effect of lime rates on the repayment period. All other variables are constant at values set in Table III.

Lime rate ($t.ha^{-1}$)	Repayment period (years)
0.5	4
1.0	4
2.0	6
3.0	7
4.0	8

repayed in four years versus a seven-year repayment period for a lime rate of 3 $t.ha^{-1}$. Longer repayment periods must be a disincentive to farmers to adopt higher rates of lime.

3. *The effects of varying livestock gross margin*

Changes in livestock prices as reflected in their gross margins significantly affected the loan repayment period and the internal rate of return but had little influence on the optimal lime rate (Table VII).

B. The effects of soil and production parameters on economic criteria

The sensitivity of the economic submodel to soil and biological input values is investigated by varying pasture responsiveness to soil pH, initial soil pH, total exchangeable cations and stocking rate.

The pH response curve selected for a model run has significant influence on the predicted repayment period, optimal lime rate and internal rate of return at the optimal lime rate. It is interesting to note that some liming can be justified even for relatively unresponsive pastures such as defined by curve number 7 of Figure 7 (Table VIII).

Table VII. Effects of changing livestock gross margins. All other variables are constant at values set in Table III.

Livestock gross margin ($\$.DSE^{-1}$)	Loan repayment period (years)	Optimal lime rate ($t.ha^{-1}$)	Internal rate of return (%)
10	7	1.5	24
12	6	2.0	25
14	5	2.0	30
16	5	2.0	35

Initial pH values (pH_i = 4.0, 4.3, 4.7 and 5.0) for 4 initial stocking rates (SR_i = 6, 7.5, 10 and 14 $DSE.ha^{-1}$) were varied on two 'soil types' with TEC values of 4 and 10 me%, representing lightly pH buffered and highly pH buffered soils. It should be noted that each pH_i/TEC combination describes a differently buffered soil.

1. Sensitivity of maximum net gross margin

a. Lime rates: The net gross margin criteria may be used to determine an optimal rate of lime. For the lightly pH buffered soils, (Fig. 9A)), optimal lime rates are relatively insensitive to SR_i at all pH_i values. Initial pH value is the critical variable, with 3.0 to 4.0 $t.ha^{-1}$ being optimal at pH_i = 4.0 down to nil lime at pH_i = 5.0. For the more strongly pH buffered soils (Fig. 9B)) optimal lime rates are approximately double those for the lightly buffered soils especially at low pH_i. Stocking rates are of greater importance in well buffered soils. At SR_i values of 10 and 14, the optimal lime rates are about 2 $t.ha^{-1}$ greater than at SR_i = 6 in the range pH_i range 4.0 to 4.7. At pH_i 5.0, lime is unprofitable while at pH_i 4.7 lime addition only increases NGM with higher stocking rates.

b. Cumulative increment ($\$.ha^{-1}.10\ years^{-1}$) earned at the optimal NGM.: A risk is only worth taking if the rewards are sufficiently attractive. For example a farmer might decide to lime only if it increases his average annual earnings by more than $\$20.ha^{-1}$ (or $\$200.ha^{-1}$ over 10 years).

Table VIII. Effect of responsiveness to lime on the repayment period, optimal lime rates and internal rate of return.

Curve no.[a]	Repayment period on 2 $t.ha^{-1}$ (years)	Optimal lime rate ($t.ha^{-1}$)	Internal rate of return (%)
1	4	4.0	37
2	4	3.0	34
3	5	2.5	28
4	6	2.0	25
5	7	1.5	24
6	8	1.0	24
7	>10	1.0	14
8	>10	0.5	11
9	>10	0.0	0

[a] Number of pH response curve. See Fig. 7.

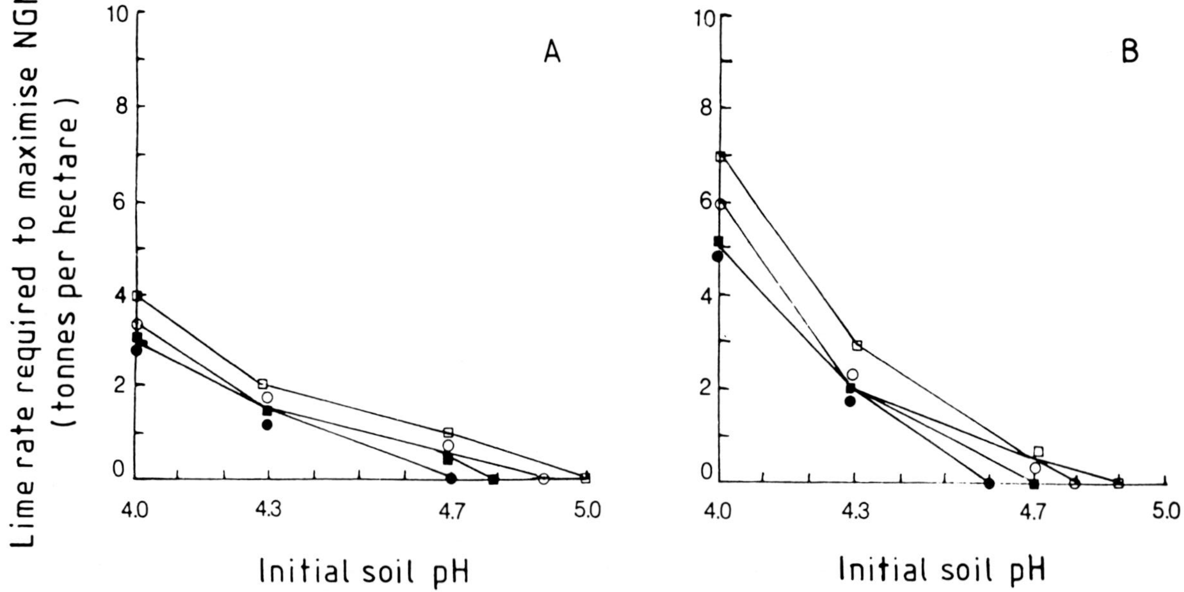

Fig. 9. Rates of lime required to maximise net gross margin (NGM) as a function of initial soil pH and carrying capacity on two soil types: (A) total exchangeable cations 4 me%; (B) total exchangeable cations 10 me%. The symbols ●, ■, ○ and □ represent stocking rates of 6, 7.5, 10 and 14 DSE.ha^{-1} respectively.

If such a farmer is on a lightly pH buffered soil with $SR_i = 14$ he would only lime paddocks with $ph_i < 4.6$. If his soil had a lower carrying capacity ($SR_i = 7.5$) he would only lime paddocks with $pH_i < 4.3$. The same farmer on a heavily pH buffered soil would lime paddocks with $pH_i < 4.5$ at $SR_i = 14$ and paddocks with $pH_i < 4.2$ at $SR_i = 7.5$ (Fig. 10).

2. *Sensitivity of internal rate of return*

A farmer who has alternative investment options with an IRR of 10% would only lime at these rates (as determined in Fig. 9) if IRR is greater than 10%. On a lightly pH buffered soil (Fig. 11A) with $SR_i = 6$, such a farmer would apply lower rates at $pH_i > 4.5$. With $SR_i = 14$, lower rates would be applied at $pH_i > 4.7$. On a more strongly pH buffered soil (Fig. 11B) with $SR_i = 6$, lower rates of lime would be applied at $pH_i > 4.3$. With a $SR_i = 14$, lower rates would be applied at $pH_i > 4.7$.

3. *Sensitivity of cash flow*

Figure 12 shows the number of years it would take a farmer to repay a loan for liming 100 ha at a rate of 2 $t.ha^{-1}$. For example a farmer who is not prepared to be in debt for more than five years would either decide against a loan or consider lower rates of lime application at $pH_i > 4.1$ for $SR_i = 6$ or at $pH_i > 4.3$ for $SR_i = 14$ on the lightly pH buffered soils. On the highly buffered soils the same considerations would apply at slightly lower pH values.

VI. CONCLUSIONS

'Lime-it' is a simple management oriented model based on field data. This model integrates several factors into a cohesive approach to decision making on use of lime in long-term pastures based on subterranean clover. It consists of four submodels; soil, pasture, livestock and economic. Each of these submodels can be independently updated in the light of new information. This information, as part of the model, can then be applied rapidly to liming strategy. We recognise the need for improvements in defining the longevity of lime effects on soil parameters, the relationship between soil parameters and pasture production, and that between pasture and animal production. Assuming that the model is a reasonable representation of the real system, then some conjecture can be made about the system on the basis of the sensitivity

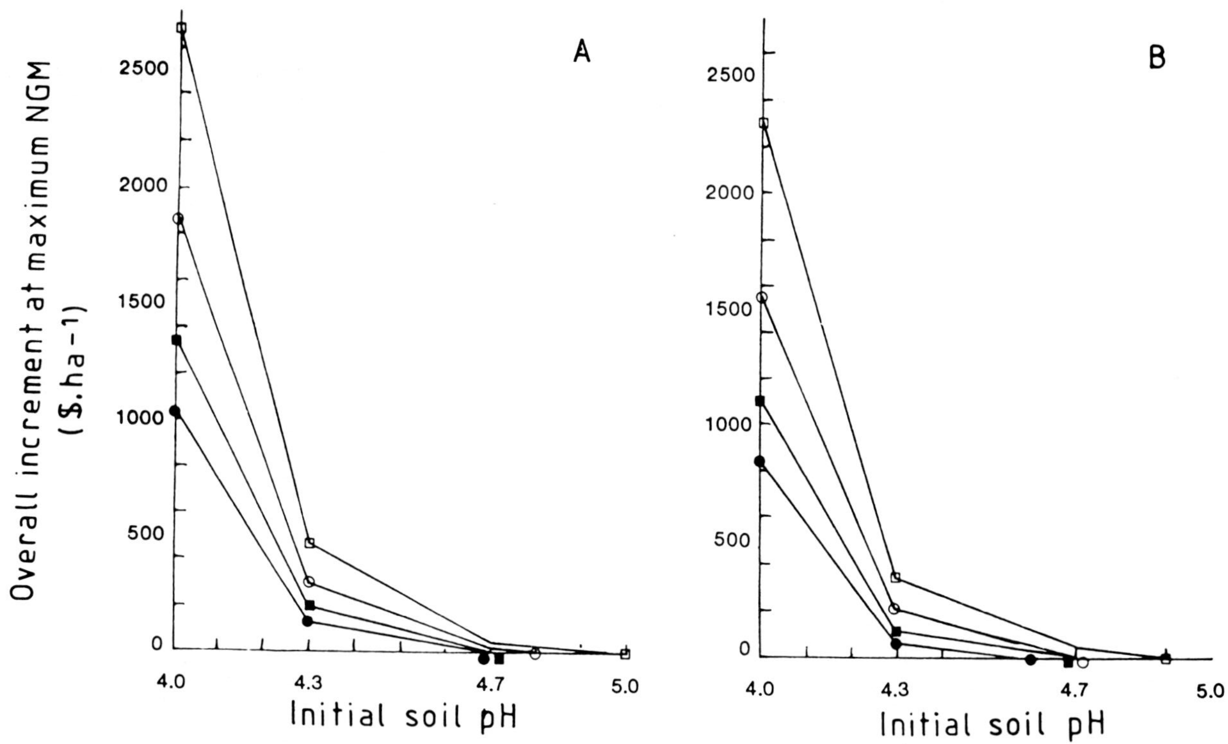

Fig. 10. The cumulative increment ($\$.ha^{-1}$) earned over 10 years at the rate of lime required to optimise net gross margin (NGM) as a function of initial soil pH and carrying capacity on two soil types. Details as in Fig. 9.

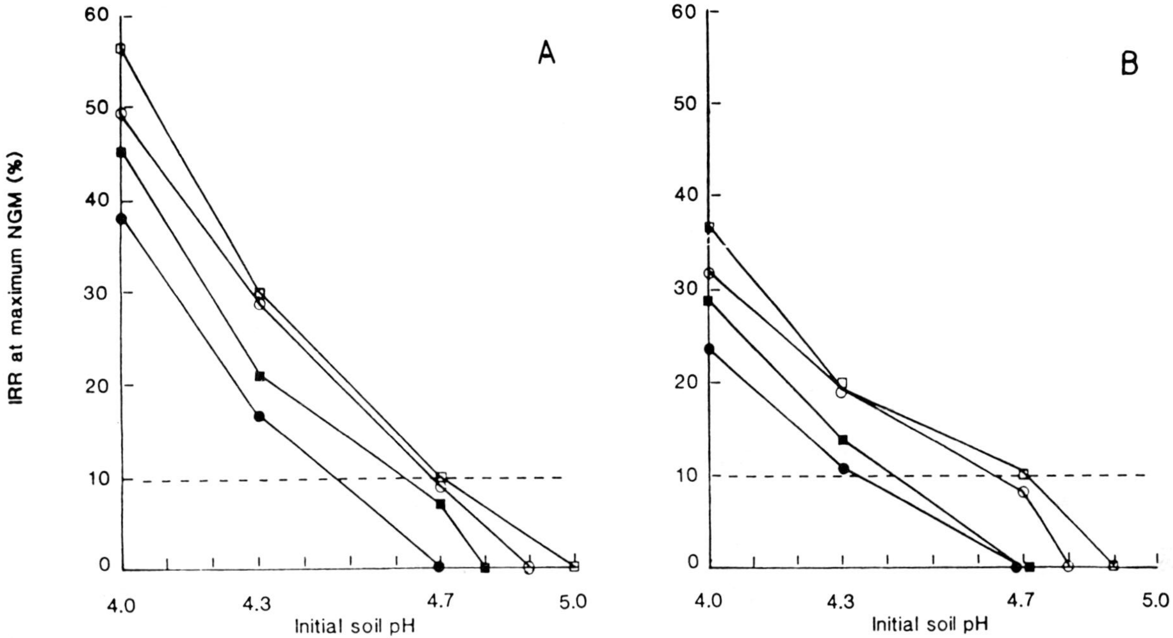

Fig. 11. Internal rates of return (IRR) to liming at rates required to maximise net gross margin (NGM) as a function of initial soil pH and carrying capacity of two soil types. Details as in Fig. 9.

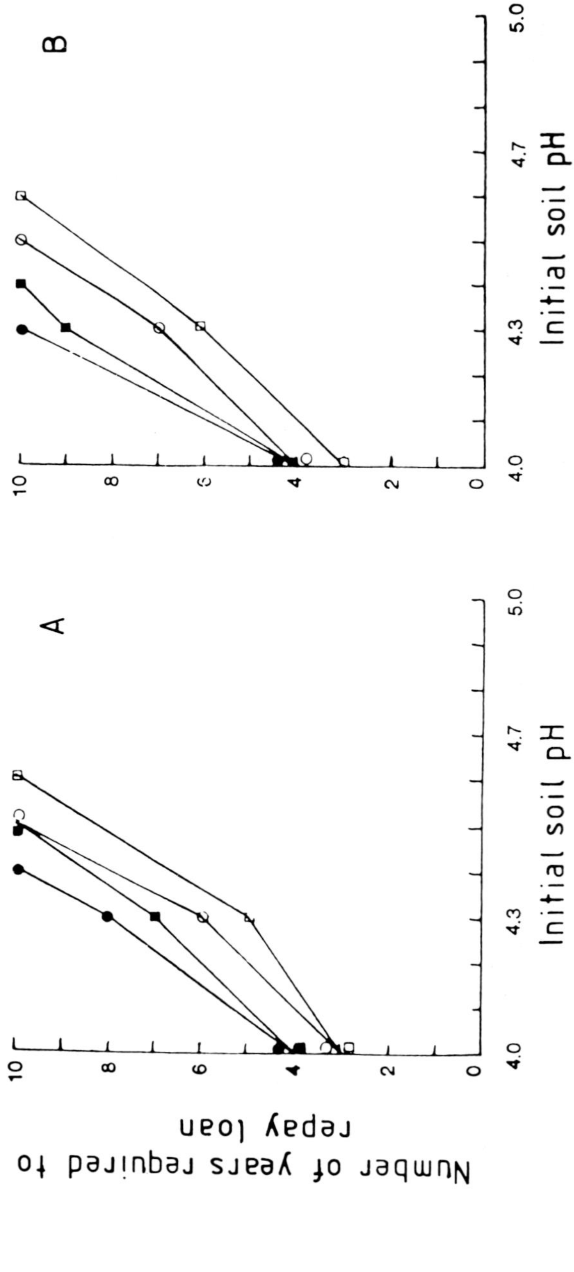

Fig. 12. The number of years required to repay a loan on costs associated with liming at 2 t.ha^{-1} as a function of initial soil pH and carrying capacity (DSE.ha^{-1}) of two soil types. Details as in Fig. 9.

analysis. The results of our simulation suggest that blanket recommendations such as liming to a target pH or a set rate of lime are inappropriate. Farmers will need advice which is tailor-made for their particular situations. Such advice should be consistent with their management constraints and objectives.

Models such as 'Lime-it' can be employed usefully by advisory and other workers required to advise on use of lime. With information on soil pH, exchangeable cations and the carrying capacity of a paddock, 'Lime-it' enables the farmer to consider his decision from several viewpoints. As well as the three economic criteria (NGM, IRR and cash flow), the implications of any liming option can also be assessed in terms of broader management goals which are not usually considered in economic analyses. For example, a farmer could consider the improvement or maintenance of soil pH and production potential over time. A farmer may accept less than optimal returns if he is more concerned with protecting the value of his asset (the soil) than with maximising shorter term income.

The sensitivity analysis also demonstrates many situations in which farmer decisions, based on an economic rationale, will result in continuing soil acidification and sub-optimal production. This observation should be an area of concern to agricultural policy makers. In this regard, 'Lime-It' can be used to study the value of a lime subsidy.

REFERENCES

Barrow, N. J. (1965). Further investigations of the use of lime on established pastures. *Australian Journal of Experimental Agriculture and Animal Husbandry* **5**, 442–449.

Bircham, J. S., Crouchley, G. and Wright, D. F. (1977). Effects of superphosphate, lime and stocking rate on pasture and animal production on the Wairarapa Plains. II. Animal production. *New Zealand Journal of Experimental Agriculture* **5**, 349–355.

Bolton, J. (1977). Changes in soil pH and exchangeable calcium in two liming experiments on contrasting soils over twelve years. *Journal of Agricultural Science*, (Camb.) **89**, 81–86.

Bromfield, S. M., Cumming, R. W., David, D. J. and Williams, C. H. (1983). Change in soil pH, manganese and aluminium under subterranean clover pasture. *Australian Journal of Experimental Agriculture and Animal Husbandry* **23**, 181–191.

Chisholm, P. H. and Dillon, J. L. (1966). Discounting and other interest rate procedures in farm management. Professional Farm Management Guidebook No. 2., Faculty of Economic Studies. University of New England, Armidale.

Coleman, N. T. and Thomas, G. W. (1967). The basic chemistry of soil acidity. *In* 'Soil Acidity and Liming' (R. W. Pearson and F. Adams, eds.), pp. 1–41. American Society of Agronomy, Madison, Wisconsin.

Dent, J. B. and Anderson, J. R. (1971). Systems, management and agriculture. *In* 'Systems Analysis in Agricultural Management' (J. B. Dent and J. R. Anderson, eds.), pp. 3–14. John Wiley and Sons, Sydney.

Edmeades, D. C., O'Connor, M. B., Saunders, W. M. H. and Toxopeus, M. R. J. (1981). The role of lime in agriculture; some recent results. Proceedings of Ruakuara Farmers Conference **32**, 45–51.

Edmeades, D. C., Pringle, R. M., Shannon, P. W. and Mansell, G. P. (1984). Effects of lime on pasture production on soils in the North Island of New Zealand. IV. Predicting lime responses. *New Zealand Journal of Agriculture Research* **27**, 371–382.

Edmeades, D. C., Pringle, R. M., Mansell, G. P., Shannon, P. W., Ritchie, J. and Stewart, K. M. (1985). Effects of lime on pasture production on soils in the North Island of New Zealand. V. Description of a lime recommendation scheme. *New Zealand Journal of Experimental Agriculture* **13**, 47–58.

Gasser, J. K. R. (1973). An assessment of the importance of some factors causing losses of lime from agricultural soils. *Experimental Husbandry* **25**, 86–95.

Gasser, J. K. R. (1985). Processes causing loss of calcium from agricultural soils. *Soil Use and Management* **1**, 14–17.

Gessa, C., Melis, P., Bellu, G. and Testini, C. (1978). Inactivation of clay pH-dependent charge in organo-mineral complexes. *Journal of Soil Science* **29**, 58–64.

Hall, H. H. (1983). Economic evaluation of crop response to lime. *American Journal of Agricultural Economics* **65**, 811–817.

Helling, C. S., Chesters, G. and Corey, R. B. (1964). Contribution of organic matter and clay to soil cation-exchange capacity as affected by the pH of the saturating solution. *Soil Science Society of America Proceedings* **28**, 517–520.

Helyar, K. R. (1978). Effects of aluminium and manganese toxicities on legume growth. *In* 'Mineral Nutrition of Legumes in Tropical and Subtropical Soils'. (C. S. Andrew and E. J. Kamprath, eds.), pp. 207–231. CSIR0, Melbourne.

Helyar, K. R. and Anderson, A. J. (1970). Responses of five pasture species to phosphorus, lime, and nitrogen on an infertile acid soil with a high phosphate sorption capacity. *Australian Journal of Agricultural Research* **21**, 677–692.

Kamprath, E. J. (1978). Lime in relation to Al toxicity in tropical soils. *In* 'Mineral Nutrition of Legumes in Tropical and Subtropical Soils' (C. S. Andrew and E. J. Kamprath, eds.), pp. 233–245. CSIRO, Melbourne.

Landbouw Economisch Instituut (1986). Landbouw Cijvers 1986, p. 123.

Lucas, R. E. (1942). Reliability of lime requirement calculations based on the rapid copper method for exchange capacity. *Soil Science Society of America Proceedings* **7**, 362–367.

McLachlan, K. D. (1980). Nutrient problems in sown pasture on an acid soil. I. Survey of the problems and suggested corrective treatments. *Australian Journal of Experimental Agriculture and Animal Husbandry* **20**, 319–326.

Metson, A. J. (1956). Methods of chemical analysis for soil survey samples. Soil Bureau Bulletin 12, DSIR, New Zealand, pp. 102–103.

Moyle, R. G. (1961). The economic interpretation of pasture responses to superphosphate. *Quarterly Review of Agricultural Economics* **14**, 79–87.

Munns, D. N. and Fox, R. L. (1977). Comparative requirements of tropical and temperate legumes. *Plant and Soil* **46**, 533–548.

O'Connor, M. B. and Edmeades, D. C. (1984). A summary of M. A. F. trials on the effects of lime on animal production. Proceedings of a Workshop on Lime in New Zealand Agriculture. (B. L. J. Jackson and D. C. Edmeades, eds.), p. 17–19. M. A. F., Wellington.

Proef Station voor de Akkerbouw en Groenteteeld in Voilegrond (1982). Kwantitatieve Informatie, Bedrijfssynthese 1982–1983, pp. 108–133.

Rhys, G. J. and Edmeades, D. C. (1984). Pasture measurement techniques and herbage consumption of limed pasture under common grazing. *New Zealand Journal of Experimental Agriculture* **12**, 225–229.

Ross, G. J., Lawton, K. and Ellis, B. G. (1964). Lime requirement related to physical and chemical properties of nine Michigan soils. *Soil Science Society of America Proceedings* **28**, 209–212.

Rowe, B. A. (1982). Effects of limestone on pasture yields and the pH of two krasnozems in north-western Tasmania. *Australian Journal of Experimental Agriculture and Animal Husbandry* **22**, 100–105.

Index

9 0 1 2 3 4 5 6 7 8
A B C D E F G H I J